S0-BCY-410

INTEGRAL EQUATIONS

INTEGRAL EQUATIONS

HARRY HOCHSTADT

Department of Mathematics,
Polytechnic Institute of Brooklyn, New York

Wiley Classics Edition Published in 1989

WILEY

A WILEY-INTERSCIENCE PUBLICATION

JOHN WILEY & SONS, New York • Chichester • Brisbane • Toronto • Singapore

Copyright © 1973, by John Wiley & Sons, Inc.

All rights reserved. Published simultaneously in Canada.

Reproduction or translation of any part of this work beyond
that permitted by Sections 107 or 108 of the 1976 United States
Copyright Act without the permission of the copyright owner
is unlawful. Requests for permission or further information
should be addressed to the Permissions Department, John
Wiley & Sons, Inc.

Library of Congress Cataloging in Publication Data:

ɻochstadt, Harry.
  Integral equations.

  (Pure and applied mathematics)
  "A Wiley-Interscience publication."
  1. Integral equations.   I. Title.

QA431.H73   515'.45   73-4230
ISBN 0-471-40165-X

ISBN 0-471-50404-1 (pbk.)

Printed in the United States of America

10–9 8

PREFACE

After having taught the subject of integral equations for a number of years I concluded that most potential texts are either too abstract, too old-fashioned, or too specialized. The classical techniques yield rather precise information regarding solvability of equations and existence and distribution of eigenvalues, but often require rather tedious and wearisome analysis. The functional analytic techniques yield certain kinds of results quite expeditiously, but often do not lead to the quantitative results provided by the classical techniques. The aim of this book is to compromise between these approaches to develop the most desirable features of each. The functional analytic approach can be developed in a Banach space setting, but to keep the treatment on a simpler level I have decided to restrict myself to Hilbert spaces.

Chapter 1 provides a heuristic discussion of integral equations based on finite difference approximations. It is assumed that the reader has some background in linear algebra, but the more specialized results of that subject that are needed later are developed here. There is also a discussion of the necessary background of Hilbert space theory.

Chapter 2 develops some elementary techniques, in particular the contraction mapping principle and its applications to integral equations.

Chapter 3 develops the theory of compact operators, which is then used to discuss a broad class of integral equations. Another long section is devoted to ordinary differential operators and their study via compact integral operators. The Fredholm alternative is discussed fully.

In Chapter 4 some applications of these techniques to boundary value problems in more than one dimension are discussed.

Chapter 5 is devoted to a complete treatment of numerous transform techniques. The Fourier transform is developed and used to develop the Laplace, Mellin, and Hankel transforms. Subsequently, the projection method is discussed and applied to Wiener-Hopf problems and to certain mixed boundary value problems. Some of the ad hoc methods for these problems are more direct than the method presented here. It is my feeling, however, that it is most desirable to discuss these problems in the framework

v

of a unified and consistent theory. All too often the student is left with the idea that such problems are solved by tricks rather than within the framework of a consistent theory.

Chapter 6 develops the classical Fredholm technique to obtain a number of results that the functional analytic techniques do not yield. Integral operators with positive kernels are also discussed.

Chapter 7 is a distinct departure from the preceding chapters. Almost all the earlier chapters are devoted to linear integral equations. The theory of nonlinear equations is not nearly as well developed as the linear theory. Here the Schauder fixed-point theorem is presented and applied to a number of nonlinear equations. A complete proof of the Schauder theorem is given, based on the Brouwer fixed-point theorem. The latter is proved by analytic rather than topological methods.

Without a doubt there are many topics that might well have been included with profit. Notable examples can be cited of such omissions. One of the most powerful and beautiful applications of the Fredholm alternative can be found in potential theory. Chapter 3 could well have been enriched by the inclusion of this topic. Because this topic is so tempting to an author, I decided to exclude it; it can be found in many other works on integral equations. Systems of integral equations, generalizations from Hilbert to Banach spaces, more general Wiener-Hopf equations, all could have been presented. No book can pretend to be complete and exhaustive. Every author has to decide at what point he is willing to stop for economic and pedagogical reasons. Experience has shown that the following material can be covered in a one-year course. Hopefully, the mixture is such that both applied scientists and mathematicians, to whom I intended to address myself, will find this material useful and interesting for its own sake.

HARRY HOCHSTADT

Department of Mathematics,
Polytechnic Institute of Brooklyn, New York
January 1973

CONTENTS

INTEGRAL EQUATIONS

1

GENERAL
INTRODUCTION

SUMMARY

This chapter is devoted to a discussion of some of the broad classes of integral equations to be discussed in more detail in later chapters. By means of examples and finite difference approximations, some motivation will be given for the differences in theory and methodology underlying these classifications and their investigation.

The necessary background in linear algebra will be sketched and some aspects of Hilbert space theory will be presented. Subsequently, we will view all integral operators to be discussed as operators acting on suitable Hilbert spaces.

1. INTRODUCTION

The theory of integral equations has close contacts with many different areas of mathematics. Foremost among these are differential equations and operator theory. Many problems in the fields of ordinary and partial differential equations can be recast as integral equations. Many existence and uniqueness results can then be derived from the corresponding results from integral equations.

Many problems of mathematical physics can be stated in the form of integral equations. Some of these will be discussed as examples and treated explicitly. To make a list of such applications would be almost impossible. Suffice it to say that there is almost no area of applied mathematics and mathematical physics where integral equations do not play a role.

In many ways one can view the subject of integral equations as an extension of linear algebra and a precursor of modern functional analysis. Especially

1

in dealing with linear integral equations the fundamental concepts of linear vector spaces, eigenvalues and eigenfunctions will play a significant role.

There are a number of classifications of linear integral equations that distinguish different kinds of equations. The following are the most frequently studied.

$$\int_a^b K(x, y)\phi(y)\, dy = f(x), \tag{1}$$

$$\phi(x) - \lambda \int_a^b K(x, y)\phi(y)\, dy = f(x), \tag{2}$$

$$a(x)\phi(x) - \lambda \int_a^b K(x, y)\phi(y)\, dy = f(x). \tag{3}$$

The above equations (1)–(3) are generally known as Fredholm equations of the first, second, and third kind, respectively. The interval (a, b) may in general be a finite interval or $(-\infty, b]$, $[a, \infty)$, or $(-\infty, \infty)$, where a and b are finite. If $a(x)$ does not vanish one can divide (3) by $a(x)$ to reduce it to (2). The functions $f(x)$, $a(x)$, and $K(x, y)$ are presumably known functions and the function $\phi(x)$ is unknown. The parameter λ could be absorbed in the function $K(x, y)$, but it is convenient to retain it in the equation. Its role will become clearer when the operators in question will be studied. The function $K(x, y)$ is generally known as the kernel of the equation.

A second class of equations are the Volterra equations of the first, second, and third kind, namely

$$\int_a^x K(x, y)\phi(y)\, dy = f(x), \tag{4}$$

$$\phi(x) - \lambda \int_a^x K(x, y)\phi(y)\, dy = f(x), \tag{5}$$

$$a(x)\phi(x) - \lambda \int_a^x K(x, y)\phi(y)\, dy = f(x). \tag{6}$$

One can view these as special cases of Fredholm equations. The latter reduce to the corresponding Volterra equations if $K(x, y) = 0$ for $y > x$. Nevertheless Volterra equations have many interesting properties that do not emerge from the general theory of Fredholm equations so that a separate study is definitely warranted.

Equations (1)–(6) have one thing in common; they are all linear equations. That is, the function ϕ enters the equations in a linear manner so that

$$\int_a^b K(x, y)[c_1\phi_1(y) + c_2\phi_2(y)]\, dy$$

$$= c_1 \int_a^b K(x, y)\phi_1(y)\, dy + c_2 \int_a^b K(x, y)\phi_2(y)\, dy.$$

If the integral were replaced by the more general

$$\int_a^b K(x, y, \phi(y))\, dy$$

one would call the equation nonlinear. Typical examples of such operators are

$$\int_a^b K(x, y)\phi^2(y)\, dy,$$

$$\int_a^b K(x, y) \sin \phi(y)\, dy.$$

2. EXAMPLES

Consider the Volterra equation

$$\phi(x) - \lambda \int_0^x y\phi(y)\, dy = f(x). \tag{7}$$

By differentiation, the above can be reduced to a first order differential equation

$$\phi'(x) - \lambda x\phi(x) = f'(x),$$

and from (7) we obtain the initial value

$$\phi(0) = f(0).$$

One can solve this equation by standard methods of differential equations to obtain

$$\phi(x) = f(x) + \lambda \int_0^x e^{\lambda/2(x^2 - y^2)} f(y)\, dy. \tag{8}$$

One can easily check that (8) satisfies the differential equation as well as the integral equation (7). To derive the differential equation we required that both $\phi(x)$ and $f(x)$ had first derivatives. The differentiation was merely an artifice that enabled us to reduce the integral equation to a differential equation, whose solution was obtainable by relatively elementary methods.

As a second example we consider the Fredholm equation

$$\phi(x) - \lambda \int_0^1 xy(y - x)\phi(y)\, dy = f(x). \tag{9}$$

The expressions $\lambda \int_0^1 y^2\phi(y)\, dy$ and $-\lambda \int_0^1 y\phi(y)\, dy$ are unknown constants depending on $\phi(y)$ and we shall denote them by a and b. Then (9) can be rewritten in the form

$$\phi(x) = f(x) + ax + bx^2. \tag{10}$$

The fact that the solution of (9) has the simple character of (10) depends on the special nature of the kernel in (9). To determine a and b we note that by multiplying (10) by λx^2 and integrating

$$\lambda \int_0^1 x^2 \phi(x)\, dx = \lambda \int_0^1 x^2 f(x)\, dx + \lambda \frac{a}{4} + \lambda \frac{b}{5} = a.$$

Similarly,

$$\lambda \int_0^1 x \phi(x)\, dx = \lambda \int_0^1 x f(x)\, dx + \lambda \frac{a}{3} + \lambda \frac{b}{4} = -b.$$

The latter are two linear algebraic equations in two unknowns a and b. These can be solved by standard methods, and inserting the solutions in (10) one finally obtains

$$\phi(x) = f(x)$$

$$+ \lambda \int_0^1 \frac{xy\{-(\lambda/5) + [1 + (\lambda/4)]y - (\lambda/3)xy - [1 - (\lambda/4)]x\} f(y)}{1 + (\lambda^2/240)}\, dy \quad (11)$$

The two examples (7) and (9) and the corresponding solutions (8) and (11) demonstrate one of the big differences between Volterra and Fredholm equations. The solution (8) exists for all finite values of λ, whereas (11) will fail to exist if its denominator $1 + (\lambda^2/240)$ vanishes. In that case (11) will in general fail to have a solution. If, however,

$$\int_0^1 x^2 f(x)\, dx = 0, \qquad \int_0^1 x f(x)\, dx = 0, \qquad 1 + \frac{\lambda^2}{240} = 0$$

(9) will have the solution

$$\phi(x) = f(x) + \frac{\lambda}{5} cx - \left(1 - \frac{\lambda}{4}\right) cx^2 \tag{12}$$

where c is completely arbitrary.

We saw that (7) could be solved by reducing it to a differential equation. In fact, the first order differential equation

$$\phi'(x) = f(x, \phi(x)) \qquad \phi(0) = \phi_0 \tag{13}$$

can be rewritten in the form of a nonlinear Volterra equation, by integrating (13).

$$\phi(x) = \phi_0 + \int_0^x f(y, \phi(y))\, dy. \tag{14}$$

In fact, the proof of the existence and uniqueness of a solution for (13) is generally based on treating the integral equation (14).

Fredholm equations also can be related to certain types of differential equations. Consider the boundary value problem

$$\phi''(x) + \lambda \phi(x) = f(x) \qquad \phi(0) = 0 \qquad \phi(1) = 0. \tag{15}$$

A double integration of (15) shows that

$$\phi(x) = c_1 + c_2 x + \int_0^x (x - y)[f(y) - \lambda\phi(y)]\,dy.$$

From $\phi(0) = 0$ we conclude that $c_1 = 0$. From $\phi(1) = 0$ we see that

$$c_2 = -\int_0^1 (1 - y)[f(y) - \lambda\phi(y)]\,dy.$$

Use of these values of c_1 and c_2 allows us to rewrite problem (15) in the form

$$\phi(x) - \lambda\int_0^1 K(x, y)\phi(y)\,dy = -\int_0^1 K(x, y)f(y)\,dy \qquad (16)$$

where

$$K(x, y) = y(1 - x), \qquad y \le x$$
$$= x(1 - y), \qquad y \ge x.$$

Equation (16) is equivalent to (15) and incorporates the boundary conditions as well.

Volterra equations of the first kind are in some ways more difficult than equations of the second kind. For example, the apparently simple equation

$$\int_0^x y\phi(y)\,dy = f(x) \qquad (17)$$

has the solution

$$\phi(x) = \frac{f'(x)}{x}. \qquad (18)$$

But this solution can make sense only if $f(x)$ satisfies certain regularity conditions. From (17) we see that $f(0) = 0$. Furthermore $f(x)$ must be differentiable for (18) to make sense. In (7) we used the differentiability of $f(x)$ merely as an artifice that could be dispensed with in (8). In (18) the differentiability is vital.

3. FINITE DIFFERENCE APPROXIMATIONS

Finding explicit solutions of integral equations is in general just as difficult as finding solutions of differential equations. Only in exceptional cases can such solutions be found. Generally, various approximate and numerical methods have to be used. Finite difference approximations are not only of great practical utility, but also provide certain insight into the nature of integral equations.

In the equation

$$\phi(x) - \lambda\int_0^1 K(x, y)\phi(y)\,dy = f(x) \qquad (19)$$

we shall replace the integral by a suitable sum:

$$\phi(x) - \lambda \sum_{i=1}^{n} \frac{1}{n} K\left(x, \frac{i}{n}\right) \phi\left(\frac{i}{n}\right) = f(x). \tag{20}$$

For large n and a continuous kernel $K(x, y)$ and continuous $\phi(x)$ the sum in (20) represents a close approximation to the integral in (19). If, furthermore, we evaluate (20) only at n discrete points

$$\phi\left(\frac{j}{n}\right) - \lambda \sum_{i=1}^{n} \frac{1}{n} K\left(\frac{j}{n}, \frac{i}{n}\right) \phi\left(\frac{i}{n}\right) = f\left(\frac{j}{n}\right), \qquad j = 1, 2, \ldots, n, \tag{21}$$

we have replaced the integral equation (19) by the algebraic system (21). We shall rewrite (21) in matrix form

$$(I - \lambda L)\Phi = F \tag{22}$$

where the general term in the matrix L is $(1/n)K(j/n, i/n)$, and Φ is a vector with components $\phi(i/n)$ and F has components $f(i/n)$.

To solve (22) we invert the matrix and find

$$\Phi = (I - \lambda L)^{-1}F. \tag{23}$$

The above inverse will exist for all λ, with the exception of at most n values. These are the roots of the characteristic determinantal equation

$$|I - \lambda L| = 0. \tag{24}$$

In (23) as in (11) we see that there may be special values of λ for which no solution exists. Such values are commonly known as eigenvalues. Every finite algebraic system (21) necessarily has eigenvalues, even though the integral equation (19) need not have eigenvalues. This is a subtle point that we shall bypass for the time being.

For the case where (19) is a Volterra equation $K(x, y) = 0$ for $y \geq x$. In that case (21) can be rewritten as

$$\phi\left(\frac{j}{n}\right) - \lambda \sum_{i=1}^{j-1} \frac{1}{n} K\left(\frac{j}{n}, \frac{i}{n}\right) \phi\left(\frac{i}{n}\right) = f\left(\frac{j}{n}\right), \qquad j = 1, 2, \ldots, n. \tag{25}$$

In this case L in (22) is in fact a triangular matrix with all diagonal elements equal to zero:

$$L = \begin{pmatrix} 0 & 0 & 0 & \cdots & 0 & 0 \\ l_{21} & 0 & 0 & \cdots & 0 & 0 \\ l_{31} & l_{32} & 0 & \cdots & 0 & 0 \\ \cdot & \cdot & \cdot & & \cdot & \cdot \\ \cdot & \cdot & \cdot & & \cdot & \cdot \\ \cdot & \cdot & \cdot & & \cdot & \cdot \\ l_{n1} & l_{n2} & l_{n3} & \cdots & l_{n,n-1} & 0 \end{pmatrix} \tag{26}$$

Inspection of (26) shows that L is a nilpotent matrix and in particular $L^n = 0$. This fact enables us to find the inverse of $I - \lambda L$ for all λ. A simple calculation shows that

$$(I - \lambda L)(I + \lambda L + \lambda^2 L^2 + \cdots + \lambda^{n-1} L^{n-1}) = I \qquad (27)$$

so that (23) can be replaced by

$$\Phi = (I + \lambda L + \lambda^2 L^2 + \cdots + \lambda^{n-1} L^{n-1}) F. \qquad (28)$$

As in (8), (28) is another heuristic demonstration that Volterra equations of the second kind will have unique solutions for all λ.

Equations of the first kind are in general more complicated. If we study the finite difference analog of a Fredholm equation of the first kind we are led to a system of the type

$$L\Phi = F. \qquad (29)$$

Now if $|L| \neq 0$, L has an inverse and (29) has a unique solution. If $|L| = 0$, (29) may or may not have a solution. But even if a solution exists it will not be unique.

For the case where (29) represents a Volterra equation, L is nilpotent, as in (26) and necessarily $|L| = 0$, so that the above remarks apply.

4. THE FREDHOLM ALTERNATIVE

The Fredholm alternative is a fundamental tool in reaching decisions regarding the solvability of certain types of integral equations. In this section the analogous theory for finite dimensional spaces will be sketched. Let X be a finite dimensional inner product space of dimension n over the field of complex numbers and L an operator on X that is represented by an $n \times n$ matrix. With L we can associate the adjoint matrix L^*, which is sometimes referred to as the hermitean transpose of L.

Let (ϕ, ψ) denote the inner product defined on X, where ϕ and ψ are arbitrary vectors in X. L and L^* are related by the expression

$$(L\phi, \psi) = (\phi, L^*\psi).$$

The above can in fact be used as the defining relationship for L^*.

By $N(L)$ we denote the nullspace of L, that is

$$N(L) = \{\phi \mid L\phi = 0\}.$$

It is easy to verify that $N(L)$ is indeed a subspace of X. By $\nu(L)$ we shall denote the dimension of $N(L)$. By $R(L)$ we shall denote the range of L, so that

$$R(L) = \{\phi \mid \phi = L\psi \text{ for some } \psi \in X\}$$

Again we can see that $R(L)$ is a subspace and its dimension is denoted by $\rho(L)$.

LEMMA 1. $\nu(L) + \rho(L) \leq n = \dim X$.

PROOF. Let $\phi_1, \phi_2, \ldots, \phi_n$ be a basis for X such that $\phi_1, \phi_2, \ldots, \phi_\nu$ is a basis for $N(L)$. Then for any $\phi \in X$ we have

$$\phi = \sum_{i=1}^{n} \alpha_i \phi_i$$

for a suitable set of scalars α_i. It follows that

$$L\phi = \sum_{i=\nu+1}^{n} \alpha_i L\phi_i$$

since $L\phi_i = 0$ for $1 \leq i \leq \nu(L)$. Since $L\phi$ is a general element in $R(L)$ and since it is a linear combination of at most $n - \nu(L)$ linearly independent vectors we see that

$$\rho(L) \leq n - \nu(L)$$

thereby proving the lemma. ■

It may be remarked that the above inequality is in fact an equality. This will be proved shortly.

If R is a subspace of X we note the set of all vectors in X that are orthogonal to R by $R^{\perp}$. $R^{\perp}$ can be shown to be a subspace of X and is called the orthogonal complement of R.

LEMMA 2. $\nu(L) + \rho(L^*) = n$.

PROOF. By $R(L^*)^{\perp}$ we denote the orthogonal complement of $R(L^*)$. We shall first prove that $N(L) \subset R(L^*)^{\perp}$. Let $\phi \in N(L)$ so that

$$0 = (L\phi, \psi) = (\phi, L^*\psi) \qquad \text{for all } \psi.$$

It follows that ϕ is orthogonal to all $L^*\psi$ so that

$$\phi \in R(L^*)^{\perp}.$$

Since ϕ is an arbitrary element of $N(L)$ we see that

$$N(L) \subset R(L^*)^{\perp}.$$

We now prove the reverse inclusion. Let $\phi \in R(L^*)^{\perp}$ so that

$$0 = (\phi, L^*\psi) = (L\phi, \psi) \qquad \text{for all } \psi.$$

From the above we see that necessarily $\phi \in N(L)$ so that

$$R(L^*)^{\perp} \subset N(L).$$

We can conclude that $N(L) = R(L^*)^\perp$. From this we conclude that $N(L)$ and $R(L^*)$ form complementary subspaces of X so that

$$X = N(L) \oplus R(L^*).$$

That is, every element of X can be uniquely decomposed into an element in $N(L)$ and an element of $R(L^*)$. It follows that

$$\nu(L) + \rho(L^*) = n. \quad \blacksquare$$

LEMMA 3. $\nu(L) = \nu(L^*)$.

PROOF. On a finite dimensional space the double adjoint $L^{**} = L$. By lemmas 1 and 2 we have

$$\nu(L^*) \leq n - \rho(L^*) = \nu(L).$$

Now we see that

$$\nu(L) = \nu(L^{**}) \leq \nu(L^*) \leq \nu(L)$$

so that $\nu(L) = \nu(L^*)$. $\quad \blacksquare$

COROLLARY. $\nu(L) + \rho(L) = n$.

PROOF. By lemmas 2 and 3

$$n = \nu(L^*) + \rho(L) = \nu(L) + \rho(L). \quad \blacksquare$$

The above shows that the inequality in lemma 1 is in fact an equality. However, $N(L)$ and $R(L)$ are not, in general, complementary subspaces as are $N(L)$ and $R(L^*)$. In the case where L is selfadjoint so that $L = L^*$ they will be, of course.

THEOREM 1. The Fredholm Alternative. The equation

$$L\phi = f \tag{30}$$

has a unique solution for all f, if and only if $\nu(L) = 0$. If $\nu(L) > 0$ the above equation has solutions only for those f, that are orthogonal to the nullspace $N(L^*)$.

PROOF. If $\nu(L) = 0$ we see from the above corollary that $\rho(L) = n$ so that the range $R(L)$ is the entire space. In that case, f necessarily belongs to $R(L)$ and can be represented by $L\phi$ for some ϕ. If there were two such solutions ϕ_1 and ϕ_2, then $L(\phi_1 - \phi_2) = 0$. But since $\nu(L) = 0$, we see that $\phi_1 = \phi_2$.

If $\nu(L) > 0$, then by lemma 2, (30) has a solution only for those $f \in R(L) = N^\perp(L^*)$. Therefore solutions exist only for those f orthogonal to $N(L^*)$. $\quad \blacksquare$

5. HADAMARD'S INEQUALITY

A familiar result from analytic geometry yields the volume of a parallelepiped in terms of a determinant. Let (a_1, a_2, a_3), (b_1, b_2, b_3), (c_1, c_2, c_3) denote the three vectors defining three edges of such a parallelepiped. Then the absolute value of the determinant

$$\begin{vmatrix} a_1 & a_2 & a_3 \\ b_1 & b_2 & b_3 \\ c_1 & c_2 & c_3 \end{vmatrix}$$

furnishes the volume V of this parallelepiped. A simple upper estimate for such a volume is given by the following inequality:

$$V \le \sqrt{(a_1{}^2 + a_2{}^2 + a_3{}^2)(b_1{}^2 + b_2{}^2 + b_3{}^2)(c_1{}^2 + c_2{}^2 + c_3{}^2)}. \tag{31}$$

Equality is attained if the three edges are mutually perpendicular. Under all other conditions (31) is a strict inequality.

THEOREM 2.   Hadamard's Inequality.   Let L be a matrix with the general element l_{ij}. An upper estimate for its determinant is given by

$$|L|^2 \le \prod_{i=1}^{n} \sum_{j=1}^{n} |l_{ij}|^2. \tag{32}$$

Equation (32) is a generalization of (31) in two ways. First of all it is a statement that applies to an n dimensional space. Here our geometric intuition no longer suffices. Secondly, we allow the terms l_{ij} to be complex numbers whereas in (31) we dealt only with real numbers. In order to prove (32) we require some preliminary results.

THEOREM 3.   On the Arithmetic-Geometric Mean.   Let $d_1, d_2, \ldots, d_n$ be a set of n non-negative numbers. Then

$$\frac{1}{n} \sum_{i=1}^{n} d_i \ge \prod_{i=1}^{n} d_i^{1/n}. \tag{33}$$

PROOF.   LET A denote the arithmetic mean $(1/n) \sum_{i=1}^{n} d_i$. If all d_i are equal the above inequality reduces to an equality. More generally, let d_M and d_m denote the greatest and smallest of the d_i respectively. Define a and b by

$$d_M = A + a, \qquad d_m = A - b.$$

We now replace d_M by A and d_m by $A + a - b$. This substitution has evidently no effect on the arithmetic mean. But in view of the fact that

$$d_M d_m = A(A + a - b) - ab \le A(A + a - b)$$

this substitution will increase the geometric mean $\prod_{i=1}^{n} d_i^{1/n}$. If a substitution of the above type is performed repeatedly, and each time the greatest of the d_i is replaced by A, we see that after at most $n - 1$ steps all d_i have been replaced by A. Each such step leaves the arithmetic mean unchanged, but increases the geometric mean. After $n - 1$ steps all d_i have been replaced by A so that the geometric mean and arithmetic mean are equal. It follows that in general (33) must be a strict inequality, unless all d_i are equal. ■

An $n \times n$ matrix L is said to be positive, if $(L\phi, \phi) \geq 0$, for all vectors ϕ in the inner product space X. For such matrices the following theorem yields an estimate that will enable us to prove the Hadamard inequality.

THEOREM 4. Let L be a selfadjoint, positive matrix. Then

$$|L| \leq \prod_{i=1}^{n} l_{ii}. \tag{34}$$

PROOF. The positivity of L implies that $(L\Phi, \Phi) \geq 0$ for all Φ. If $|L| = 0$ (34) is trivial. If $|L| \neq 0$, $(L\Phi, \Phi) > 0$ for all $\Phi \neq 0$. Now let Φ_i be the ith canonical unit vector, namely $\Phi_i = (\delta_{1i}, \delta_{2i}, \ldots, \delta_{ni})$. Here δ_{ij} denotes the Kronecker δ defined by

$$\delta_{ij} = 0 \qquad i \neq j$$
$$= 1 \qquad i = j.$$

Then

$$(L\Phi_i, \Phi_i) = l_{ii} > 0.$$

Let D denote the diagonal matrix with diagonal terms $1/\sqrt{l_{11}}, 1/\sqrt{l_{22}}, \ldots,$ $1/\sqrt{l_{nn}}$. Then we consider the matrix $T = DLD$. First of all T is selfadjoint since

$$T^* = D^*L^*D^* = DLD = T.$$

It is also positive, since

$$(T\Phi, \Phi) = (DLD\Phi, \Phi) = (LD\Phi, D\Phi) \geq 0.$$

It is also evident that the diagonal terms of T are all equal to 1. We also see that

$$|T| = |D|^2 |L| = |L| \prod_{i=1}^{n} \frac{1}{l_{ii}}. \tag{35}$$

We shall show that $|T| \leq 1$, so that (35) yields the desired result (34).

To establish the fact that $|T| \leq 1$, we denote the eigenvalues of T by λ_i. It follows that

$$\text{trace } T = n = \sum_{i=1}^{n} \lambda_i.$$

$$|T| = \prod_{i=1}^{n} \lambda_i.$$

We now have, by theorem 3

$$1 = \frac{1}{n}\sum_{i=1}^{n} \lambda_i \geq \left[\prod_{i=1}^{n} \lambda_i\right]^{1/n} = |T|^{1/n}$$

so that $|T| \leq 1$, which implies (34). ■

PROOF OF HADAMARD'S INEQUALITY.   Let L be a general matrix. Then LL^* is clearly selfadjoint and positive, since

$$(LL^*\Phi, \Phi) = (L^*\Phi, L^*\Phi) = \|L^*\Phi\|^2 \geq 0.$$

The general term of LL^* is $\sum_{k=1}^{n} l_{ik}\bar{l}_{jk}$ and an application of (34) to LL^* yields

$$|LL^*| = |L|^2 \leq \prod_{i=1}^{n}\sum_{j=1}^{n}|l_{ij}|^2. ■$$

COROLLARY.   Let $L = (l_{ij})$ and $|l_{ij}| \leq l$. Then

$$|L| \leq l^n n^{n/2}.$$

PROOF.   Use of (34) shows that

$$|L|^2 \leq \prod_{i=1}^{n} nl^2 = n^n l^{2n}.$$

Taking square roots yields the result. ■

6. HILBERT SPACES

So far we have not really addressed ourselves to the question of what we mean by a solution of an integral equation. Basically, of course, a solution of any equation must reduce the equation to an identity. But often one may impose additional restrictions on the solution, such as demanding that it should belong to a particular class of functions. For these purposes it will prove to be convenient to work in so-called Hilbert spaces.

DEFINITION.   A linear space over a field F (in our work F will invariably be the field of complex numbers) is a collection X of elements with two defined operations. The first of these is addition of elements in X and the second multiplication of elements in X by scalars in F. In addition we stipulate the following conditions:

1. X forms a commutative group under the additive operation. That is, if $f, g, h, \ldots$ belong to X, then

(a) the operation is closed so that $f + g$ belongs to X for all f and g in X.
(b) the operation is associative: $(f + g) + h = f + (g + h)$.

(c) there exists an identity 0 for which $f + 0 = f$ for all f in X.
(d) for every f there exists an inverse element denoted by $(-f)$ such that

$$f + (-f) = 0.$$

(e) the operation is commutative

$$f + g = g + f \quad \text{for all } f, g \text{ in } X.$$

2. Multiplication by scalars is closed. That is,

(a) $1 \cdot f = f$ for all f in X,
(b) αf is in X for all f in X and α in F.
(c) for all α, β in F and f in X

$$\alpha(\beta f) = (\alpha\beta)f.$$

3. The following distributive laws hold:

(a) $\alpha(f + g) = \alpha f + \alpha g$ for all α in F and f, g in X.
(b) $(\alpha + \beta)f = \alpha f + \beta f$ for all α, β in F and f in X.

EXAMPLE 1. Consider the set of all vectors of the form $f = (a_1, a_2, \ldots, a_n, \ldots)$ where all a_i are complex numbers, and only a finite number of a_i do not vanish. If $g = (b_1, b_2, \ldots, b_n, \ldots)$ is a second such vector we define

$$\alpha f = (\alpha a_1, \alpha a_2, \ldots, \alpha a_n, \ldots)$$

where α is any complex number, and

$$f + g = (a_1 + b_1, a_2 + b_2, \ldots, a_n + b_n, \ldots).$$

It is a simple matter to verify that the above set of vectors forms a linear space.

EXAMPLE 2. Consider the set of all continuous functions $f(x)$ defined on the closed interval $[0, 1]$, that take complex values. We denote this set by $C[0, 1]$. For any complex scalars $(\alpha f)(x) = \alpha \cdot f(x)$ in $C[0, 1]$ if $f(x)$ is in $C[0, 1]$. Addition is defined pointwise, that is

$$f + g = f(x) + g(x).$$

Again, one can verify that these form a linear space.

The linear spaces do not have enough structure to be useful in the area of integral equations. But the inner product spaces prove to be much more useful.

DEFINITION. A linear space X is said to be an inner product space if an inner product is defined on it. Such an inner product assigns to every pair

f and g in X a complex number denoted by (f, g). By definition, such an inner product has the following properties:

1. $(f, g) = \overline{(g, f)}$
2. $(\alpha f + \beta g, h) = \alpha(f, h) + \beta(g, h)$
3. $(f, f) \geq 0$ and $(f, f) = 0$ if and only if $f = 0$.

N.B. (f, f) is real, since by property 1 $(f, f) = \overline{(f, f)}$. We let $(f, f)^{\frac{1}{2}} = \|f\|$ and call it the norm of f.

THEOREM 5.   Cauchy-Schwarz Inequality.   Let f and g belong to an inner product space. Then

$$|(f, g)| \leq \|f\| \, \|g\| \tag{35}$$

Equality is achieved if and only if f and g are linearly dependent; that is for suitable scalars α and β, $\alpha f + \beta g = 0$.

PROOF.   We assume that $f \neq 0$, because if $f = 0$, (35) is trivially true. Let

$$(f, g) = |(f, g)| \, e^{i\theta}, \qquad \alpha = \frac{|(f, g)| \, e^{-i\theta}}{\|f\|^2}$$

so that

$$(\alpha f - g, \alpha f - g) \|f\|^2 = \|\alpha f - g\|^2 \|f\|^2 \geq 0.$$

Using the properties of inner products we can expand the left side of this inequality to obtain

$$[|\alpha|^2 \|f\|^2 - \alpha(f, g) - \overline{\alpha(f, g)} + \|g\|^2] \|f\|^2 \geq 0.$$

So far no use has been made of the particular value of α. If that value is inserted in the above inequality, we obtain

$$\|g\|^2 \|f\|^2 - |(f, g)|^2 \geq 0$$

which is equivalent to (35). To complete the proof we note that if f and g are linearly independent

$$\|\alpha f - g\| > 0$$

so that we get a strict inequality. Equality will occur if $g = \alpha f$ for a suitable α.   ∎

THEOREM 6.   The norm $\|f\|$ has the following three properties:

1. $\|f\| \geqslant 0$ and $\|f\| = 0$ if and only if $f = 0$
2. $\|\alpha f\| = |\alpha| \, \|f\|$
3. $\|f + g\| \leq \|f\| + \|g\|$

The last property is known as the triangle inequality.

PROOF.   Equations (1) and (2) are immediate consequences of the definition of an inner product. To prove (3) we use the Cauchy–Schwarz inequality in

$$\|f + g\|^2 = (f + g, f + g) = \|f\|^2 + (f, g) + \overline{(f, g)} + \|g\|^2$$
$$\leq \|f\|^2 + 2\,|(f, g)| + \|g\|^2 \leq \|f\|^2 + 2\,\|f\|\,\|g\| + \|g\|^2$$
$$\leq [\|f\| + \|g\|]^2.$$

By taking square roots the result follows. One can also see that (3) reduces to an equality only if f and g are linearly dependent.   ■

EXAMPLE 3.   Consider the linear space X of Example 1. Let

$$(f, g) = \sum_{i=1}^{\infty} a_i \bar{b}_i.$$

It is easy to verify that the above forms an inner product. Convergence is really no problem, since only a finite number of terms in the above sum do not vanish.

EXAMPLE 4.   Consider the linear space of Example 2. One can define an inner product on this space by

$$(f, g) = \int_0^1 f(x)\overline{g(x)}\, dx.$$

DEFINITION.   Let H be an inner product space and $\{f_n\}$ a Cauchy sequence in H. Such a sequence has the property that for every $\epsilon > 0$ we can find an $N(\epsilon)$ such that

$$\|f_n - f_m\| < \epsilon \qquad \text{for} \quad n, m > N(\epsilon).$$

In other words

$$\lim_{n, m \to \infty} \|f_n - f_m\| = 0.$$

H is said to be a Hilbert space if every Cauchy sequence converges to an element in H.

It is easy to see that if such a Cauchy sequence converges it must converge to a unique element. Suppose

$$\lim_{n_k \to \infty} f_{n_k} = g$$

$$\lim_{m_k \to \infty} f_{m_k} = h$$

that is, we have two subsequences of $\{f_n\}$ such that each converges to a different element. Then

$$\|g - h\| = \|g - f_{n_k} + f_{n_k} - f_{m_k} + f_{m_k} - h\|$$
$$\leq \|g - f_{n_k}\| + \|f_{n_k} - f_{m_k}\| + \|f_{m_k} - h\|.$$

For n_k and m_k sufficiently large we have

$$\|g - f_{n_k}\| < \epsilon$$
$$\|f_{n_k} - f_{m_k}\| < \epsilon$$
$$\|f_{m_k} - h\| < \epsilon$$

so that

$$\|g - h\| < 3\epsilon.$$

Since g and h are independent of n_k and m_k and ϵ is arbitrary, we see that necessarily

$$\|g - h\| = 0$$

so that $g = h$.

In general an inner product space need not be a Hilbert space. For example, the space treated in Example 3 is not a Hilbert space. To see this we consider the following Cauchy sequence $\{f_n\}$ where

$$f_n = \left(1, \frac{1}{2}, \frac{1}{3}, \ldots, \frac{1}{n}, 0, 0, \ldots\right).$$

For $n > m$ we have

$$\|f_n - f_m\| = \left[\sum_{k=m+1}^{n} \frac{1}{k^2}\right]^{\frac{1}{2}}$$

and it follows that

$$\lim_{n, m \to \infty} \|f_n - f_m\| = 0.$$

Nevertheless, the sequence does not converge to an element in the space. Clearly, in a purely formal manner,

$$\lim_{n \to \infty} f_n = (1, \tfrac{1}{2}, \tfrac{1}{3}, \ldots) \tag{36}$$

and none of the components of the above vanish. Since in the space in question only a finite number of components were allowed not to vanish (36) is not in the space.

We can, however, enlarge this space by adding to it all limits of Cauchy sequences. For example, we consider the space H consisting of all $f = (a_1, a_2, a_3, \ldots)$ such that $\sum_{n=1}^{\infty} |a_n|^2 < \infty$. Clearly, if $f_n = (a_1, a_2, \ldots, a_n, 0, 0, \ldots)$ is an element of the space X in Example 3 then

$$\lim_{n \to \infty} f_n = f.$$

Also, the set $\{f_n\}$ forms a Cauchy sequence, since

$$\|f_n - f_m\| = \left[\sum_{k=m+1}^{n} |a_k|^2\right]^{\frac{1}{2}} \quad \text{for} \quad n > m$$

and

$$\lim_{n, m \to \infty} \|f_n - f_m\| = 0.$$

In this fashion we see that X can be completed to a Hilbert space H.

A standard theorem in functional analysis guarantees that every inner product space X can be completed to form a Hilbert space H. Such a Hilbert space H is said to be the completion of X.

We now turn to the space $C[0, 1]$ of Example 4. Here again we can verify that we are not dealing with a Hilbert space. Consider the sequence of functions defined by

$$f_n(x) = 1, \qquad\qquad 0 \leq x \leq \tfrac{1}{2}$$

$$= 1 - 2n(x - \tfrac{1}{2}), \quad \tfrac{1}{2} \leq x \leq \frac{1 + n}{2n}$$

$$= 0, \qquad\qquad \frac{1 + n}{2n} \leq x \leq 1$$

$$n = 1, 2, 3, \ldots.$$

Clearly, each $f_n(x)$ is continuous and a simple calculation shows that

$$\|f_n - f_m\| \leq \left[\frac{1}{n} + \frac{1}{m}\right]^{1/2}.$$

The upper estimate on the right is excessively large, but it does show that the sequence is a Cauchy sequence. Nevertheless

$$\lim_{n \to \infty} f_n(x) = 1, \qquad 0 \leq x \leq \tfrac{1}{2}.$$

$$= 0, \qquad \tfrac{1}{2} < x \leq 1.$$

Clearly $f_n(x)$ has a limit, but the limiting function is not in the space $C[0, 1]$. The limit function is no longer continuous.

In accordance with the general theorem this space has a completion denoted by $L_2[0, 1]$. It consists of all functions $f(x)$ for which we can find Cauchy sequences in $C[0, 1]$, $\{f_n(x)\}$ such that

$$\lim_{n \to \infty} \int_0^1 |f(x) - f_n(x)|^2 \, dx = 0.$$

Such functions will not be continuous in general and in fact will not even be integrable in the Riemann sense. To define their integrals over any interval in $[0, 1]$, say $[a, b]$; we can, however, proceed as follows. By definition

$$\int_a^b f(x) \, dx = \lim_{n \to \infty} \int_a^b f_n(x) \, dx.$$

That the limit on the right exists follows from the fact that

$$\left| \int_a^b f_n(x)\, dx - \int_a^b f_m(x)\, dx \right| \leq \int_a^b |f_n(x) - f_m(x)|\, dx$$

$$\leq \int_0^1 |f_n(x) - f_m(x)|\, dx$$

$$\leq \left[\int_0^1 |f_n(x) - f_m(x)|^2\, dx \right]^{\frac{1}{2}}$$

$$= \|f_n - f_m\|$$

by the Cauchy–Schwarz inequality. It follows that the sequence $\{\int_a^b f_n(x)\, dx\}$ is a Cauchy sequence in the complex plane.

A word of caution is in order here. To ask for the $\lim f_n(x)$ makes no sense. The $\lim \int_a^b f_n(x)\, dx$ is well defined however for all intervals $[a, b]$ in the domain of definition $[0, 1]$. We cannot conclude that the completion process assigns to every Cauchy sequence of continuous functions a unique function. But it does assign unique integrals. The difference of two different limits, say $f(x)$ and $g(x)$ must be such that

$$\int_0^1 |f(x) - g(x)|^2\, dx = 0.$$

But this does not imply that $f(x) = g(x)$. For example, consider the function

$$h(x) = 0, \qquad x \neq \tfrac{1}{2}$$
$$= 1, \qquad x = \tfrac{1}{2}.$$

Clearly

$$\int_0^1 |h(x)|^2\, dx = 0$$

but $h(x) \neq 0$. Similarly, for the function

$$k(x) = 0, \qquad x \text{ irrational}$$
$$= 1, \qquad x \text{ rational}$$

$$\int_0^1 |k(x)|^2\, dx = 0$$

but $k(x) \neq 0$. A function with this property is known as a null function. If two functions $f(x)$ and $g(x)$ differ by a null function then

$$\int_a^b f(x)\, dx = \int_a^b g(x)\, dx \tag{37}$$

for all $0 \leq a \leq b \leq 1$. Our completion process will give us not functions, but equivalence classes of functions that differ by a null function. Any two functions $f(x)$ and $g(x)$ satisfying (37) for all a, b will be considered identical in H. Any two functions that differ by a null function will be said to be equal almost everywhere.

A similar argument can be used to discuss the space $L_2(-\infty, \infty)$. We start with the space $C_0[-\infty, \infty]$. By the latter we denote the set of all continuous complex valued functions, that vanish outside some finite interval. This interval will, of course, depend on the function. The space becomes a linear vector space under the ordinary addition of functions and multiplication by complex scalars. We can introduce the following inner product

$$(f, g) = \int_{-\infty}^{\infty} f(x)\overline{g(x)} \, dx,$$

and thus obtain an inner product space. There is no problem regarding the convergence of the above integral, since $f(x)$ and $g(x)$ vanish for sufficiently large $|x|$. The space is certainly not complete. Consider the following sequence.

$$f_n(x) = e^{-|x|} - e^{-n}, \qquad |x| \leq n$$
$$= 0, \qquad\qquad |x| > n.$$

Obviously $f_n(x)$ is continuous and vanishes outside some finite interval. Furthermore for $n > m$

$$\|f_n(x) - f_m(x)\|^2 = \int_{-\infty}^{\infty} |f_n(x) - f_m(x)|^2 \, dx \leq \int_{-n}^{n} |e^{-n} - e^{-m}|^2 \, dx \leq 8ne^{-2m}$$

and it follows that we have a Cauchy sequence. But

$$\lim_{n \to \infty} f_n(x) = e^{-|x|}$$

which is not in $C_0[-\infty, \infty]$. However, we form the completion of that space and thus obtain a Hilbert space denoted by $L_2[-\infty, \infty]$.

In the study of integral equations, the notion of an operator and specifically an integral operator is fundamental. Such an operator assigns to an element f in H a new element, say Kf in H. If the operator satisfies the condition

$$K(\alpha f + \beta g) = \alpha Kf + \beta Kg \tag{38}$$

we say that K is a linear operator. An example of a linear operator is

$$Kf = \int_0^1 K(x, y)f(y) \, dy,$$

where $f \in L_2[0, 1]$ and $K(x, y)$ is continuous in x and y. Such an operator may or may not be defined on the whole space H.

EXAMPLE. Consider the space $L_2[0, 1]$ and the kernel $K(x, y) = \sin xy$. If $f(x)$ is an element in the space, then

$$Kf = \int_0^1 \sin (xy) f(y) \, dy \tag{39}$$

is also in the space since

$$\|Kf\| = \left[\int_0^1 \left|\int_0^1 \sin xy f(y) \, dy\right|^2 dx\right]^{\frac{1}{2}}$$

$$\leq \left[\int_0^1 \int_0^1 \sin^2 (xy) \, dy \int_0^1 |f(y)|^2 \, dy \, dx\right]^{\frac{1}{2}}$$

$$\leq \|f\|. \tag{40}$$

The Cauchy–Schwarz inequality was used in the above. It follows that for f in $L_2[0, 1]$, Kf is also in $L_2[0, 1]$.

EXAMPLE. Consider the operator

$$Kf = \int_0^1 \frac{\sin (xy)}{y^2} f(y) \, dy. \tag{41}$$

For an arbitrary element $f(y)$ in $L_2[0, 1]$ (41) may not be defined. It will certainly be well defined for all $f(y)$ such that

$$\int_0^1 \left|\frac{f(y)}{y}\right|^2 dy < \infty. \tag{42}$$

In the case of (41) K is not defined on all of $L_2[0, 1]$, but on a subset of $L_2[0, 1]$. This subset is said to be the domain of K. For linear operators the domain will be a subspace of the Hilbert space.

Operators that do not satisfy (38) are said to be nonlinear. A general nonlinear integral operator is of the form

$$Kf = \int_0^1 K(x, y, f(y)) \, dy.$$

Examples of such operators are

$$Kf = \int_0^1 K(x, y) f^2(y) \, dy$$

$$Kf = \int_0^1 \sin (xf(y)) \, dy.$$

An important class of operators are the so-called bounded operators.

DEFINITION. An operator K is said to be bounded if for some constant M we have

$$\|Kf\| \leq M \|f\| \tag{43}$$

for all f in a Hilbert space H. The greatest lower bound of all M for which (43) holds is called the norm of K and denoted by $\|K\|$. Another way of defining $\|K\|$ is by

$$\|K\| = \sup_{f \neq 0} \frac{\|Kf\|}{\|f\|}. \tag{44}$$

Note that if $f(x)$ is in $L_2[0, 1]$ and K is an operator, then $\|f\|$ and $\|K\|$ are different expressions. Yet there should be no confusion in their usage, since within a given context it will always be clear when we are dealing with a function or an operator. Equation (39) defines a bounded operator and from (40) it follows that $\|K\| \leq 1$. One can show that (41) defines an unbounded operator.

THEOREM 7. If K_1 and K_2 are two bounded operators, then their product is also bounded.

PROOF. Let $\|K_1 f\| \leq M_1 \|f\|$
$\|K_2 f\| \leq M_2 \|f\|$.
Then

$$\|K_1 K_2 f\| \leq M_1 \|K_2 f\| \leq M_1 M_2 \|f\|. \quad \blacksquare$$

In general, it is not a simple matter to decide when a given integral operator is bounded. We shall describe two sufficient conditions for the boundedness of such an operator.

THEOREM 8. Consider $L_2[a, b]$, where $[a, b]$ is a finite interval. Suppose $K(x, y)$ is continuous for all x, y in $[a, b]$. Then the operator

$$Kf = \int_a^b K(x, y) f(y) \, dy$$

is bounded.

PROOF. Since $K(x, y)$ is continuous on a closed and bounded set, it must be bounded. Then, if $|K(x, y)| \leq M$

$$|Kf| \leq M \int_a^b |f(y)| \, dy \leq M \left[(b - a) \int_a^b |f(y)|^2 \, dy \right]^{1/2} = M(b - a)^{1/2} \|f\|$$

and it follows that

$$\|Kf\| = \left[\int_a^b |Kf|^2 \, dx \right]^{1/2} \leq M(b - a) \|f\|.$$

It is now evident that

$$\|K\| \leq M(b - a). \quad \blacksquare$$

THEOREM 9.   Consider $L_2[a, b]$, where the interval $[a, b]$ may be infinite. If

$$\int_a^b \int_a^b |K(x, y)|^2 \, dx \, dy = M^2 < \infty$$

the operator

$$Kf = \int_a^b K(x, y)f(y) \, dy$$

is bounded.

PROOF.   By the Cauchy–Schwarz inequality

$$|Kf| \leq \left[\int_a^b |K(x, y)|^2 \, dy \int_a^b |f(y)|^2 \, dy \right]^{\frac{1}{2}}.$$

Then

$$\|Kf\| \leq \left[\int_a^b \int_a^b |K(x, y)|^2 \, dx \, dy \, \|f\|^2 \right]^{\frac{1}{2}} = M \, \|f\|. \quad \blacksquare$$

For finite intervals the second theorem includes the first as a special case. For an infinite interval the conditions of the first theorem no longer guarantee that K is bounded.

Integral operators, just like matrices on finite dimensional spaces, have adjoints. The fundamental property of such an adjoint is described by the following relationship.

$$(Kf, g) = (f, K^*g) \tag{45}$$

for all f, g in the space. Now

$$(Kf, g) = \int_a^b \int_a^b K(x, y)f(y) \, dy \overline{g(x)} \, dx,$$

and by changing the order of integration we find

$$(Kf, g) = \int_a^b f(y) \int_a^b \overline{K(x, y)} g(x) \, dx \, dy.$$

If we define

$$K^*g = \int_a^b \overline{K(x, y)} g(x) \, dx \tag{46}$$

we see that (45) is satisfied. Equation (46) therefore shows that the adjoint of an integral operator is again an integral operator, whose kernel is simply related to the kernel of K.

Evidently, when

$$K(x, y) = \overline{K(y, x)} \tag{47}$$

we have $K = K^*$, or equivalently

$$(Kf, g) = (f, Kg)$$

and in analogy to the terminology for matrices with the above property we shall call an integral operator with this property selfadjoint. A bounded selfadjoint integral operator has a kernel that satisfies (47).

EXERCISES

1. Verify in detail the steps leading from (15) to (16).
2. Solve [a] $u(x) = x + \int_0^x (t - x)u(t)\, dt$
   [b] $u(x) = 1 + \int_0^x (t - x)u(t)\, dt$.
3. Solve $u(x) = 1 + \int_0^x u(t)\, dt$.
4. Solve $u(x) = x + \int_0^{1/2} u(t)\, dt$.
5. Solve $u(x) = \sin x - x/4 + \frac{1}{4}\int_0^{\pi/2} txu(t)\, dt$.
6. Form the integral equation corresponding to the following differential equation:

$$\frac{dy}{dx} - y = 0,$$

$$y(0) = 1.$$

7. Form the integral equation corresponding to the following differential equation:

$$\frac{d^2 y}{dx^2} + y = 0,$$

$$y(0) = 0, \qquad y'(0) = 1.$$

8. In proving the theorem of the arithmetic–geometric mean

$$\frac{1}{n}\sum_{i=1}^{n} d_i \geq \prod_{i=1}^{n} d_i^{1/n}$$

we used the averaging process. Replace d_M by S, d_m by $S + a - b$. Here

$$S = \frac{1}{n}\sum_{i=1}^{n} d_i, \qquad d_M = \max d_i = S + a$$

$$d_m = \min d_i = S - b.$$

Show that after n such steps all d_i have been replaced by S.

9. Let X be a finite dimensional vector space and L an operator acting on it. Let $N(L)$ and $R(L)$ denote the null space and range of L respectively. Show that $N(L)$ and $R(L)$ are subspaces.

10. Verify that the vector space mentioned in Example 1 of Section 6 is indeed a vector space. Do the same for Example 2.
11. Verify that for the set of vectors in Example 1 of Section 6 $(f, g) = \sum_{i=1}^{\infty} a_i \bar{b}_i$ is an inner product.
12. Verify that for the set of vectors in Example 2 of Section 6 $(f, g) = \int_0^1 f(x)\overline{g(x)}\, dx$ is an inner product.
13. Let X be an inner product space and $\{x_n\}$ a sequence of elements of X. If $\|x_n\| \to \|x\|$ for some x and $(x_n, x) \to (x, x)$, show that $x_n \to x$.
14. Let X be the linear space of polynomials with the inner product

$$(f, g) = \int_0^1 f(x)\overline{g(x)}\, dx$$

Let $f(x) = x^2 + 2$, $g(x) = 3x$. Find (f, g) and $\|f\|$.
15. Suppose $|(u, v)| = \|u\| \, \|v\|$. Show that u and v are linearly dependent.
16. Show that the operator

$$Kf = \int_0^1 \frac{\sin (xy)}{y} f(y)\, dy$$

acting on $L_2[0, 1]$ is unbounded.

2
BASIC
EXISTENCE THEOREMS

SUMMARY

There are a number of integral equations that can be studied by some relatively elementary methods. Such cases will be discussed in this chapter.

Some of the methods to be examined are topological in nature. In particular if our operators are small (in a sense that will be made more precise) existence and uniqueness theorems can be established by a certain fundamental fixed point principle. These techniques can be shown to apply even if the kernels of our integral operator have so-called weak singularities.

Finally, degenerate operators will be examined. These can be discussed using strictly the techniques of linear algebra.

1. FIXED POINT THEOREMS

In this section we shall develop some properties of a class of so-called contraction operators. These shall enable us to derive a number of basic existence and uniqueness theorems for integral equations in a rather expeditious way. Basically this technique is topological in nature. These topological methods are most useful in connection with nonlinear operators, where analytical methods often fail. For linear equations the analytical methods often lead to strong results that are not obtainable by topological methods.

DEFINITION. Let H be a Hilbert space and T a bounded operator on H. T is not necessarily a linear operator. T is said to be a contraction operator if there exists a positive constant $\alpha < 1$ such that

$$\|Tf_1 - Tf_2\| \leq \alpha \|f_1 - f_2\|$$

for all f_1, f_2 in H.

THEOREM 1.   Let T be a contraction operator on H. The equation

$$Tf = f \tag{1}$$

has a unique solution f in H. Such a solution is said to be a fixed point of T.

PROOF.   Suppose there are two fixed points f and g so that

$$Tf = f$$
$$Tg = g$$

Then

$$\|f - g\| = \|Tf - Tg\| \leq \alpha \|f - g\|$$

and

$$(1 - \alpha) \|f - g\| \leq 0.$$

Since $\|f - g\|$ is necessarily non-negative we see that

$$\|f - g\| = 0$$

so that $f = g$. It follows that if (1) does have a solution it must be unique.

To show that (1) has a solution we shall set up an iteration procedure. Select any f_0 and then construct a sequence $\{f_n\}$ defined by

$$f_{n+1} = Tf_n, \qquad n = 0, 1, 2, \ldots.$$

We shall first show that this sequence is a Cauchy sequence, and then that its limit is indeed a solution of (1). That it has a limit will follow from the fact that a Cauchy sequence must have a unique limit in a Hilbert space. The limit will be independent of the initial choice f_0, since it will be a solution of (1), which must be unique.

First we note that

$$\|f_{n+1} - f_n\| = \|Tf_n - Tf_{n-1}\| \leq \alpha \|f_n - f_{n-1}\|.$$

By a successive application of the above we have

$$\|f_{n+1} - f_n\| \leq \alpha \|f_n - f_{n-1}\| \leq \alpha^2 \|f_{n-1} - f_{n-2}\| \leq \cdots \leq \alpha^n \|f_1 - f_0\|.$$

More generally we have, if $n > m$,

$$\|f_n - f_m\| = \|(f_n - f_{n-1}) + (f_{n-1} - f_{n-2}) + \cdots + (f_{m+1} - f_m)\|$$
$$\leq \|f_n - f_{n-1}\| + \|f_{n-1} - f_{n-2}\| + \cdots + \|f_{m+1} - f_m\|$$
$$\leq (\alpha^{n-1} + \alpha^{n-2} + \cdots + \alpha^m) \|f_1 - f_0\|$$
$$\leq (\alpha^m + \alpha^{m+2} + \cdots) \|f_1 - f_0\| = \frac{\alpha^m}{1 - \alpha} \|f_1 - f_0\|$$

so that

$$\lim_{n, m \to \infty} \|f_n - f_m\| = 0.$$

It follows that $\{f_n\}$ is a Cauchy sequence, and we denote its limit by f.

We shall have to show that the limit f is a solution of (1). In view of the fact that T is a continuous operator, we have

$$Tf = T(\lim f_n) = \lim Tf_n = \lim f_{n+1} = f$$

and $Tf = f$. ■

N.B.   It follows that $f = \lim_{n \to \infty} T^n f_0$.

There is a generalization of the preceding theorem that will prove to be particularly convenient for Volterra operators.

THEOREM 2.   Let T be an operator on H, such that the nth power of T, namely T^n is a contraction operator. Then the equation

$$Tf = f$$

has a unique solution f in H.

PROOF.   By the previous theorem we can assert that the equation

$$T^n f = f$$

has a unique solution. In fact, we can obtain the solution by finding

$$\lim_{k \to \infty} T^{kn} f_0 = f,$$

for an arbitrary initial function f_0. In particular, we see that, by letting $f_0 = Tf$

$$\lim_{k \to \infty} T^{kn} Tf = f.$$

But since $T^n f = f$ we also have $T^{kn} f = f$ so that

$$\lim_{k \to \infty} T^{kn} Tf = \lim_{k \to \infty} T T^{kn} f = \lim_{k \to \infty} Tf = Tf$$

so that $Tf = f$.

To show that this solution is unique we note that if

$$Tf = f, \qquad Tg = g$$

then we also have

$$T^n f = f, \qquad T^n g = g$$

and since T^n is a contraction operator with a unique fixed point $f = g$. ■

2. ELEMENTARY EXISTENCE THEOREMS

In this section we shall be concerned with equations of the type

$$\phi - \lambda K \phi = f \tag{2}$$

where f is in a Hilbert space H, and K is a bounded operator with the property

$$\|K\phi_1 - K\phi_2\| \leq M \|\phi_1 - \phi_2\|. \tag{3}$$

In the applications that will concern us the operator K will be an integral operator of the type described in Chapter 1. It may be linear or nonlinear. For a linear operator (3) is, of course, equivalent to the statement that K is bounded. For a nonlinear operator (3) is not equivalent to K being bounded and represents an additional hypothesis.

We can rewrite (2) in the form

$$T\phi = \phi \tag{4}$$

where

$$T\phi = f + \lambda K\phi. \tag{5}$$

T is a nonlinear operator if $f \neq 0$, even if K is linear.

THEOREM 3. Equation (2) has a unique solution for all f and sufficiently small $|\lambda|$, provided K is a bounded operator, that also satisfies (3).

PROOF. For T, as defined in (5), we find by virtue of (3), that

$$\| T\phi_1 - T\phi_2 \| = |\lambda| \, \| K\phi_1 - K\phi_2 \| \leq |\lambda| \, M \, \| \phi_1 - \phi_2 \|.$$

For $|\lambda| \, M < 1$, T is a contraction operator, so that (4), and equivalently (2), has a unique solution for all f. ■

The above theorem can be applied immediately to the Fredholm integral equation

$$\phi(x) - \lambda \int_a^b K(x, y)\phi(y) \, dy = f(x) \tag{6}$$

if $f(x)$ is in $L_2[a, b]$ and the integral operator is bounded. In a later section some sufficient conditions on $K(x, y)$ for the associated integral operator K to be bounded will be discussed. In accordance with the proof of the existence of a fixed point the solution is given by

$$\phi(x) = \lim_{n \to \infty} T^n f_0(x) \tag{7}$$

where $f_0(x)$ is an arbitrary initial function. We have

$$T f_0 = f + \lambda K f_0$$
$$T^2 f_0 = T[f + \lambda K f_0] = f + \lambda K f + \lambda^2 K^2 f_0$$

$$\cdot$$
$$\cdot$$
$$\cdot$$

$$T^n f_0 = f + \lambda K f + \lambda^2 K^2 f + \cdots + \lambda^{n-1} K^{n-1} f + \lambda^n K^n f_0.$$

and finally

$$\phi = f + \lambda K f + \lambda^2 K^2 f + \cdots + \lambda^n K^n f + \cdots. \tag{8}$$

Formally (8) could be obtained from (2) by expanding $(I - \lambda K)^{-1}$ in a geometric series in λK. That this procedure is valid is a consequence of the fact that T is a contraction operator for $|\lambda|$ small enough.

A few words are in order on the meaning of K^n, where K is an integral operator.

$$Kf(x) = \int_a^b K(x, y)f(y)\, dy$$

$$K^2 f(x) = K\left[\int_a^b K(x, y)f(y)\, dy\right]$$

$$= \int_a^b K(x, z)\int_a^b K(z, y)f(y)\, dy\, dz$$

$$= \int_a^b \left[\int_a^b K(x, z)K(z, y)\, dz\right]f(y)\, dy.$$

It follows that K^2 is an integral operator, whose kernel is given by

$$\int_a^b K(x, z)K(z, y)\, dz.$$

More generally it is easy to show that

$$K^n f(x) = \int_a^b K_n(x, y)f(y)\, dy$$

where $K_n(x, y)$ can be defined recursively by

$$K_n(x, y) = \int_a^b K(x, z)K_{n-1}(z, y)\, dz, \qquad n = 2, 3, \ldots. \tag{9}$$

$$K_1(x, y) = K(x, y)$$

In the same fashion one can show that

$$K_{m+n}(x, y) = \int_a^b K_m(x, z)K_n(z, y)\, dz. \tag{10}$$

The conclusions of theorem 3 can also be applied to the more general nonlinear integral equation

$$\phi(x) - \lambda \int_a^b K(x, y, \phi(y))\, dy = f(x) \tag{11}$$

provided the function $K(x, y, z)$ satisfies suitable hypotheses.

THEOREM 4. Suppose

$$\left\|\int_a^b K(x, y, \phi(y))\, dy\right\| \leq M\, \|\phi(y)\|$$

and that

$$|K(x, y, z_1) - K(x, y, z_2)| \leq N(x, y) |z_1 - z_2|$$

where

$$\int_a^b \int_a^b |N(x, y)|^2 \, dx \, dy = P^2 < \infty.$$

If $f(x) \in L_2[a, b]$, (11) has a unique solution in $L_2[a, b]$, if $|\lambda| \, P < 1$.

PROOF.   We consider the operator

$$T\phi = f + \lambda K\phi,$$

where K is defined in (11)

$$\|T\phi_1 - T\phi_2\| = |\lambda| \left\| \int_a^b [K(x, y, \phi_1(y)) - K(x, y, \phi_2(y))] \, dy \right\|$$

$$\leq |\lambda| \left\{ \int_a^b \left[\int_a^b |K(x, y, \phi_1(y)) - K(x, y, \phi_2(y))| \, dy \right]^2 dx \right\}^{1/2}$$

$$\leq |\lambda| \left\{ \int_a^b \left[\int_a^b N(x, y) |\phi_1(y) - \phi_2(y)| \, dy \right]^2 dx \right\}^{1/2}$$

$$\leq |\lambda| \, P \, \|\phi_1 - \phi_2\|.$$

It follows that for $|\lambda| \, P < 1$, T is a contraction operator so that it has a unique fixed point. That fixed point is a solution of (11).   ∎

Equation (6) is a special case of (11) where $K(x, y)\phi(y) = K(x, y, \phi(y))$ so that the above theorem covers the linear case as well.

As an application of this theorem we consider the following boundary value problem.

$$\phi''(x) + \lambda L(x, \phi(x)) = f(x) \tag{12}$$

$$\phi(0) = \phi(1) = 0.$$

Furthermore, we assume that

$$\|L(x, \phi(x))\| \leq M \, \|\phi(x)\|$$

$$|L(x, \phi_1(x)) - L(x, \phi_2(x))| \leq N(x) |\phi_1(x) - \phi_2(x)|, \tag{13}$$

where

$$\int_0^1 |N(y)|^2 \, dy = P^2 < \infty.$$

By using the result of Chapter 1, (16) we can rewrite (12) in the form

$$\phi(x) - \lambda \int_0^1 K_1(x, y)L(y, \phi(y)) \, dy = - \int_0^1 K_1(x, y)f(y) \, dy. \tag{14}$$

where

$$K_1(x, y) = y(1 - x), \qquad y \le x.$$
$$= x(1 - y), \qquad y \ge x.$$

Since $|K_1(x, y)| \le \frac{1}{4}$ and $L(x, \phi(x))$ satisfies (13) the conditions of the last theorem are satisfied. It follows that if

$$\tfrac{1}{4} |\lambda| P < 1$$

(12) will have a unique solution for all $f(x)$ in $L_2[0, 1]$.

3. VOLTERRA EQUATIONS

The results of the previous section apply to Volterra equations as well. As was pointed out earlier the equation

$$\phi(x) - \lambda \int_a^b K(x, y)\phi(y) \, dy = f(x)$$

reduces to a Volterra equation if $K(x, y) = 0$ for $y > x$. In that case however, we can strengthen the results of the previous section considerably.

THEOREM 5. Let $f(x) \in L_2[0, 1]$ (we consider a finite interval and without loss of generality let it be $[0, 1]$) and suppose that $K(x, y)$ is continuous for $x, y \in [0, 1]$ and therefore uniformly bounded, say $|K(x, y)| \le M$. Then the equation

$$\phi(x) - \lambda \int_0^x K(x, y)\phi(y) \, dy = f(x) \tag{15}$$

has a unique solution $\phi(x)$, for all λ and $f(x)$ in $L_2[0, 1]$.
N.B. The requirement that $K(x, y)$ be continuous is too strong. In theorem 6 it will be replaced by a weaker condition.

PROOF. We consider the operator

$$T\phi = f(x) + \lambda \int_0^x K(x, y)\phi(y) \, dy.$$

If $T\phi$ has a fixed point, such a fixed point must be a solution of (15). To show that such a fixed point exists we will show that T^n, for some n, will be a contraction operator. By theorem 2 T will then have a unique fixed point. Now

$$T^n\phi = f + \lambda K f + \cdots + \lambda^{n-1} K^{n-1} f + \lambda^n K^n \phi$$

where

$$K^n\phi = \int_0^x K_n(x, y)\phi(y) \, dy,$$

so that

$$\| T^n \phi_1 - T^n \phi_2 \| = |\lambda|^n \left\| \int_0^x K_n(x, y)(\phi_1(y) - \phi_2(y))\, dy \right\|.$$

To determine $K_n(x, y)$ we can use (9) specialized to Volterra kernels.

$$K_1(x, y) = K(x, y)$$

$$K_n(x, y) = \int_y^x K(x, z) K_{n-1}(z, y)\, dz, \qquad n = 2, 3, \ldots.$$

By hypothesis $|K_1(x, y)| \leq M$ and one can then show inductively that

$$|K_n(x, y)| \leq \frac{M^n (x - y)^{n-1}}{(n - 1)!}, \qquad 0 \leq y \leq x.$$

For $n = 1$, the above is obviously valid. If it is true for n, then

$$|K_{n+1}(x, y)| \leq \int_y^x |K(x, z)|\, |K_n(z, y)|\, dz$$

$$\leq \frac{M^{n+1}}{(n - 1)!} \int_y^x (z - y)^{n-1}\, dz$$

$$= \frac{M^{n+1}(x - y)^n}{n!}.$$

We have, therefore,

$$\| T^n \phi_1 - T^n \phi_2 \| \leq \frac{|\lambda|^n M^n}{(n - 1)!} \left\| \int_0^x (\phi_1(y) - \phi_2(y))\, dy \right\|$$

$$\leq \frac{|\lambda|^n M^n}{(n - 1)!} \| \phi_1 - \phi_2 \|.$$

For n sufficiently large

$$\frac{|\lambda|^n M^n}{(n - 1)!} < 1$$

so that T^n is a contraction operator, and (15) has theref
tion. ∎

Theorems 3 and 5 both apply to (15). The additio:
$K(x, y)$ in theorem 5 allow us to show, however, that a un
for all finite λ. For the more general Fredholm equation, t
condition encountered in theorem 3, however, is vital.

EXAMPLE. $\phi(x) - \lambda\int_0^1 xy\phi(y)\,dy = 1$.

As in (8) we find immediately that

$$\phi(x) = 1 + \frac{\lambda x}{2}\sum_{n=0}^{\infty}\frac{\lambda^n}{3^n}$$

which converges only for $|\lambda| < 3$. For $\lambda = 3$ one can check that no solution exists. For $|\lambda| > 3$, and more generally for $\lambda \neq 3$ the solution is

$$\phi(x) = 1 + \frac{\lambda x}{2[1 - (\lambda/3)]}.$$

It is also worth noting that our existence and uniqueness theorems apply only to the Hilbert spaces in which we are working. If we chose to work in other spaces, these results need not hold any longer.

EXAMPLE. Consider

$$\phi(x) - \int_0^x y^{x-y}\phi(y)\,dy = 0. \tag{16}$$

The above has the general solution

$$\phi(x) = cx^{x-1} \tag{17}$$

where c is an arbitrary constant.

It is easy to verify that the preceding theory applies to (16). Solution (17) however is not in $L_2[0, 1]$, and our general theorem merely asserts that in that space (16) has a unique solution, given by $\phi(x) = 0$. The above example is a classical one and emphasizes the fact that the space in which one chooses to work is significant in determining the resultant theory.

Theorem 5 still demands that $K(x, y)$ be continuous. The proof of the results can be modified so as to yield similar results for more general kernels.

THEOREM 6. Let $f(x) \in L_2[0, 1]$, and suppose that $K(x, y)$ is such that

$$\int_0^1\int_0^1 |K(x, y)|^2\,dx\,dy < \infty.$$

Then

$$\phi(x) - \lambda\int_0^x K(x, y)\phi(y)\,dy = f(x) \tag{18}$$

has a unique solution for all λ, in $L_2[0, 1]$.

PROOF. We let

$$A^2(x) = \int_0^x |K(x, y)|^2\,dy, \qquad B^2(y) = \int_y^1 |K(x, y)|^2\,dx$$

and by hypothesis both $A^2(x)$ and $B^2(y)$ are integrable. Let N be such that

$$\int_0^1 A^2(x)\,dx \leq N, \qquad \int_0^1 B^2(y)\,dy \leq N.$$

Furthermore, we define the function $\rho(x)$ by

$$\rho(x) = \int_0^x A^2(y)\,dy \qquad \text{so that} \quad \rho(1) \leq N.$$

As in the proof of theorem 5 we consider instead of (18) the equivalent equation

$$\phi(x) = f(x) + \lambda Kf + \lambda^2 K^2 f + \cdots + \lambda^{n-1} K^{n-1} f + \lambda^n K^n \phi, \qquad (19)$$

$$K^n \phi = \int_0^x K_n(x, y)\phi(y)\,dy.$$

To estimate $\|K^n\|$ we examine $K_n(x, y)$. Now

$$K_2(x, y) = \int_y^x K(x, z)K(z, y)\,dz$$

and by the Cauchy–Schwarz inequality

$$|K_2(x, y)|^2 \leq \int_y^x |K(x, z)|^2\,dz \int_y^x |K(z, y)|^2\,dz \leq A^2(x)B^2(y).$$

Similarly,

$$K_3(x, y) = \int_y^x K(x, z)K_2(z, y)\,dy$$

so that

$$|K_3(x, y)|^2 \leq \int_y^x |K(x, z)|^2\,dz \int_y^x |K_2(z, y)|^2\,dz$$

$$\leq A^2(x)B^2(y) \int_y^x A^2(z)\,dz = A^2(x)B^2(y)[\rho(x) - \rho(y)].$$

It is easy to carry through an inductive argument to show that

$$|K_n(x, y)|^2 \leq A^2(x)B^2(y) \frac{[\rho(x) - \rho(y)]^{n-2}}{(n-2)!}, \qquad n \geq 2.$$

We can then rewrite (19) in the form

$$\phi = T^n \phi, \qquad \text{where} \quad T\phi = f + \lambda K\phi$$

and then we can show that for n sufficiently large T^n is a contraction operator.

$$|T^n\phi_1 - T^n\phi_2|^2 = \left|\int_0^x K_n(x, y)[\phi_1(y) - \phi_2(y)] \, dy\right|^2$$

$$\leq \int_0^x \frac{A^2(x)B^2(y)[\rho(x) - \rho(y)]^{n-2}}{(n - 2)!} \, dy$$

$$\times \int_0^x |\phi_1(y) - \phi_2(y)|^2 \, dy$$

$$\leq \frac{A^2(x)[\rho(x)]^{n-2}}{(n - 2)!} \int_0^1 B^2(y) \, dy \, \|\phi_1 - \phi_2\|^2.$$

By an integration we then find

$$\|T^n\phi_1 - T^n\phi_2\|^2 \leq \frac{[\rho(1)]^{n-1}N}{(n - 1)!} \|\phi_1 - \phi_2\|^2 \leq \frac{N^n}{(n - 1)!} \|\phi_1 - \phi_2\|^2,$$

so that T is a contraction operator if $[N^n/(n - 1)!] < 1$. For large n that will be the case so that (19) and therefore (18) as well will have a unique solution in $L_2[0, 1]$. ∎

4. KERNELS WITH WEAK SINGULARITIES

In general, integral equations with singular kernels are very difficult to handle. For the moment, we shall not offer a precise definition for a singular kernel. Loosely speaking, these are kernels that do not satisfy the previously discussed conditions. There is one type, however, that is amenable to a rather simple discussion.

THEOREM 7.   Consider

$$\phi(x) - \lambda \int_0^x K(x, y)\phi(y) \, dy = f(x) \tag{20}$$

where $K(x, y) = H(x, y)/|x - y|^\alpha$, and $H(x, y), f(x)$ are continuous in both variables and $0 \leq \alpha < 1$. Under these conditions (20) has a unique solution for all λ.

PROOF.   For $0 \leq \alpha < \frac{1}{2}$, $K(x, y)$ is square-integrable and theorem 6 can be applied. We shall therefore suppose that $\frac{1}{2} \leq \alpha < 1$. Then we consider

$$K_2(x, y) = \int_y^x \frac{H(x, z)}{(x - z)^\alpha} \cdot \frac{H(z, y)}{(z - y)^\alpha} \, dz.$$

Let $z = y + (x - y)u$ and suppose $|H(x, y)| \leq M$, so that

$$|K_2(x, y)| \leq \frac{M^2}{(x - y)^{2\alpha-1}} \int_0^1 \frac{du}{[u - u^2]^\alpha}.$$

The integral in the above exists and we see that if $\frac{1}{2} \leq \alpha < \frac{3}{4}$, then $0 \leq 2\alpha - 1 < \frac{1}{2}$, so that $K_2(x, y)$ is square-integrable. Finally, by repeating the process

$$|K_{2^n}(x, y)| \leq \frac{M^n}{(x - y)^{2^n\alpha+1-2u}}$$

and for $\alpha < 1 - (1/2^{n+1})$, $K_{2^n}(x, y)$ is square-integrable. It follows that, if we consider (19), for sufficiently large n, theorem 6 can be applied. ■

It is not absolutely necessary to assume that $f(x)$ is continuous. What is required is that for a suitable n, $K^n f(x) \in L_2[0, 1]$. If $f(x)$ is continuous, $K^n f(x) \in L_2[0, 1]$ for all n.

A similar theorem can be developed for Fredholm equations.

THEOREM 8. Consider

$$\phi(x) - \lambda \int_0^1 K(x, y)\phi(y)\, dy = f(x) \tag{21}$$

where

$$K(x, y) = \frac{H(x, y)}{|x - y|^\alpha},$$

and $H(x, y), f(x)$ are continuous functions for $0 \leq x, y \leq 1$, and $0 \leq \alpha < 1$. For sufficiently small $|\lambda|$ (21) will have a unique solution.

PROOF. Instead of (21) we consider the equation

$$\phi = f + \lambda K f + \cdots + \lambda^{n-1}K^{n-1}f + \lambda^n K^n \phi. \tag{22}$$

Equation (22) is in the form $\phi = T^n\phi$, and if the latter has a unique fixed point it follows that $\phi = T\phi$ does also.

As in the preceding theorem we shall study the structure of the iterated kernels. We assume that

$$|H(x, y)| \leq M$$

so that

$$K_2(x, y) = \int_0^1 \frac{H(x, z)H(z, y)}{|x - z|^\alpha |z - y|^\alpha}\, dz$$

and

$$|K_2(x, y)| \leq M^2 \int_0^1 \frac{dz}{|x - z|^\alpha |z - y|^\alpha}.$$

Now let $z = xu + y(1 - u)$, and, without loss of generality, assume $x > y$. Then the integral in the above inequality reduces to

$$\frac{1}{|x - y|^{2\alpha-1}} \int_{-y/x-y}^{1-y/x-y} \frac{du}{|u - u^2|^\alpha} \leq \frac{1}{|x - y|^{2\alpha-1}} \int_{-\infty}^{\infty} \frac{du}{|u - u^2|^\alpha}.$$

Now for $0 \leq \alpha < \frac{1}{2}$, $K(x, y)$ is already square integrable, and therefore (21) has a unique solution for small $|\lambda|$. The above estimate shows that for $\frac{1}{2} < \alpha < \frac{3}{4}$, $K_2(x, y)$ is square integrable by the above estimates, so that theorem 2 can be applied to T^2. For $\alpha = \frac{1}{2}$ we note that

$$|K_2(x, y)| \leq M^2 \int_{-y/x-y}^{1-y/x-y} \frac{du}{|u - u^2|^{1/2}}.$$

From the above we note that at worst $K_2(x, y)$ will have a logarithmic singularity of the type $\log |x - y|$, which is still square-integrable so that theorem 2 applies. For $\frac{3}{4} \leq \alpha < 1$ one can use T^n for sufficiently large n as in theorem 7. ■

5. DEGENERATE KERNELS

There is a class of Fredholm equations that can be treated by the standard methods of linear algebra. These are equations of the second kind.

$$\phi(x) - \lambda \int_a^b K(x, y)\phi(y)\, dy = f(x) \tag{23}$$

where

$$K(x, y) = \sum_{j=1}^n a_j(x)b_j(y). \tag{24}$$

Kernels of the type shown in (24) are said to be degenerate. We assume that the functions $a_j(x)$ and $b_j(x)$ belong to $L_2[a, b]$, so that $K(x, y)$ is square-integrable. Furthermore, we shall assume that the set of functions $\{a_j(x)\}$ is linearly independent. That is

$$\sum_{j=1}^n \alpha_j a_j(x) = 0 \quad \text{almost everywhere}$$

if and only if all α_j vanish. If that were not the case, one of the $a_j(x)$ could be expressed in terms of a linear combination of the rest of the $a_j(x)$ and consequently the number of terms in (24) could be reduced from n to $n - 1$. Similar remarks apply to the set $\{b_i(y)\}$.

If (24) is inserted in (23) we obtain

$$\phi(x) - \lambda \sum_{j=1}^n a_j(x) \int_a^b b_j(y)\phi(y)\, dy = f(x). \tag{25}$$

If we denote $\int_a^b b_j(y)\phi(y)\, dy$ by z_j, multiply (25) by $b_i(x)$, and integrate we obtain

$$z_i - \lambda \sum_{j=1}^n a_{ij} z_j = f_i, \qquad i = 1, 2, \ldots, n \tag{26}$$

where $a_{ij} = \int_a^b a_j(x) b_i(x)\, dx$, and $\int_a^b f(x) b_i(x)\, dx = f_i$ (26) is a system of n linear algebraic equations in n unknowns $z_1, z_2, \ldots, z_n$. If this system has a solution, one can immediately solve (25) and one finds

$$\phi(x) = f(x) + \lambda \sum_{i=1}^n z_i a_i(x). \tag{27}$$

Equation (25) will have a unique solution for all $f(x)$ if and only if (26) has a unique solution for all $\{f_i\}$. That will happen only if

$$|I - \lambda A| \neq 0 \tag{28}$$

where the matrix $A = (a_{ij})$. Since $|I - \lambda A|$ is a polynomial of degree n, there will be at most n values of λ for which (25) and (26) fail to have unique solutions.

To discuss the situation, where (28) is not satisfied, we examine the adjoint homogeneous equation associated with (25), namely

$$\psi(x) - \bar{\lambda} \sum_{j=1}^n \overline{b_j(x)} \int_a^b \overline{a_j(y)} \psi(y)\, dy = 0. \tag{29}$$

As before, we can reduce this equation to an algebraic system. We let $\int_a^b \psi(x) a_i(x)\, dx = w_i$. Multiply (29) by $\overline{a_i(x)}$ and integrate so that

$$w_i - \bar{\lambda} \sum_{j=1}^n \overline{a_{ji}} w_j = 0, \qquad i = 1, 2, \ldots, n. \tag{30}$$

If (30) has solutions, other than the trivial set where all w_j vanish, we see that

$$\psi(x) = \bar{\lambda} \sum_{i=1}^n w_i \overline{b_i(x)} \tag{31}$$

satisfies (29).

We are now in a position to apply the Fredholm alternative to (26) and (30). If $|I - \lambda A| = 0$ for some value λ, (29) will have nontrivial solutions. Then (26) will have (nonunique) solutions only for those f_i satisfying

$$\sum_{i=1}^n f_i \bar{w}_i = 0, \tag{32}$$

where the w_i satisfy (30). We now consider the inner product

$$(f, \psi) = \int_a^b f(x) \overline{\psi(x)}\, dx = \lambda \sum_{i=1}^n \bar{w}_i \int_a^b f(x) b_i(x)\, dx = \lambda \sum_{i=1}^n f_i \bar{w}_i.$$

The case where $|I - \lambda A| = 0$ certainly cannot occur for $\lambda = 0$, so that (32) is equivalent to

$$(f, \psi) = 0 \tag{33}$$

or all $\psi(x)$ satisfying (29).

Similarly, one can show that if $f(x) = 0$, (25) and (29) or equivalently (26) with all $f_i = 0$ and (30) have the same number of linearly independent solutions.

These results can be summarized as follows.

The Fredholm Alternative

The integral equation

$$\phi(x) - \lambda \int_a^b K(x, y)\phi(y) \, dy = f(x) \tag{34}$$

where $K(x, y) = \sum_{i=1}^n a_i(x)b_i(y)$ and all $a_i(x)$, $b_i(x)$, $f(x)$ are in $L_2[a, b]$, has a unique solution for all $f(x)$ if and only if, the associated homogeneous equation

$$\phi(x) - \lambda \int_a^b K(x, y)\phi(y) \, dy = 0 \tag{35}$$

has only $\phi(x) = 0$ as a solution. The adjoint equation

$$\psi(x) - \bar{\lambda} \int_a^b \overline{K(y, x)}\psi(y) \, dy = 0 \tag{36}$$

has the same number of linearly independent solutions as (35). If this number of solutions of (35) and (36) is positive, (34) will have a (nonunique) solution, if and only if

$$(f, \psi) = \int_a^b f(x)\overline{\psi(x)} \, dx = 0$$

for all solutions $\psi(x)$ of (36).

We see that the Fredholm alternative holds for linear algebraic systems, as well as integral equations with degenerate kernels. In the next chapter we shall see how it can be generalized to a broader class of integral operators.

EXAMPLE.

$$\phi(x) - \lambda \left[\pi x \int_0^1 \sin \pi y \phi(y) \, dy + 2\pi x^2 \int_0^1 \sin 2\pi y \phi(y) \, dy \right] = f(x).$$

Let

$$\int_0^1 \sin \pi y \phi(y) \, dy = z_1, \qquad \int_0^1 \sin \pi y f(y) \, dy = f_1$$

$$\int_0^1 \sin 2\pi y \phi(y) \, dy = z_2, \qquad \int_0^1 \sin 2\pi y f(y) \, dy = f_2$$

so that the integral equation can be reduced to

$$(1 - \lambda)z_1 - 2\left(1 - \frac{4}{\pi^2}\right)\lambda z_2 = f_1$$

$$\frac{\lambda}{2}z_1 + (1 - \lambda)z_2 = f_2$$

If $\lambda^2 \neq (\pi^2/4)$, we find

$$z_1 = \frac{(1 + \lambda)f_1 + 2[1 - (4/\pi^2)]\lambda f_2}{1 - (4\lambda^2/\pi^2)}, \qquad z_2 = \frac{(-\lambda/2)f_1 + (1 - \lambda)f_2}{1 - (4\lambda^2/\pi^2)}$$

so that

$$\phi(x) = f(x) + \pi\lambda z_1 x + 2\pi\lambda z_2 x^2.$$

If $\lambda = (\pi/2)$ we consider the adjoint system

$$\left(1 - \frac{\pi}{2}\right)\zeta_1 + \frac{\pi}{4}\zeta_2 = 0, \qquad -\pi\left(1 - \frac{4}{\pi^2}\right)\zeta_1 + \left(1 + \frac{\pi}{2}\right)\zeta_2 = 0$$

with solutions $\zeta_1 = (\pi/4)C$, $\zeta_2 = -[1 - (\pi/2)]C$, where C is arbitrary. In order for a solution to exist we require that

$$\frac{\pi}{4}f_1 - \left(1 - \frac{\pi}{2}\right)f_2 = 0.$$

If the above condition holds we obtain as our solution

$$\phi(x) = f(x) + \frac{\pi}{2}\left\{\frac{f_1}{1 - (\pi/2)} - 2\left(1 + \frac{2}{\pi}\right)Cx + \pi Cx^2\right\}$$

where C is arbitrary, so that a nonunique solution exists.

6. VOLTERRA EQUATIONS OF THE FIRST KIND

In general, it is exceedingly difficult to deal with Volterra equations of the first kind. We shall examine a number of cases that can be discussed in terms of the results derived so far. Consider

$$\int_0^x K(x, y)\phi(y)\, dy = f(x) \tag{37}$$

and suppose that $K(x, y)$ and $f(x)$ are sufficiently differentiable. Then by differentiating both sides of (37) with respect to x one finds,

$$K(x, x)\phi(x) + \int_0^x \frac{\partial}{\partial x} K(x, y)\phi(y)\, dy = f'(x).$$

If $K(x, x) \neq 0$, the above can be reduced to an equation of the second type

$$\phi(x) + \int_0^x \frac{(\partial/\partial x)K(x, y)}{K(x, x)} \, \phi(y) \, dy = \frac{f'(x)}{K(x, x)} . \tag{38}$$

If the kernel in (38) is square-integrable, and $[f'(x)/K(x, x)] \in L_2[0, 1]$, (38) will have a unique solution in $L_2[0, 1]$, by the preceding theory.

There is a second method, whereby (37) can be reduced to an equation of the second type. Let

$$\psi(x) = \int_0^x \phi(y) \, dy$$

and perform an integration by parts in (37). Then

$$K(x, x)\psi(x) - \int_0^x \frac{\partial}{\partial y} K(x, y)\psi(y) \, dy = f(x)$$

and if $K(x, x) \neq 0$

$$\psi(x) - \int_0^x \frac{(\partial/\partial y)K(x, y)}{K(x, x)} \, \psi(y) \, dy = \frac{f(x)}{K(x, x)} . \tag{39}$$

Equation (39) will have a unique solution in $L_2[0, 1]$, provided its kernel is square-integrable and $[f(x)/K(x, x)] \in L_2[0, 1]$.

An equation that can be solved explicitly is Abel's equation

$$\int_0^x \frac{\phi(y)}{(x - y)^\alpha} \, dy = f(x) \tag{40}$$

where $0 \leq \alpha < 1, f(x)$ is continuous and $f(0) = 0$. If we denote the kernel in (40) by $K_\alpha(x, y)$, and apply to both sides of (40) the Volterra operator associated with $K_\beta(x, y)$ and then interchange the order of integration on the left we obtain

$$\int_0^x \left[\int_0^x \frac{dz}{(x - z)^\beta (z - y)^\alpha} \right] \phi(y) \, dy = \int_0^x \frac{f(y)}{(x - y)^\beta} \, dy.$$

Letting $z = y + (x - y)u$ we see that

$$\int_0^x \frac{dz}{(x - z)^\beta (z - y)^\alpha} = (x - y)^{1-\alpha-\beta} \int_0^1 \frac{du}{(1 - u)^\beta u^\alpha}$$

$$= (x - y)^{1-\alpha-\beta} \frac{\Gamma(1 - \alpha)\Gamma(1 - \beta)}{\Gamma(2 - \alpha - \beta)}$$

where $\Gamma(\rho)$ denotes the gamma function. The above integral equation now takes the form

$$\int_0^x \frac{\phi(y)}{(x-y)^{\alpha+\beta-1}}\,dy = \frac{\Gamma(2-\alpha-\beta)}{\Gamma(1-\alpha)\Gamma(1-\beta)}\int_0^x \frac{f(y)}{(x-y)^\beta}\,dy.$$

In particular for $\beta = 1 - \alpha$ it takes the simple form

$$\int_0^x \phi(y)\,dy = \frac{\Gamma(1)}{\Gamma(\alpha)\Gamma(1-\alpha)}\int_0^x \frac{f(y)}{(x-y)^{1-\alpha}}\,dy = \frac{\sin\pi\alpha}{\pi}\int_0^x \frac{f(y)}{(x-y)^{1-\alpha}}\,dy$$

where the facts $\Gamma(1) = 1$ and $\Gamma(\alpha)\Gamma(1-\alpha) = (\pi/\sin\pi\alpha)$ were used. If the right side has a derivative we finally obtain

$$\phi(x) = \frac{\sin\pi\alpha}{\pi}\frac{d}{dx}\int_0^x \frac{f(y)}{(x-y)^{1-\alpha}}\,dy \tag{41}$$

as the solution of (40).

The case where $\alpha = \frac{1}{2}$ leads to the equation

$$\int_0^x \frac{\phi(y)}{(x-y)^{1/2}}\,dy = f(x) \tag{42}$$

and its solution

$$\phi(x) = \frac{1}{\pi}\frac{d}{dx}\int_0^x \frac{f(y)}{(x-y)^{1/2}}\,dy. \tag{43}$$

This case was first treated by Abel in 1823 in connection with the problem of the tautochrone. In that problem a particle slides down an unknown curve in a prescribed time, depending on the initial elevation. The resultant equation then becomes

$$t(x) = \frac{1}{\sqrt{2g}}\int_0^x \frac{s'(y)\,dy}{(x-y)^{1/2}}. \tag{44}$$

Here $t(x)$ denotes the time of descent from elevation x to $x = 0$, $x(s)$ the arclength from $x = 0$ to elevation x along the curve, and g is the gravitational constant. Clearly (44) is of type (42) and $s'(x)$ can be found via (43). Then $s(x)$ can be found and the curve is thus determined.

EXERCISES

1. Verify that existence and uniqueness results apply to

$$\phi(x) - \int_0^x y^{x-y}\phi(y)\,dy = 0.$$

2. Show that $\phi(x) = Cx^{x-1}$ satisfies the above equation and that $\int_0^1 \phi^2(x)\,dx$ does not exist.

3. Complete the inductive argument used in the proof of theorem 6.
4. Complete the proof of theorem 7 by showing in detail that

$$|K_{2^n}(x, y)| \leq \frac{M^n}{(x - y)^{2^n \alpha + 1 - 2^n}} \cdots$$

5. Consider the two kernels

$$K_1(x, y) = \frac{H_1(x, y)}{(x - y)^\alpha}, \qquad K_2(x, y) = \frac{H_2(x, y)}{(x - y)^\beta}$$

that define Volterra operators. Construct the product operator K_1, K_2, show that it is a Volterra operator and find its kernel. $H_i(x, y)$ is assumed to be continuous and $0 < \alpha, \beta < 1$.

6. Let $H(x, y)$ be continuous and show that

$$\phi(x) - \lambda \int_0^x H(x, y) \log (x - y)\phi(y) \, dy = f(x)$$

has a solution.

7. Solve $\phi(x) - \lambda \int_0^x a(x)b(y)\phi(y) \, dy = f(x)$.
8. Solve $u(x) = \sec^3 x + \lambda \int_0^1 u(t) \, dt$.
9. Solve $u(x) = e^x + \lambda \int_0^{10} u(t) \, dt$.
10. Solve $u(x) = e^x + \lambda \int_0^1 2e^x e^t u(t) \, dt$.
11. Solve $u(x) = \frac{1}{2} \int_0^\pi \sin xu(t) \, dt$.
12. Consider

$$\int_a^b K(x, y)\phi(y) \, dy = f(x).$$

where K is a degenerate operator. What can one say about the solvability and uniqueness, if solutions exist?

3

INTEGRAL EQUATIONS
WITH L_2 KERNELS

SUMMARY

In the first part of this chapter, the so-called compact operators are defined and their principal properties are derived. For selfadjoint compact operators one can obtain more detailed information regarding their eigenvalues, eigenfunctions and the associated expansion theorems. These results are then applied to compact integral operators. It will be shown that all regular ordinary differential operators have inverses that are compact. The general Sturm-Liouville problem can be treated by these methods.

The above results can be refined when dealing with positive operators and self-adjoint integral operators that have continuous kernels.

A general theory of integral equations with compact integral operators can be developed. The Fredholm Alternative will be proved for these cases. All these results can be generalized to weighted integral operators.

1. INTRODUCTION

We shall be concerned with integral equations of the type

$$\phi(x) - \lambda \int_a^b K(x, y)\phi(y)\, dy = f(x) \tag{1}$$

in the Hilbert space $L_2[a, b]$. The function $f(x)$ will be assumed to belong to $L_2[a, b]$ and the kernel will be assumed to be square-integrable, so that

$$\int_a^b \int_a^b |K(x, y)|^2\, dx\, dy < \infty$$

and we shall refer to such a kernel as an L_2 kernel for short. If for some particular value of λ the homogeneous equation

$$\phi(x) - \lambda \int_a^b K(x, y)\phi(y)\, dy = 0 \tag{2}$$

has solutions other than $\phi(x) = 0$, we shall call λ an eigenvalue of K and the solutions of (2) eigenfunctions. It will be shown that if $K(x, y)$ is an L_2 kernel, then for every eigenvalue (2) has at most a finite number of solutions.

In particular if the operator K is self-adjoint all eigenvalues will be shown to be real and the associated eigenfunctions will be orthogonal. That is, if $\phi(x)$ and $\psi(x)$ are two distinct eigenfunctions

$$(\phi, \psi) = \int_a^b \phi(x)\overline{\psi(x)}\, dx = 0.$$

We will derive suitable expansion theorems in terms of these eigenfunctions that will enable us to solve not only (1) but also the equation of the first kind

$$\int_a^b K(x, y)\phi(y)\, dy = f(x). \tag{3}$$

The eigenvalues will also be characterized in terms of certain extremal properties, that are useful in providing numerical estimates.

Finally we shall be able to derive a Fredholm alternative for (1) even when K is not self-adjoint.

2. COMPACT OPERATORS

We shall show that L_2 kernels define integral operators that are special cases of so-called compact operators.

DEFINITION. Let K be a bounded, linear operator on a Hilbert space H. Let $\{f_n\}$ be an infinite uniformly bounded sequence in H; that is for some M we have $\|f_n\| \leq M$ for all n. K will be said to be a compact operator if from the sequence $\{Kf_n\}$ one can extract a subsequence $\{Kf_{n_k}\}$ that is a Cauchy sequence. The sequence $\{Kf_{n_k}\}$ converges, of course, since H is a Hilbert space.

THEOREM 1. Let H be $L_2[a, b]$, and K a degenerate integral operator

$$Kf = \sum_{i=1}^m a_i(x) \int_a^b b_i(y)f(y)\, dy.$$

K is a compact operator, if $a_i(x)$ and $b_i(x)$ belong to $L_2[a, b]$ for all i.

PROOF. Consider a uniformly bounded set of functions $\{f_n\}$.

$$Kf_n = \sum_{i=1}^{m} a_i(x) \int_a^b b_i(y)f_n(y) \, dy.$$

Clearly

$$\left| \int_a^b b_i(y)f_n(y) \, dy \right| \leq \|b_i(y)\| \, \|f_n\| \leq M \, \|b_i(y)\|$$

so that the sequence $\{\int_a^b b_i(y)f_n(y) \, dy\}$ is a uniformly bounded sequence of complex numbers. It must therefore have a point of accumulation, and a suitable subsequence will converge to that point of accumulation.

Let $\{f_{n(1)}\}$ be a subsequence of $\{f_n\}$ such that $\{\int_a^b b_1(y)f_{n(1)}(y) \, dy\}$ converges to some complex number, say c_1. By the same argument we can now extract a subsequence $\{f_{n(2)}\}$ of $\{f_{n(1)}\}$ such that $\{\int_a^b b_2(y)f_{n(2)}(y) \, dy\}$ converges to, say c_2. By extracting successive subsequences we finally arrive at $\{f_{n(m)}\}$ with property that

$$\lim_{n(m) \to \infty} \int_a^b b_i(y)f_{n(m)}(y) \, dy = c_i, \qquad i = 1, 2, \ldots, m.$$

Since we have the inclusion

$$\{f_n\} \supset \{f_{n(1)}\} \supset \{f_{n(2)}\} \supset \cdots \supset \{f_{n(m)}\}$$

we see that the above limit exists for all i.

Finally we note that the sequence $\{Kf_{n(m)}\}$ is a subsequence of $\{Kf_n\}$, and is also a Cauchy sequence. In fact

$$\lim_{n(m) \to \infty} Kf_{n(m)} = \sum_{i=1}^{m} c_i a_i(x),$$

so that K is a compact operator. ∎

There are a number of properties of compact operators that are especially useful. Some of these will be formulated as theorems.

THEOREM 2. Let K be a compact operator on the Hilbert space H, and L a bounded operator. Then both KL and LK are compact operators.

PROOF. Let $\{f_n\}$ be a uniformly bounded sequence in H. Since K is compact it contains a subsequence $\{f_{n'}\}$ such that $\{Kf_{n'}\}$ is a Cauchy sequence. Since L is bounded, we see that

$$\|LKf_{n'} - LKf_{m'}\| \leq \|L\| \, \|Kf_{n'} - Kf_{m'}\|$$

so that $\{LKf_{n'}\}$ is also a Cauchy sequence and LK is compact.

Now we consider the operator KL. If $\{f_n\}$ is uniformly bounded, and L is bounded, then $\{Lf_n\}$ is again uniformly bounded. Since K is compact there exists a subsequence such that $\{KLf_{n'}\}$ is a Cauchy sequence, so that KL is also compact. ■

THEOREM 3. Let K and L be two compact operators on a Hilbert space H. Then $K + L$ is again compact, and so is αK for any scalar α.

PROOF. Let $\{f_n\}$ be a uniformly bounded sequence in H. It certainly contains a subsequence $\{f_{n'}\}$ such that $\{Kf_{n'}\}$ is a Cauchy sequence. If α is any scalar then clearly $\{\alpha Kf_{n'}\}$ will again be a Cauchy sequence so that αK is compact.

Now we can extract from $\{f_{n'}\}$ a subsequence $\{f_{n''}\}$ such that both $\{Kf_{n''}\}$ and $\{Lf_{n''}\}$ are Cauchy sequences. In that case $\{(K + L)f_{n''}\}$ is a Cauchy sequence, so that $K + L$ is also compact. ■

COROLLARY. Let $\{K_k\}$ be a finite sequence of n compact operators on a Hilbert space H, and $\{\alpha_k\}$ a set of scalars. Then $\sum_{k=1}^{n} \alpha_k K_k$ is compact.

PROOF. The result follows by an immediate application of theorem 3. ■

THEOREM 4. Let K be a compact operator on a Hilbert space H, and let $\{f_n\}$ be a linearly independent sequence of eigenfunctions corresponding to some nonzero eigenvalue μ, that is $Kf_n = \mu f_n$ for all n. Then $\{f_n\}$ contains a finite number of elements.

PROOF. We shall replace the set $\{f_n\}$ by an equivalent set $\{g_n\}$, where all g_n are again eigenfunctions, but are orthonormal, so that

$$(g_n, g_m) = 0, \qquad n \neq m$$
$$= 1, \qquad n = m.$$

To accomplish this we use the well-known Gram-Schmidt process. Let

$$g_1 = \frac{f_1}{\|f_1\|}$$

$$g_2 = \frac{f_2 - (f_2, g_1)g_1}{\|f_2 - (f_2, g_1)g_1\|}$$

.

.

.

$$g_n = \frac{f_n - \sum_{k=1}^{n-1} (f_n, g_k)g_k}{\|f_n - \sum_{k=1}^{n-1} (f_n, g_k)g_k\|}$$

.

.

.

and it is easy to verify that both sets $\{f_1, f_2, \ldots, f_n\}$ and $\{g_1, g_2, \ldots, g_n\}$ span the same n dimensional subspace of H. Furthermore, a simple inductive argument shows that if $(g_1, g_2, \ldots, g_{j-1})$ is orthonormal, then for $i < j$

$$(g_j, g_i) = \frac{(f_j, g_i) - \sum_{k=1}^{j-1} (f_j, g_k)(g_k, g_i)}{\|f_j - \sum_{k=1}^{j-1} (f_j, g_k)g_k\|} = 0$$

so that the set $\{g_n\}$ is orthonormal. Note that none of the denominators can vanish; otherwise the $\{f_n\}$ would not be linearly independent. One can verify that

$$\|g_n - g_m\|^2 = \|g_n\|^2 - (g_n, g_m) - (g_m, g_n) + \|g_m\|^2 = 2$$

if $n \neq m$.

Clearly, the set of eigenfunctions $\{g_n\}$ is uniformly bounded. If it were infinite we could select a subsequence such that $\{Kg_{n'}\}$ is a Cauchy sequence. Then

$$\|Kg_{n'} - Kg_{m'}\| = \|\mu g_{n'} - \mu g_{m'}\| = |\mu| \sqrt{2},$$

because the set $\{g_n\}$ is orthonormal. Since $\mu \neq 0$ the above cannot be a Cauchy sequence; otherwise $\|Kg_{n'} - Kg_{m'}\|$ could be made smaller than $|\mu| \sqrt{2}$ for sufficiently large n', m'. It follows that the number of linearly independent eigenfunctions must be finite. The requirement that $\mu \neq 0$ is clearly vital in the proof. For a degenerate operator it is easy to verify that $\mu = 0$ is an eigenvalue with infinite multiplicity. ∎

The next theorem will prove to be of great importance in the sequel. It is often rather difficult to prove directly that an operator is compact. This theorem often facilitates such proof.

THEOREM 5. Let $\{K_n\}$ be a sequence of compact operators on a Hilbert space H, such that for some operator K we have

$$\lim_{n \to \infty} \|K - K_n\| = 0.$$

Then K is also compact.

PROOF. Let $\{f_n\}$ be any infinite, uniformly bounded sequence. We can select a subsequence $\{f_{n(1)}\}$ such that $\{K_1 f_{n(1)}\}$ is a Cauchy sequence. From the sequence $\{f_{n(1)}\}$ we now extract a subsequence $\{f_{n(2)}\}$ such that $\{K_2 f_{n(2)}\}$ is a Cauchy sequence. Proceeding in this fashion we obtain a succession of subsequences

$$\{f_n\} \supset \{f_{n(1)}\} \supset \{f_{n(2)}\} \supset \cdots \supset \{f_{n(k)}\} \supset \cdots$$

such that $\{K_k f_{n(i)}\}$ is a Cauchy sequence for $k = 1, 2, \ldots, i$.

Finally we select the sequence $\{f_{n(n)}\}$, which is a subsequence of every $\{f_{n(k)}\}$, except possibly for a finite number of terms so that $\{K_k f_{n(n)}\}$ will be a

Cauchy sequence for all k. We shall now show that $\{Kf_{n^{(n)}}\}$ is also a Cauchy sequence from which we can conclude that K is compact. Now

$$
\|Kf_{n^{(n)}} - Kf_{m^{(m)}}\|
$$
$$
= \|(K - K_k)f_{n^{(n)}} + K_kf_{n^{(n)}} - K_kf_{m^{(m)}} + (K_k - K)f_{m^{(m)}}\|
$$
$$
\leq \|K - K_k\| \, \|f_{n^{(n)}}\| + \|K_kf_{n^{(n)}} - K_kf_{m^{(m)}}\| + \|K_k - K\| \, \|f_{m^{(m)}}\|.
$$

The first and third term can be made smaller than $M\epsilon$, if $\|f_{n^{(n)}}\| \leq M$, and $k > k(\epsilon)$, and the middle term will also be smaller than ϵ if $n^{(n)}$ and $m^{(m)}$ are larger than $N(\epsilon)$, so that

$$
\|Kf_{n^{(n)}} - Kf_{m^{(m)}}\| \leq (2M + 1)\epsilon, \qquad k > k(\epsilon), \quad n^{(n)}, m^{(m)} > N(\epsilon).
$$

It follows that $\{Kf_{n^{(n)}}\}$ is a Cauchy sequence so that K is compact. ∎

This last theorem will help us in the following manner. First we shall prove that, if $K(x, y)$ is continuous for $0 \leq x, y \leq 1$, then the associated integral operator on $L_2[0, 1]$ is compact. We showed earlier that the space $L_2[0, 1]$ can be viewed as the completion of $C[0, 1]$, with respect to the appropriate norm. In a similar manner we can view L_2 kernels as suitable limits. Then we have

$$
\lim_{n \to \infty} \int_0^1 \int_0^1 |K(x, y) - K_n(x, y)|^2 \, dx \, dy = 0,
$$

where $K(x, y)$ is an L_2 kernel and all $K_n(x, y)$ are continuous. If we knew that the associated integral operators K_n are compact, then the above statement is equivalent to

$$
\lim_{n \to \infty} \|K - K_n\| = 0
$$

so that K would also be compact.

THEOREM 6. Let $K(x, y)$ be continuous for all $0 \leq x, y \leq 1$. The associated integral operator

$$
Kf = \int_0^1 K(x, y)f(y) \, dy
$$

is a compact operator on $L_2[0, 1]$.

PROOF. Since $|K(x, y)| \leq M$, the above operator is clearly bounded

$$
\|Kf\| \leq M \, \|f\|
$$

and without loss of generality we can assume that $\|K\| = 1$. We now take any uniformly bounded sequence $\{f_n(x)\}$, such that (again without loss of generality) $\|f_n\| \leq 1$. Consider now the set of functions $\{g_n(x)\}$ where $g_n = Kf_n$. We shall show that there is a convergent subsequence $\{g_{n'}(x)\}$ that converges to a function in $L_2[0, 1]$.

To accomplish this result we select a dense countable set of points on the interval $[0, 1]$. The rational numbers are certainly acceptable. We shall denote these by $\{r_k\}$. Now we consider the set $\{g_n(r_1)\}$. From the properties of $K(x, y)$ and $\{f_n(x)\}$ it follows that

$$|g_n(x)| \leq M \|f_n\| = M.$$

In particular we have

$$|g_n(r_1)| \leq M$$

so that the set $\{g_n(r_1)\}$ has a point of accumulation in the interval $[-M, M]$. We can now select a subsequence $\{g_{n^{(1)}}(r_1)\}$ that converges to, say, b_1. Next we consider the sequence of functions $\{g_{n^{(1)}}(x)\}$. At $x = r_1$, it converges to b_1. Now we examine $\{g_{n^{(1)}}(r_2)\}$ and select a subsequence $\{g_{n^{(2)}}(r_2)\}$ that converges to, say, b_2. We repeat this process, ad infinitum, at all rationals. Finally we consider the set of functions $\{g_{n^{(n)}}(x)\}$. This sequence converges at all rationals by virtue of its construction.

For convenience we shall reindex the sequence $\{g_{n^{(n)}}(x)\}$ so that we call it $\{g_n(x)\}$. We now define a function $g(x)$ by

$$\lim_{n \to \infty} g_n(r_k) = g(r_k)$$

so that $g(x)$ is only defined for rational x. We can also show that on the rationals $g(x)$ is continuous. Let x_1 and x_2 be any two rational numbers. Then

$$|g(x_1) - g(x_2)| = \lim_{n \to \infty} |g_n(x_1) - g_n(x_2)|$$

$$= \lim_{n \to \infty} \left| \int_0^1 [K(x_1, y) - K(x_2, y)] f_n(y) \, dy \right|$$

$$\leq \lim_{n \to \infty} \left[\int_0^1 |K(x_1, y) - K(x_2, y)|^2 \, dy \right]^{1/2} \|f_n\|$$

$$\leq \left[\int_0^1 |K(x_1, y) - K(x_2, y)|^2 \, dy \right]^{1/2}$$

and since $K(x, y)$ is continuous so is $g(x)$. Now we can in fact define $g(x)$ for all $0 \leq x \leq 1$. Let x be arbitrary, and x_k a sequence of rational numbers such that

$$\lim_{k \to \infty} x_k = x.$$

Then

$$g(x) = \lim_{k \to \infty} g(x_k).$$

The above $g(x)$ is well-defined. To see this suppose that $\{\tilde{x}_k\}$ is a second sequence of rationals converging to x and suppose that $\lim_{k \to \infty} g(\tilde{x}_k) = \tilde{g}(x)$. Then

$$|g(x) - \tilde{g}(x)| \leq |g(x) - g(x_k)| + |g(x_k) - g(\tilde{x}_k)| + |g(\tilde{x}_k) - \tilde{g}(x)|.$$

Now for $k > k(\epsilon)$ the first and third terms on the right are less than ϵ. Since $g(x)$ is continuous on the rationals the middle term is also less than ϵ and it follows that $g(x) = \tilde{g}(x)$. Now it follows that $g(x)$ is well defined as a continuous function for all $0 \leq x \leq 1$. Being continuous on a closed and bounded interval it is also uniformly continuous and thus certainly belongs to $L_2[0, 1]$. It follows that the operator K is indeed compact. ∎

The preceding argument applies to $L_2[a, b]$, where $[a, b]$ is any finite interval, since by a simple change of variable such an interval can be normalized to $[0, 1]$.

In the course of proving theorem 6 we have in fact succeeded in proving another standard theorem of analysis.

Arzelà's Theorem. Let $\{g_n(x)\}$ be a set of continuous functions defined on $[0, 1]$, that is uniformly bounded and equicontinuous. That is

$$|g_n(x)| \leq M \qquad \text{for all } n$$
$$|g_n(x_1) - g_n(x_2)| < \epsilon \qquad \text{for } |x_1 - x_2| < \delta(\epsilon)$$

where M and $\delta(\epsilon)$ are independent of n. One can then extract a subsequence $\{g_{n_k}(x)\}$ that converges uniformly to a continuous function $g(x)$.

The proof of Arzelà's theorem is implicitly contained in the proof of theorem 6. There we took the set $\{f_n(x)\}$ to be in $L_2[0, 1]$, such that $\|f_n\| \leq 1$. Then the set $\{g_n(x)\}$ was defined by

$$g_n(x) = \int_0^1 K(x, y) f_n(y)\, dy$$

so that

$$|g_n(x)| \leq \left\{ \int_0^1 |K(x, y)|^2\, dy \int_0^1 |f_n(y)|^2\, dy \right\}^{\frac{1}{2}} \leq k\, \|f_n\| \leq kM$$

where $|K(x, y)| \leq k$. Clearly $\{g_n(x)\}$ is uniformly bounded. Since $K(x, y)$ is continuous we have

$$|g_n(x_1) - g_n(x_2)| \leq \left\{ \int_0^1 |K(x_1, y) - K(x_2, y)|^2\, dy \right\}^{\frac{1}{2}} \|f_n\| < \epsilon$$

$$\text{if } |x_1 - x_2| < \delta(\epsilon)$$

so that $\{g_n(x)\}$ is also equicontinuous.

We can now extend the results of theorem 6 from continuous kernels to L_2 kernels.

THEOREM 7. Let $K(x, y)$ be such that

$$\int_0^1 \int_0^1 |K(x, y)|^2\, dx\, dy < \infty.$$

Then the operator

$$Kf = \int_0^1 K(x, y) f(y) \, dy$$

is a compact operator on $L_2[0, 1]$.

PROOF. As was remarked earlier, we consider an L_2 kernel $K(x, y)$ as a limit of continuous kernels, such that

$$\lim_{n \to \infty} \int_0^1 \int_0^1 |K(x, y) - K_n(x, y)|^2 \, dx \, dy = 0$$

and we define

$$\int_0^1 \int_0^1 |K(x, y)|^2 \, dx \, dy = \lim_{n \to \infty} \int_0^1 \int_0^1 |K_n(x, y)|^2 \, dx \, dy.$$

Using the Cauchy-Schwarz inequality we see that

$$\|(K - K_n)f\| = \left\| \int_0^1 [K(x, y) - K_n(x, y)] f(y) \, dy \right\|$$

$$\leq \left\| \left\{ \int_0^1 |K(x, y) - K_n(x, y)|^2 \, dy \int_0^1 |f(y)|^2 \, dy \right\}^{\frac{1}{2}} \right\|$$

$$\leq \left\{ \int_0^1 \int_0^1 |K(x, y) - K_n(x, y)|^2 \, dx \, dy \right\}^{\frac{1}{2}} \|f\|$$

so that

$$\|K - K_n\| \leq \left\{ \int_0^1 \int_0^1 |K(x, y) - K_n(x, y)|^2 \, dx \, dy \right\}^{\frac{1}{2}}$$

and

$$\lim_{n \to \infty} \|K - K_n\| = 0.$$

Since each K_n is a compact operator, by theorem 6, it follows from theorem 5 that K is also compact. ■

To complete our discussion we still have to consider compact operators on $L_2[0, \infty]$ and $L_2[-\infty, \infty]$. Here again theorem 5 will prove to be a vital tool.

THEOREM 8. Let $K(x, y)$ be an L_2 kernel on $L_2[0, \infty]$ $(L_2[-\infty, \infty])$. Then the operator

$$Kf = \int_0^\infty K(x, y) f(y) \, dy \left(\int_{-\infty}^\infty K(x, y) f(y) \, dy \right)$$

is compact.

PROOF. We shall consider only the case $L_2[0, \infty]$, but the case $L_2[-\infty, \infty]$ is so similar that no additional discussion is really necessary. In order for

$K(x, y)$ to be an L_2 kernel we require that

$$\int_0^\infty \int_0^\infty |K(x, y)|^2 \, dx \, dy < \infty.$$

So far we have discussed integrals over finite domains. Integrals over infinite domains have to be treated as limits. We require that

$$\lim_{n \to \infty} \int_0^n \int_0^n |K(x, y)|^2 \, dx \, dy < \infty$$

and then this limit defines the integral over an infinite square.

Now we define the kernels

$$K_n(x, y) = K(x, y), \qquad 0 \leq x, y \leq n$$
$$= 0, \qquad\qquad \text{otherwise.}$$

Evidently

$$\lim_{n \to \infty} \| K - K_n \|^2 \leq \lim_{n \to \infty} \int_0^\infty \int_0^\infty |K(x, y) - K_n(x, y)|^2 \, dx \, dy = 0.$$

Inspection of K_n shows that we can consider it as an operator on $L_2[0, n]$, and by theorem 7 it will be compact. Consequently, it will also be compact on $L_2[0, \infty]$, and by theorem 5 K will be compact on $L_2[0, \infty]$. ■

The preceding theorems summarize the most important facts regarding compact operators, that we shall require. To go further we shall have to specialize to an even more restricted class, the self-adjoint operators.

3. SELF-ADJOINT COMPACT OPERATORS

We first recall two theorems regarding any self-adjoint operators.

THEOREM 9. Let K be a self-adjoint operator on a Hilbert space H. All eigenvalues of K are real.

PROOF. Suppose

$$Kf = \mu f.$$

Then by taking inner products of Kf and f we find

$$(Kf, f) = \mu(f, f)$$
$$(f, Kf) = \bar{\mu}(f, f).$$

But since K is self-adjoint

$$(Kf, f) = (f, Kf)$$

so that

$$\mu(f, f) = \bar{\mu}(f, f)$$

and since $(f, f) > 0$ we have $\mu = \bar{\mu}$ so that μ is real. ■

THEOREM 10. Let K be a self-adjoint operator on a Hilbert space H. All eigenfunctions corresponding to distinct eigenvalues are orthogonal. A set of eigenfunctions that is finite or countable corresponding to the same eigenvalue can be orthogonalized.

PROOF. Suppose $\mu_1 \neq \mu_2$ and $Kf_1 = \mu_1 f_1$, $Kf_2 = \mu_2 f_2$. Then

$$(Kf_1, f_2) = \mu_1(f_1, f_2)$$

and also

$$(Kf_1, f_2) = (f_1, Kf_2) = \mu_2(f_1, f_2)$$

so that

$$\mu_1(f_1, f_2) = \mu_2(f_1, f_2).$$

Since $\mu_1 \neq \mu_2$ it follows that $(f_1, f_2) = 0$.

Suppose that corresponding to a given eigenvalue μ, K has more than one eigenfunction. If such a set $\{f_n\}$ is finite or countable it can be replaced by an equivalent, but orthonormal, set $\{g_n\}$ by the Gram-Schmidt process (see the proof of theorem 4). ■

N.B. We shall be concerned exclusively with so-called separable Hilbert spaces. The space $L_2[a, b]$ is separable and for these we can show that all orthogonal sets are finite or countable.

Earlier the norm of a bounded operator was defined by

$$\|K\| = \sup_{f \neq 0} \frac{\|Kf\|}{\|f\|} = \sup_{\|f\|=1} \|Kf\|.$$

We shall show that if K is self-adjoint and compact one of $\pm \|K\|$ must indeed be an eigenvalue.

THEOREM 11. Let K be a compact, self-adjoint operator on a Hilbert space H. At least one of $\pm \|K\|$ must be an eigenvalue.

PROOF. Let $\{f_n\}$ be a sequence such that $\|f_n\| = 1$ and $\|Kf_n\|$ converges to $\|K\|$. Now we have

$$0 \leq \|K^2 f_n - \|Kf_n\|^2 f_n\|^2 = \|K^2 f_n\|^2 - 2\|Kf_n\|^2 (K^2 f_n, f_n) + \|Kf_n\|^4$$

$$= \|K^2 f_n\|^2 - \|Kf_n\|^4$$

$$\leq \|K\|^2 \|Kf_n\|^2 - \|Kf_n\|^4.$$

Since $\|Kf_n\|$ converges to $\|K\|$ we see that

$$\lim_{n \to \infty} \|K^2 f_n - \|Kf_n\|^2 f_n\| = 0. \tag{4}$$

K^2 as the product of compact operators is compact and we can extract a subsequence of $\{f_n\}$, say $\{f_{n'}\}$, such that $\{K^2 f_{n'}\}$ converges to an element, say $\|K\|^2 g$. Using this in (4) we see that

$$\lim_{n \to \infty} \| \|K\|^2 g - \|K\|^2 f_{n'} \| = 0$$

so that $f_{n'}$ converges to g and

$$K^2 g = \|K\|^2 g.$$

We have now shown that K^2 has an eigenvalue $\|K\|^2$. To complete the proof we observe that

$$(K - \|K\|)(K + \|K\|)g = 0.$$

If $h = (K + \|K\|)g \neq 0$, $(K - \|K\|)h = 0$ so that $\|K\|$ is an eigenvalue of K. If $h = 0$, $-\|K\|$ is an eigenvalue of K. ■

This theorem guarantees the existence of at least one nonzero eigenvalue, but no more in general. It is easy to give examples of operators with precisely one such eigenvalue.

EXAMPLE. Let $K(x, y)$ be defined by

$$K(x, y) = a(x)\overline{a(y)}$$

where $a(x) \in L_2[a, b]$. Then

$$Kf = \int_a^b K(x, y) f(y) \, dy$$

is a compact, self-adjoint operator on $L_2[a, b]$. Clearly

$$Ka(x) = \int_a^b a(x)\overline{a(y)}a(y) \, dy = \left(\int_a^b |a(y)|^2 \, dy \right) a(x)$$

so that $a(x)$ is an eigenfunction of K, with eigenvalue $\int_a^b |a(y)|^2 \, dy = \|a\|^2$. One can also verify that $\|K\| = \|a\|^2$. In fact $a(x)$ is the only linearly independent eigenfunction with a nonzero eigenvalue. If $b(x)$ were any other eigenfunction

$$Kb(x) = \left(\int_a^b \overline{a(y)}b(y) \, dy \right) a(x) = \alpha b(x)$$

so that either $\alpha = 0$ or $b(x)$ is linearly dependent on $a(x)$.

A few remarks are in order here regarding our terminology. In terms of operator theory K is said to have an eigenvalue μ, if there exists a nonzero element f such that

$$Kf = \mu f. \tag{5}$$

In integral equation theory we deal with operators of the form I-λK, and λ is said to be an eigenvalue if for some f

$$(I - \lambda K)f = 0 \qquad (6)$$

or equivalently,

$$Kf = \frac{1}{\lambda} f. \qquad (7)$$

Inspection of (5) and (6), (7) shows a lack of consistency in our terminology. In fact according to (5), $1/\lambda$ is an eigenvalue in (7). Within a given context, it will generally be clear whether the terminology of (5) or (7) is being used. Furthermore, we shall reserve the symbol μ for eigenvalues as defined in (5), and λ as defined in (6) and (7).

The requirement of selfadjointness in theorem 11 is a vital one. In Chapter 3 it was seen that the Volterra operators, whether compact or merely bounded, had no eigenvalue, since $(I - \lambda K)f = 0$ had only the solution $f = 0$ for all λ. It is easy to check that a Volterra operator is not self-adjoint.

The proof of theorem 11 shows that $\pm \|K\|$ must be an eigenvalue. Since the quantity $\|K\|$ is characterized by an extremal property one can always estimate $\|K\|$ by calculating $\|Kf\|/\|f\|$ for a number of different f and thereby obtain a suitable numerical approximation.

We shall now examine the structure of K in terms of all of its eigenfunctions. Let μ_1 denote the eigenvalue guaranteed by theorem 11 and ϕ_1 a corresponding eigenfunction. We shall also assume that $\|\phi_1\| = 1$. Now we shall construct a new operator K_1 defined by

$$K_1 f = Kf - \mu_1 \phi_1 (f, \phi_1). \qquad (8)$$

If we view K as an integral operator its kernel is given by

$$K_1(x, y) = K(x, y) - \mu_1 \phi_1(x)\overline{\phi_1(y)}.$$

The operator K_1 is again compact and self-adjoint, since it is the sum of two such operators. There are now two possibilities; either $\|K_1\| = 0$ so that $K(x, y) = \mu_1 \phi_1(x)\overline{\phi_1(y)}$, or else $\|K_1\| \neq 0$.

If $\|K_1\| \neq 0$, then it is possible that K has more than one independent eigenfunction associated with the eigenvalue $\mu_1 = \|K\|$. Suppose there are k such independent eigenfunctions. In accordance with theorem 4 there is at most a finite number of such independent eigenfunctions. By use of the Gram-Schmidt process we can assume that the set of eigenfunctions ϕ_1, $\phi_2, \ldots, \phi_k$ is orthonormal. Following the proof of theorem 11 it is of course possible that $-\|K\|$ is also an eigenvalue and we can associate with it the independent eigenfunctions $\phi_{k+1}, \phi_{k+2}, \ldots, \phi_n$, that are also orthonormal.

In analogy to (8) we now construct the operator

$$K_n f = Kf - \sum_{i=1}^{n} \mu_i \phi_i(f, \phi_i), \qquad (9)$$

which is an integral operator with the kernel

$$K_n(x, y) = K(x, y) - \sum_{i=1}^{n} \mu_i \phi_i(x)\overline{\phi_i(y)}. \qquad (10)$$

In the above we used the convention that

$$\mu_1 = \mu_2 = \cdots = \mu_k = \|K\|$$

$$\mu_{k+1} = \mu_{k+2} = \cdots = \mu_n = -\|K\|$$

showing that $\|K\|$ is an eigenvalue of multiplicity k, and $-\|K\|$ one of multiplicity $n - k$. The operator K_n is again a compact, self-adjoint operator. Here again we have two possibilities. Either $\|K_n\| = 0$ or $\|K_n\| \neq 0$. In the former case we see from (9) that K is a degenerate operator. We shall now study the structure of K_n, if $\|K_n\| \neq 0$.

Let f be any element in H. We now write

$$f = \sum_{i=1}^{n} (f, \phi_i)\phi_i + f^{\perp} \qquad (11)$$

and (11) defines $f^{\perp}$. Since

$$(f, \phi_j) = (f, \phi_j) + (f^{\perp}, \phi_j)$$

it follows that

$$(f^{\perp}, \phi_j) = 0, \qquad j = 1, 2, \ldots, n. \qquad (12)$$

Let H_n be the Hilbert space spanned by the eigenfunctions $\phi_1, \phi_2, \ldots, \phi_n$, and let $H_n^{\perp}$ denote its orthogonal complement. By the latter is meant the space of all elements in H orthogonal to H_n. Equation (11) shows that every element in f can be written uniquely as a sum of two terms, one in H_n and the other in $H_n^{\perp}$. We then say that

$$H = H_n \oplus H_n^{\perp}. \qquad (13)$$

(The above is read as: H is the orthogonal sum of H_n and $H_n^{\perp}$.)

DEFINITION. A subspace H' of a Hilbert space H is said to be invariant under an operator K if K applied to any element of H' yields an element in H'.

We are now in a position to prove the following lemma.

LEMMA. H_n and $H_n^{\perp}$ are invariant under K.

PROOF.   If $f \in H_n$ we have

$$f = \sum_{i=1}^{n} (f, \phi_i)\phi_i$$

and

$$Kf = \sum_{i=1}^{n} (f, \phi_i)\mu_i\phi_i$$

so that $Kf \in H_n$.

If $f \in H_n^{\perp}$, then

$$(f, \phi_i) = 0, \qquad i = 1, 2, \ldots, n.$$

Now

$$(Kf, \phi_i) = (f, K\phi_i) = \mu_i(f, \phi_i) = 0$$

so that $Kf \in H_n^{\perp}$. ■

A simple calculation using (9) and (11) now shows that

$$K_n f = Kf^{\perp}. \tag{14}$$

Equation (14) shows that K_n maps any element f into $H_n^{\perp}$. We also have

$$\|K_n\| = \sup_{\|f\|=1} \|K_n f\| = \sup_{\|f^{\perp}\|=1} \|Kf^{\perp}\| \leq \|K\|,$$

since the sup over a subspace cannot exceed the sup over the full space. From the above we can conclude in fact that

$$\|K_n\| < \|K\|. \tag{15}$$

Otherwise

$$\sup_{\|f\|=1, f \in H_n^{\perp}} \|Kf\| = \|K\|$$

so that K must have an eigenvector in $H_n^{\perp}$, that is associated with an eigenvalue $\pm\|K\|$. But all such eigenvectors have already been included in H_n so that (15) must hold.

Using theorem 11 we can now conclude that K, viewed as a compact self-adjoint operator on $H_n^{\perp}$ has norm $\|K_n\|$ and hence must have an eigenfunction in $H_n^{\perp}$ associated with an eigenvalue $\pm\|K_n\|$. In analogy to (9) we can now write

$$K_m^f = Kf - \sum_{i=1}^{m} \mu_i\phi_i(f, \phi_i) \tag{16}$$

where the sequence $\mu_1, \mu_2, \ldots, \mu_m$ contains all eigenvalues $\pm\|K\|, \pm\|K_n\|$ appearing in accordance with their multiplicity. If $\|K_m\| = 0$ we see that K is a degenerate operator. Otherwise we repeat the preceding calculation. Either the process stops after a finite number of steps, in which case K is

We see that

$$\|K_n f - K_m f\|^2 = \sum_{i=m+1}^{n} \mu_i^2 |(f, \phi_i)|^2 \leq \mu_{m+1}^2 \sum_{i=m+1}^{n} |(f, \phi_i)|^2$$

$$\leq \mu_{m+1}^2 \|f\|^2$$

by Bessel's inequality. Since $\mu_m \to 0$, it follows that $\{K_n f\}$ is a Cauchy sequence and therefore converges to an element in H. ∎

We have concluded that the sequence $\{K_n f\}$ converges. In fact we can show that $K_n f \to 0$. We define the sequence $\{f_n\}$ by

$$f_n = f - \sum_{i=1}^{n} (f, \phi_i)\phi_i$$

so that

$$Kf_n = Kf - \sum_{i=1}^{n} \mu_i(f, \phi_i)\phi_i = K_n f.$$

Clearly

$$(Kf_n, \phi_i) = 0, \qquad i = 1, 2, \ldots, n$$

so that H_n is a subspace on which the largest eigenvalue of K, in absolute value, is $|\mu_{n+1}|$. Then

$$\|Kf_n\| \leq |\mu_{n+1}| \|f_n\| \leq |\mu_{n+1}| \left[\|f\| + \left\| \sum_{i=1}^{n} (f, \phi_i)\phi_i \right\| \right]$$

$$\leq |\mu_{n+1}| \left[\|f\| + \left(\sum_{i=1}^{n} |(f, \phi_i)|^2 \right)^{1/2} \right]$$

$$\leq 2 |\mu_{n+1}| \|f\|$$

by Bessel's inequality. Since $\mu_n \to 0$ we see that $Kf_n = K_n f \to 0$ so that

$$Kf = \sum \mu_i(f, \phi_i)\phi_i. \tag{22}$$

If K is degenerate the sum in (22) is finite; otherwise the sum is infinite, but converges to Kf.

We can now state and prove another key theorem regarding compact, self-adjoint operators.

THEOREM 13. Let K be a compact, self-adjoint operator on a Hilbert space H and $\{\phi_i\}$ the set of all orthonormal eigenfunctions associated with nonzero eigenvalues of K. Denote the corresponding nonzero eigenvalues by $\{\mu_i\}$. Let f be any element in H. Then f can be represented in the form

$$f = \sum_i (f, \phi_i)\phi_i + f_0 \tag{23}$$

where f_0 is a suitable element in the nullspace of K (i.e., $Kf_0 = 0$).

PROOF.   We shall denote the subspace of H that is spanned by the set $\{\phi_i\}$ by H_0. One can verify quite easily that H_0 is in fact a Hilbert space. That it is an inner product space is immediate. To show that it is complete we take a Cauchy sequence $\{f_n\}$ in H_0. Clearly

$$f_n = \sum_{i=1}^{\infty} \alpha_i^{(n)} \phi_i$$

and

$$\|f_n - f_m\| = \sum_{i=1}^{\infty} |\alpha_i^{(n)} - \alpha_i^{(m)}|^2.$$

From the above we see that if $\{f_n\}$ is a Cauchy sequence in H_0, $\{\alpha_i^{(n)}\}$ is a Cauchy sequence in the complex plane for every i. Then

$$\lim_{n \to \infty} \alpha_i^{(n)} = \alpha_i, \qquad i = 1, 2, 3, \ldots.$$

Since every Cauchy sequence is uniformly bounded

$$\|f_n\|^2 = \sum_{i=1}^{n} |\alpha_i^{(n)}|^2 \le M$$

so that

$$\sum_{i=1}^{\infty} |\alpha_i|^2 \le M$$

and finally

$$f = \lim_{n \to \infty} f_n = \sum_{i=1}^{\infty} \alpha_i \phi_i \in H_0.$$

We now denote the set of all elements in H, that are orthogonal to H_0 by $H_0^{\perp}$, the orthogonal complement of H_0. It is easy to see that both are invariant under K and in fact $H_0^{\perp}$ is the nullspace of K. Suppose $f \in H_0$ so that

$$f = \sum_i (f, \phi_i)\phi_i$$

and

$$Kf = \sum_i \mu_i(f, \phi_i)\phi_i \in H_0.$$

Now if $f \in H_0^{\perp}$ so that $(f, \phi_i) = 0$ for all i, then

$$(Kf, \phi_i) = (f, K\phi_i) = \mu_i(f, \phi_i) = 0$$

so that $Kf \in H_0^{\perp}$. We see that every element in H can be decomposed into a sum of two elements, one in H_0 and one in $H_0^{\perp}$, and we can write

$$H = H_0 \oplus H_0^{\perp}.$$

If $f \in H$ we can write

$$f = f_1 + f_2$$

where $f_1 \in H_0$ and $f_2 \in H_0{}^\perp$. For f_1 we have

$$f_1 = \sum_i (f_1, \phi_i)\phi_i$$

and since $(f_2, \phi_i) = 0$

$$f_1 = \sum_i (f, \phi_i)\phi_i$$

so that

$$f = \sum_i (f, \phi_i)\phi_i + f_2$$

which is equivalent to (23). ∎

As a final result in this section we consider the following theorem.

THEOREM 14. Let K be a compact, self-adjoint integral operator with the L_2 kernel $K(x, y)$. If as before we denote by $\{\mu_i\}$ the nonzero eigenvalues and by $\{\phi_i\}$ the corresponding eigenvectors associated with K, and the Hilbert space is $L_2[a, b]$, then

$$\lim_{n \to \infty} \int_a^b \int_a^b \left| K(x, y) - \sum_{i=1}^n \mu_i\phi_i(x)\overline{\phi_i(y)} \right|^2 dx\, dy = 0 \qquad (24)$$

so that

$$K(x, y) = \sum_{i=1}^\infty \mu_i\phi_i(x)\overline{\phi_i(y)}$$

and convergence is in the mean (in the sense of (24)).

PROOF. Since

$$\int_a^b \int_a^b |K(x, y)|^2\, dx\, dy < \infty$$

the function $k(x) = \int_a^b |K(x, y)|^2\, dy$ as a function of x is certainly integrable on the interval $[a, b]$, so that

$$\int_a^b k(x)\, dx < \infty.$$

It follows that $k(x)$ is finite for almost all x on $[a, b]$. Let x be such a point. For all $f \in H_0{}^\perp$ (the nullspace of K) we have

$$\int_a^b K(x, y)f(y)\, dy = 0$$

so that $\overline{K(x, y)}$, as a function of y, belongs to H_0, that is, it is orthogonal to f for fixed x. Accordingly, we have

$$K(x, y) = \sum_{i=1}^\infty \alpha_i\overline{\phi_i(y)}.$$

It is evident that if $\{\phi_i(x)\}$ is a complete, orthonormal set on $[a, b]$, so is $\{\overline{\phi_i(x)}\}$. It follows that

$$\alpha_i = \int_a^b K(x, y)\phi_i(y)\, dy.$$

But $\phi_i(y)$ is an eigenfunction of K so that

$$\alpha_i = \mu_i \phi_i(x)$$

and

$$K(x, y) = \sum_{i=1}^{\infty} \mu_i \phi_i(x)\overline{\phi_i(y)}.$$

We now use Parseval's equality (21) to obtain

$$\int_a^b |K(x, y)|^2\, dy = \sum_{i=1}^{\infty} \mu_i^2 |\phi_i(x)|^2.$$

So far x was held fixed, but the above holds for almost all x, and the left side is integrable so that

$$\int_a^b \int_a^b |K(x, y)|^2\, dx\, dy = \sum_{i=1}^{\infty} \mu_i^2. \tag{25}$$

Using these results we find that

$$\int_a^b \int_a^b \left| K(x, y) - \sum_{i=1}^{n} \mu_i \phi_i(x)\phi_i(y) \right|^2 dy\, dx$$

$$= \int_a^b \left\{ \int_a^b |K(x, y)|^2\, dy - \sum_{i=1}^{n} \mu_i^2 |\phi_i(x)|^2 \right\} dx$$

$$= \int_a^b \int_a^b |K(x, y)|^2\, dx\, dy - \sum_{i=1}^{n} \mu_i^2.$$

The above coupled with (25) yields the desired result. ■

COROLLARY. Let K be a compact self-adjoint integral operator with an L_2 kernel $K(x, y)$, and $\{\mu_i\}$ the set of eigenvalues. Then

$$\int_a^b \int_a^b |K(x, y)|^2\, dx\, dy = \sum_{i=1}^{\infty} \mu_i^2.$$

PROOF. Immediate by (25). ■

Up to theorem 14 we worked with compact operators, that were not necessarily integral operators. In theorem 14 we used that fact that $K(x, y)$ is an L_2 kernel which is sufficient but not necessary in order for the operator K to be compact. In theorem 14 and its corollary the fact that $K(x, y)$ is square-integrable became vital. For arbitrary compact operators $\sum_{i=1}^{\infty} \mu_i^2$ need not be finite. Operators for which $K(x, y)$ is an L_2 kernel are frequently referred to as Hilbert-Schmidt operators.

4. APPLICATIONS TO DIFFERENTIAL EQUATIONS

In this section we shall consider boundary value problems associated with second order differential equations. It will be shown that so-called regular differential operators have inverses that are integral operators, that are also compact. If the differential operators are self-adjoint then the integral inverse operator will also be self-adjoint so that the results of the preceding section can be applied.

EXAMPLE 1. Consider the operator

$$Lu = -u'' \qquad (26)$$

(where primes denote differentiation) with the boundary conditions

$$u(0) = u(1) = 0. \qquad (27)$$

L is not defined on the whole space $L_2[0, 1]$, but only on those functions, that are sufficiently differentiable and satisfy the boundary conditions. For those functions we see that

$$(Lu, v) = (u, Lv) \qquad (28)$$

since

$$(Lu, v) - (u, Lv) = \int_0^1 (-u''\bar{v} + u\bar{v}'') \, dt = -u'\bar{v} + u\bar{v}' \, |_0^1 = 0$$

by applying an integration by parts and use of the boundary conditions.

We now seek an integral operator K such that

$$Ku = \int_0^1 K(x, y)u(y) \, dy, \qquad u \in L_2[0, 1], \qquad (29)$$

and

$$LKu = u. \qquad (30)$$

To satisfy the boundary conditions we require that

$$K(0, y) = K(1, y) = 0. \qquad (31)$$

Proceeding formally, we see that for K to satisfy (30) we require that

$$LKu = \int_0^1 -K_{xx}(x, y)u(y) \, dy = u(x) \qquad (32)$$

where the subscript in $K_x(x, y)$ denotes the partial derivative with respect to x.

In order to be able to construct suitable kernels by a convenient computational device we shall now introduce so-called delta or impulse functions.

It should be emphasized they are mere computational aids. As will be seen later they do not detract from a rigorous development of the subject. A kernel that has the properties required in (32) is the delta function $\delta(x - y)$. It is formally defined by

$$\delta(x - y) = 0, \qquad x \neq y$$

$$\int_0^1 \delta(x - y)\, dy = 1. \tag{33}$$

If $u(x)$ is a continuous function we have

$$\int_0^1 \delta(x - y)u(y)\, dy = \int_{x-\epsilon}^{x+\epsilon} \delta(x - y)u(y)\, dy, \qquad \epsilon > 0$$

since the integrand vanishes outside every interval $[x - \epsilon, x + \epsilon]$. We also have

$$\int_0^1 \delta(x - y)u(x)\, dy = u(x) \int_0^1 \delta(x - y)\, dy = u(x).$$

Since $u(x)$ is continuous

$$|u(x) - u(y)| < \eta(\epsilon) \qquad \text{if} \quad |x - y| < \epsilon$$

where $\lim_{\epsilon \to 0} \eta(\epsilon) = 0$. It follows that

$$\left| u(x) - \int_0^1 \delta(x - y)u(y)\, dy \right| = \left| \int_{x-\epsilon}^{x+\epsilon} \delta(x - y)(u(x) - u(y))\, dy \right|$$

$$< \eta(\epsilon) \int_{x-\epsilon}^{x+\epsilon} \delta(x - y)\, dy = \eta(\epsilon).$$

In the limit as $\epsilon \to 0$ we have

$$\int_0^1 \delta(x - y)u(y)\, dy = u(x)$$

so that we can let

$$-K_{xx}(x, y) = \delta(x - y) \tag{34}$$

thereby satisfying (32). This discussion was rather heuristic, but it will be shown, that the function $K(x, y)$ to be determined has the appropriate properties.

We shall now try to solve (34) subject to the boundary conditions (31). For $x < y$

$$-K_{xx}(x, y) = 0$$

so that

$$K(x, y) = C_1(y) + C_2(y)x$$

showing that $K(x, y)$ is linear in x, and that the constants of integration may depend on y. Since $K(0, y) = 0$ we see that $C_1(y) = 0$. Similarly we find that for $y > x$

$$K(x, y) = C_3(y) + C_4(y)x$$

and since $K(1, y) = 0$, $C_4(y) = -C_3(y)$ so that finally

$$K(x, y) = C_2(y)x, \qquad x < y$$
$$= C_3(y)(1 - x), \qquad x > y.$$

Now we shall require that $K(x, y)$ be continuous in x for $0 \leq x \leq 1$, and in particular near $x = y$. This leads to

$$C_2(y)y = C_3(y)(1 - y)$$

so that

$$K(x, y) = C_3(y)(1 - y)\frac{x}{y}, \qquad x < y$$
$$= C_3(y)(1 - x), \qquad x > y. \tag{35}$$

Finally to determine $C_3(y)$ we return to (34). By an integration

$$\int_{y-\epsilon}^{y+\epsilon} -K_{xx}(x, y)\, dx = \int_{y-\epsilon}^{y+\epsilon} \delta(x - y)\, dx = 1$$

for all $\epsilon > 0$. The fact that we integrated over y in (33) and over x above should cause no concern, since $\delta(x - y)$ is symmetric. Then

$$-K_x(y + \epsilon, y) + K_x(y - \epsilon, y) = 1$$

and using (35) we find

$$C_3(y) + \frac{C_3(y)(1 - y)}{y} = 1$$

so that

$$C_3(y) = y$$

and

$$K(x, y) = x(1 - y), \qquad x < y$$
$$= y(1 - x), \qquad x > y. \tag{36}$$

A more convenient notation for the above is the following

$$K(x, y) = x_<(1 - x_>) \tag{37}$$

where

$$x_< = \min (x, y)$$
$$x_> = \max (x, y).$$

Using (37) we see that

$$v(x) \equiv Ku = \int_0^1 x_<(1 - x_>)u(y)\, dy$$

$$= (1 - x) \int_0^x yu(y)\, dy + x \int_x^1 (1 - y)u(y)\, dy,$$

so that

$$v(0) = v(1) = 0.$$

Clearly $v = Ku$ satisfies the boundary conditions. By successive differentiation we see that

$$v'(x) = -\int_0^x yu(y)\, dy + \int_x^1 (1 - y)u(y)\, dy$$

$$-v''(x) = xu(x) + (1 - x)u(x) = u(x)$$

so that

$$LKu = u.$$

It should be emphasized at this point that $u(x) \in L_2[0, 1]$. For continuous functions $u(x)$ all the above differentiations are legitimate for all $x \in [0, 1]$. When operating on the Hilbert space $L_2[0, 1]$ such operations are more delicate. It was pointed out earlier in Chapter 1, that two functions that differ by a null function are considered as one element in $L_2[0, 1]$ and are said to be equal almost everywhere. One can show that all the above differentiations are legitimate almost everywhere.

We have now verified that the integral operator K associated with the kernel (37) has the properties required in (32). The introduction of the delta function was merely a crutch that aided us in finding $K(x, y)$. This function is customarily known as the Green's function for the operator L defined in (26) and (27).

A kernel with the symmetry property

$$K(x, y) = \overline{K(y, x)}$$

defines a self-adjoint operator. Since $K(x, y)$ in (37) has this property, and is also an L_2 kernel, K defines a compact self-adjoint operator on $L_2[0, 1]$.

We shall now apply theorem 13 to this operator. If K has an eigenfunction $\phi_n(x)$ we have

$$K\phi_n = \mu_n \phi_n$$

and if we apply L, as defined in (26) we find

$$L\phi_n = \frac{1}{\mu_n} \phi_n$$

so that ϕ_n satisfies the differential equation

$$\phi_n'' + \lambda_n \phi_n = 0 \tag{38}$$

where $\lambda_n = 1/\mu_n$ and ϕ_n also satisfies the boundary conditions

$$\phi_n(0) = \phi_n(1) = 0.$$

A general solution of (38) is given by

$$\phi_n(x) = C_1 \sin \sqrt{\lambda_n}\, x + C_2 \cos \sqrt{\lambda_n}\, x.$$

Since $\phi_n(0) = 0$ we require that $C_2 = 0$. From

$$\phi_n(1) = C_1 \sin \sqrt{\lambda_n} = 0$$

it follows that necessarily $\sqrt{\lambda_n}$ is an integral multiple of π, say $n\pi$. To normalize $\phi_n(x)$ we require that

$$\|\phi_n(x)\|^2 = \int_0^1 C_1^{\,2} \sin^2 n\pi x \, dx = \frac{C_1^{\,2}}{2} = 1$$

so that finally $\{\sqrt{2} \sin n\pi x\}$ is the set of orthonormal eigenfunctions of K and $\{n^2\pi^2\}$ is the corresponding set of eigenvalues. To apply theorem 13 we also have to investigate the null space of K. Suppose

$$Kf = 0,$$

then

$$\int_0^1 x_<(1 - x_>)f(y)\, dy = (1 - x)\int_0^x yf(y)\, dy + x\int_x^1 (1 - y)f(y)\, dy = 0.$$

A differentiation of the previous result shows that

$$-\int_0^x yf(y)\, dy + \int_x^1 (1 - y)f(y)\, dy = -\int_0^1 yf(y)\, dy + \int_x^1 f(y)\, dy = 0.$$

and for $x = 1$ we have

$$\int_0^1 yf(y)\, dy = 0.$$

A combination of the last two expressions now shows that

$$\int_x^1 f(y)\, dy = 0 \qquad \text{for all} \quad 0 \le x \le 1$$

so that $f(x)$ is a null function. It follows that the null space of K is empty.

Theorem 13 now tells us that the set of eigenfunctions is a complete orthonormal set so that

$$f(x) = \sum_{n=1}^{\infty} \left(\int_0^1 \sqrt{2} \sin n\pi y f(y)\, dy \right) \sqrt{2} \sin n\pi x, \tag{39}$$

for all $f(x)$ in $L_2[0, 1]$. Convergence of (39) is in the mean so that

$$\lim_{N \to \infty} \int_0^1 \left| f(x) - \sum_{n=1}^N \left(\int_0^1 \sqrt{2} \sin n\pi y f(y) \, dy \right) \sqrt{2} \sin n\pi x \right|^2 dx = 0. \quad (40)$$

In general theorem 13 tells us no more about the convergence of (39) than that it converges in norm in the sense of (40). Under certain conditions one can do even better.

THEOREM 15. Let $K(x,y)$ be an L_2 kernel on $[a, b]$ such that $\int_a^b |K(x, y)|^2 \, dy \leq M^2$ and suppose that $K(x, y) = \overline{K(y, x)}$. Let $\{\phi_n\}$ denote the set of eigenfunctions of the associated compact, self-adjoint operator K. Let $f(x) \in L_2[a, b]$, such that f is in the range of K, that is, there exists $g(x) \in L_2[a, b]$ such that $f = Kg$. Then the Fourier series

$$f = \sum_{n=1}^\infty (f, \phi_n)\phi_n \quad (41)$$

converges uniformly and absolutely to $f(x)$ on $[a, b]$.

PROOF. To prove the result it is sufficient to show that

$$\sum_{i=n}^m |(f, \phi_i)\phi_i| < \epsilon \quad \text{for} \quad n, m > N(\epsilon)$$

where $N(\epsilon)$ is independent of x. Now

$$\sum_{i=n}^m |(f, \phi_i)\phi_i| = \sum_{i=n}^m |(Kg, \phi_i)\phi_i| = \sum_{i=n}^m |(g, K\phi_i)\phi_i| = \sum_{i=n}^m |\mu_i(g, \phi_i)\phi_i|$$

$$\leq \left\{ \sum_{i=n}^m |(g, \phi_i)|^2 \sum_{i=1}^\infty \mu_i^2 |\phi_i(x)|^2 \right\}^{1/2}$$

by the Cauchy-Schwarz inequality. But

$$\sum_{i=1}^\infty \mu_i^2 |\phi_i(x)|^2 = \int_a^b |K(x, y)|^2 \, dy \leq M^2$$

(see the proof of theorem 14) and the series $\sum_{i=1}^\infty |(g, \phi_i)|^2$ converges so that finally

$$\sum_{i=n}^m |(f, \phi_i)\phi_i| \leq M \left\{ \sum_{i=n}^m |(g, \phi_i)|^2 \right\}^{1/2} < \epsilon, \quad n, m > N(\epsilon).$$

Therefore (41) converges uniformly and absolutely. In view of the fact that the series converges to $f(x)$ in the mean it must converge to $f(x)$ uniformly and absolutely. ∎

EXAMPLE 2. Let $f(x)$ be such that $f''(x) \in L_2[0, 1]$ and $f(0) = f(1) = 0$. Then the Fourier series of $f(x)$

$$f(x) = \sum_{n=1}^{\infty} \left(\int_0^1 \sqrt{2} \sin n\pi y f(y) \, dy \right) \sqrt{2} \sin n\pi x$$

converges to $f(x)$ absolutely and uniformly. Since $f = -Kf''$, where K is defined by the kernel (37), we can apply theorem 15.

EXAMPLE 3. Consider the differential operator

$$Lu = -(p(x)u')' + q(x)u \tag{42}$$

subject to the boundary conditions

$$\begin{aligned} a_0 u(0) + a_1 u'(0) &= 0, \qquad |a_0| + |a_1| \neq 0 \\ b_0 u(1) + b_1 u'(1) &= 0. \qquad |b_0| + |b_1| \neq 0. \end{aligned} \tag{43}$$

It will be assumed that $q(x), p(x), a_0, a_1, b_0, b_1$ are real, $p(x)$ positive on $[0, 1]$ and $q(x)$ and $p'(x)$ are continuous. By virtue of the boundary conditions, one can show that

$$(Lu, v) = (u, Lv)$$

for those u and v for which Lu and Lv is defined and in $L_2[0, 1]$.

As in Example 1 we shall find an integral operator

$$Ku = \int_0^1 K(x, y)u(y) \, dy \tag{44}$$

defined on $L_2[0, 1]$ such that Ku satisfies the boundary conditions and

$$LKu = u. \tag{45}$$

To do so we shall seek a Green's function $K(x, y)$ such that

$$-(p(x)K_x(x, y))_x + q(x)K(x, y) = \delta(x - y) \tag{46}$$

as in (34). For $x < y$ we require that

$$\begin{aligned} -(p(x)K_x(x, y))_x + q(x)K(x, y) &= 0 \\ a_0 K(0, y) + a_1 K_x(0, y) &= 0 \end{aligned}$$

That such a solution exists, is well known from the theory of differential equations. Alternatively one can rewrite the above initial value problem as a Volterra integral equation

$$K(x, y) = K(0, y) + p(0)K_x(0, y) \int_0^x \frac{dt}{p(t)} + \int_0^x q(t) \int_t^x \frac{dz}{p(z)} K(t, y) \, dt \tag{47}$$

$K(0, y)$ and $K_x(0, y)$ are assigned any values that also satisfy the initial condition at 0. Such a set is not unique, but will still involve a parameter that can be assigned arbitrarily. For example, if $a_1 \neq 0$, let $K(0, y) = 1$ so that $K_x(0, y) = -(a_0/a_1)$. With these assignments the integral equation (47) has a unique solution. Denote it by $k_1(x)$.

Similarly we find $k_2(x)$ such that

$$-(p(x)k_2')' + q(x)k_2 = 0$$
$$b_0 k_2(1) + b_1 k_2'(1) = 0$$

Now we let

$$K(x, y) = C_1 k_1(x), \qquad x < y$$
$$= C_2 k_2(x), \qquad x > y \qquad (48)$$

where C_1 and C_2 are arbitrary. Equation (48) satisfies (46) for $x < y$ and $x > y$ and also the boundary conditions. We also require that $K(x, y)$ be continuous at $x = y$ so that

$$C_1 k_1(y) = C_2 k_2(y).$$

In addition by integrating (46) over the interval $[y - \epsilon, y + \epsilon]$ and then letting $\epsilon \to 0$ we find

$$-p(y)C_2 k_2'(y) + p(y)C_1 k_1'(y) = 1$$

so that C_1 and C_2 are fully determined. Finally

$$K(x, y) = \frac{k_1(x_<)k_2(x_>)}{p(y)[k_2(y)k_1'(y) - k_1(y)k_2'(y)]}$$

where the notation in (37) is being used again.

The denominator in $K(x, y)$ appears to depend on y, but we shall see that it is in fact a constant, independent of y. To do so we differentiate the denominator,

$$\{p(y)[k_2(y)k_1'(y) - k_1(y)k_2'(y)]\}' = k_2(y)(p(y)k_1'(y))' - k_1(y)(p(y)k_2'(y))'$$
$$= k_2(y)(q(y)k_1(y)) - k_1(y)(q(y)k_2(y)) = 0$$

and use the fact that both $k_1(y)$ and $k_2(y)$ satisfy the same differential equation. Since the derivative of the denominator vanishes, it must be a constant, and we denote the constant by w. There are now two possibilities, either $w = 0$ or $w \neq 0$. If $w = 0$ we see that

$$k_2(y)k_1'(y) - k_1(y)k_2'(y) = 0$$

so that $k_1(y)$ is a multiple of $k_2(y)$, so that they are linearly dependent solutions of the differential equation. In that case our search for an integral operator (44) breaks down.

We shall now assume that $w \neq 0$ (the case where $w = 0$ will be investigated separately) so that

$$K(x, y) = \frac{k_1(x_<)k_2(x_>)}{w} . \tag{49}$$

Clearly $K(x, y)$ is continuous for $0 \leq x, y \leq 1$, and also $K(x, y) = \overline{K(y, x)}$, so that (43) is a compact self-adjoint operator. One can show as in Example 1 that if $Kf = 0$, it follows that $f = 0$. If $\{\phi_i(x)\}$ and $\{\mu_i\}$ represent the orthonormal eigenfunctions and eigenvalues of K and $f \in L_2[0, 1]$ we have by theorem 13.

$$f = \sum_{i=1}^{\infty} (f, \phi_i)\phi_i.$$

By theorem 15 if $f = Kg$ for some g, the convergence in the above is absolute and uniform.

In this example as in Example 1 the search for eigenfunctions and eigenvalues can be reduced to solving the differential equation

$$Lu = \lambda u$$

subject to the boundary conditions (43).

EXAMPLE 4. The corollary to theorem 14 tells us that

$$\int_a^b \int_a^b |K(x, y)|^2 \, dx \, dy = \sum_{i=1}^{\infty} \mu_i^2.$$

If we apply this result to Example 1 we find

$$\int_0^1 \int_0^1 [x_<(1 - x_>)]^2 \, dx \, dy = \sum_{n=1}^{\infty} \frac{1}{n^4\pi^4} .$$

Direct integration now shows that

$$\sum_{n=1}^{\infty} \frac{1}{n^4\pi^4} = \frac{1}{90} .$$

Example 3 was not general in two ways. First of all the boundary conditions (43) are said to be separated, that is, one refers to the end point $x = 0$, and the other to $x = 1$. One can take more general mixed conditions. Secondly, the constant w in (49) was assumed to be nonzero. We shall shortly consider an example with mixed boundary conditions where $w = 0$. In fact this condition can occur if and only if the differential equation

$$Lu = 0$$

with the boundary conditions assigned to u, has a nontrivial solution. In this case L has the eigenvalue $\lambda = 0$. But since the eigenvalues of K are reciprocals of those of L, K can clearly not exist. We formulate this result as a theorem.

THEOREM 16. Let L be a differential operator

$$Lu = -(p(x)u'(x))' + q(x)u(x)$$

defined on those functions in $L_2[0, 1]$, that have two derivatives. Suppose the boundary conditions are such that

$$(Lu, v) = (u, Lv),$$

for all u, v satisfying the boundary conditions and sufficiently differentiable so that Lu and Lv belong to $L_2[0, 1]$. We also assume all functions $p(x)$ and $q(x)$ to be real and such that $q(x)$ and $p'(x)$ are continuous and $p(x)$ is positive. Then an inverse integral operator Ku exists such that $LKu = u$ if and only if the operator L does not have $\lambda = 0$ as an eigenvalue.

PROOF. We omit details, but one simply traces carefully the construction of $K(x, y)$ as in Example 3. ∎

EXAMPLE 5. Consider

$$Lu = -u'' \tag{50}$$

with the boundary conditions

$$u(0) - u(1) = 0$$
$$u'(0) - u'(1) = 0. \tag{51}$$

We first consider

$$Lu = \lambda u$$

so that

$$u'' + \lambda u = 0 \tag{52}$$

and

$$u = C_1 \sin \sqrt{\lambda}\, x + C_2 \cos \sqrt{\lambda}\, x.$$

Use of (51) shows that

$$C_2(1 - \cos \sqrt{\lambda}) - C_1 \sin \sqrt{\lambda} = 0$$
$$\sqrt{\lambda}\, C_2 \sin \sqrt{\lambda} + \sqrt{\lambda}\, C_1(1 - \cos \sqrt{\lambda}) = 0.$$

Nontrivial solutions for C_1 and C_2 will exist if and only if the determinant of the above system vanishes. This leads to

$$2\sqrt{\lambda}\, (1 - \cos \sqrt{\lambda}) = 0$$

and the possible eigenvalues are

$$\lambda_0 = 0, \qquad \lambda_n = 4n^2\pi^2, \qquad n = 1, 2, \ldots.$$

For $\lambda_0 = 0$ we find $\phi_0(x) = 1$ satisfies (51) and (52). For $\lambda_n = 4n^2\pi^2$ we find that

$$u = C_1 \sin 2n\pi x + C_2 \cos 2n\pi x$$

satisfies (51) and (52). In this case we see that we can find two linearly independent eigenfunctions associated with one eigenvalue. These can be orthonormalized so that

$$\phi_{n,e} = \sqrt{2} \cos 2n\pi x, \qquad \phi_{n,0} = \sqrt{2} \sin 2n\pi x, \qquad \phi_0(x) = 1. \quad (53)$$

We note that $\lambda_0 = 0$ is the only eigenvalue for which there is only one eigenfunction.

The expansions associated with (53) are the standard Fourier series. To prove their completeness we shall use our integral operator approach. In the light of theorem 16, the approach used in Examples 1 and 3 is bound to break down. There are two modifications that enable us to find a suitable integral operator. The first is to introduce a shifted operator L_ρ defined by

$$L_\rho u = -u'' + \rho u \quad (54)$$

subject to the boundary conditions (51). The eigenvalues associated with (52) are given by

$$\lambda_n = \rho + 4n^2\pi^2 \quad (55)$$

and one can certainly select ρ so that no λ_n vanishes. The eigenfunctions of L_ρ and L will be the same. To see this we note that if

$$L\phi_n = 4n^2\pi^2\phi_n$$

then

$$L_\rho\phi_n = (\rho + 4n^2\pi^2)\phi_n.$$

Now the operator L_ρ has only nonzero eigenvalues and an inverse integral operator can be found by the methods that were discussed previously.

The above approach has two disadvantages. In the first place we have to know a priori that such a ρ exists. In a more complicated case this may not be obvious. Secondly, to find the Green's function $K(x, y)$ we need to solve the equation

$$LK(x, y) = \delta(x - y)$$

and the functions required in the solution may be more complicated than those associated in the case $\rho = 0$. An alternative procedure will therefore be described.

Let H_0 be the space spanned by all eigenfunctions associated with $\lambda = 0$, and find its orthogonal complement $H_0^\perp$ such that

$$L_2[0, 1] = H_0 \oplus H_0^\perp.$$

For differential operators H_0 will be finite dimensional. Suppose the orthonormal set $\phi_{0,1}(x), \ldots, \phi_{0,n}(x)$ spans H_0 and consider

$$\delta^*(x - y) = \delta(x - y) - \sum_{k=1}^{n} \phi_{0,k}(x)\overline{\phi_{0,k}(y)}.$$

It is easy to see that

$$\int_0^1 \delta^*(x - y)\overline{\phi_{0,k}(x)} \, dx = 0$$

so that $\delta^*(x - y)$ is orthogonal to H_0. Then solve

$$LK(x, y) = \delta^*(x - y)$$

subject to the boundary conditions and the conditions $\int_0^1 K(x, y)\phi_{0,k}(x) \, dx = 0$, $k = 1, 2, \ldots, n$. The resultant integral operator is then an operator on $H_0^{\perp}$.

We shall now apply this to (50). In this case H_0 is spanned by $\phi_0(x) = 1$; in other words H_0 is a one-dimensional space. $H_0^{\perp}$ is the subspace of $L_2[0, 1]$ orthogonal to H_0, so that if $f(x) \in H_0^{\perp}$, then $\int_0^1 f(x) \, dx = 0$. Now

$$\delta^*(x - y) = \delta(x - y) - 1$$

and

$$-K_{xx}(x, y) = \delta(x - y) - 1. \tag{56}$$

Using the standard techniques and requiring in addition that $\int_0^1 K(x, y) \, dx = 0$ we find

$$K(x, y) = \tfrac{1}{12} - \tfrac{1}{2}(x_> - x_<) + \tfrac{1}{2}(x - y)^2 \tag{57}$$

The operator K, associated with (57), is a compact self-adjoint operator on $L_2[0, 1]$. Its null space is H_0 since

$$K1 = \int_0^1 K(x, y) \, dx = 0.$$

Its eigenfunctions are given by (53), and using theorem 13 we find for any $f(x) \in L_2[0, 1]$.

$$f(x) = \sum_{n=1}^{\infty} \left(\int_0^1 f(y)\sqrt{2} \sin 2n\pi y \, dy \sqrt{2} \sin 2n\pi x \right.$$
$$\left. + \int_0^1 f(y)\sqrt{2} \cos 2n\pi y \, dy \sqrt{2} \cos 2n\pi x \right) + f_0. \tag{58}$$

f_0 has to be an element in the nullspace of K. In this case it is easy to see that

$$f_0 = \int_0^1 f(x) \, dx.$$

One can, with a slight modification, apply theorem 15 and conclude that if $f''(x) \in L_2[0, 1]$ and $f(0) = f(1), f'(0) = f'(1)$, then (58) converges to $f(x)$ uniformly and absolutely.

The situation described in Example 3 can be generalized in another direction also. This will be discussed in the next example.

EXAMPLE 6.   We shall now consider the operator

$$Lu = \frac{1}{r(x)} [-(p(x)u')' + q(x)u] \tag{59}$$

defined for those $u \in L_2[0, 1]$ that are twice differentiable and subject to a set of boundary conditions:

$$a_0u(0) + a_1u'(0) + a_2u(1) + a_3u'(1) = 0$$
$$b_0u(0) + b_1u'(0) + b_2u(1) + b_3u'(1) = 0. \tag{60}$$

$$\sum_{i=0}^{3} |a_i| \neq 0, \qquad \sum_{i=0}^{3} |b_i| \neq 0.$$

If $a_2 = a_3 = b_0 = b_1 = 0$ the boundary conditions are said to be separated; the first condition applies to the end point $x = 0$, and the second to the end point $x = 1$. More general conditions are said to be mixed. We suppose that all a_i, b_i, $p(x)$, $r(x)$, $q(x)$ are real, that $p'(x)$, $q(x)$, $r(x)$ are continuous and that $p(x)$, $r(x)$ are positive on $[0, 1]$. In particular we let r_0, r_1 be two positive constants such that

$$0 < r_0 \leq r(x) \leq r_1.$$

The above operator will not satisfy the symmetry property

$$(Lu, v) = (u, Lv)$$

for those u, v satisfying (60), and such that Lu and Lv are in $L_2[0, 1]$. In other words it will not be possible to find an inverse integral operator that is compact as well as self-adjoint. The situation can be saved however by working not on $L_2[0, 1]$, but on a different Hilbert space.

Consider the space $L_2([0, 1], r)$ defined as follows. We define the inner product

$$(f, g)_r = \int_0^1 r(x)f(x)\overline{g(x)} \, dx \tag{61}$$

from which we obtain the norm

$$\|f\|_r = \left\{ \int_0^1 r(x) |f(x)|^2 \, dx \right\}^{1/2}. \tag{62}$$

As with $L_2[0, 1]$, we can start with continuous functions and view $L_2([0, 1], r)$ as the completion of the space of continuous functions with respect to the norm (62). In fact $L_2[0, 1]$ and $L_2([0, 1], r)$ contain the same functions, but have different inner products and norms. It is obvious that

$$r_0^{\frac{1}{2}} \|f\| \leq \|f\|_r \leq r_1^{\frac{1}{2}} \|f\|. \tag{63}$$

Norms satisfying (63) are said to be equivalent. It is evident from (63) that if $\{f_n\}$ is a Cauchy sequence with respect to $\| \ \|$ it will also be one with respect to $\| \ \|_r$ and vice versa.

Suppose that $a_2 = a_3 = b_0 = b_1 = 0$ in (60). Then one can show that

$$(Lu, v)_r = (u, Lv)_r.$$

We now construct a Green's function $K(x, y)$, as before by solving

$$-(p(x)K_x(x, y))_x + q(x)K(x, y) = \delta(x - y) \tag{64}$$

with the boundary conditions

$$a_0K(0, y) + a_1K_x(0, y) = 0$$
$$b_2K(1, y) + b_3K_x(1, y) = 0.$$

With this kernel we associate the integral operator

$$Kf = \int_0^1 r(y)K(x, y)f(y) \, dy. \tag{65}$$

Using (59) we see that (64) is equivalent to the equation

$$Lu = \frac{\delta(x - y)}{r(x)}$$

and a calculation in (65) shows that

$$LKf = \int_0^1 r(y) \frac{\delta(x - y)}{r(x)} f(y) \, dy = f(x)$$

so that K is the inverse operator to L. We also can verify that

$$(Kf, g)_r = (f, Kg)_r$$

so that K is in fact a self-adjoint operator with respect to the inner product $(f, g)_r$.

We shall show that K is also compact. Let $\{f_n\}$ be a bounded sequence in $L_2([0, 1], r)$, and consider, using (65), the expression

$$\|Kf\|_r = \left\{ \int_0^1 r(x) \left| \int_0^1 r(y)K(x, y)f(y) \, dy \right|^2 dx \right\}^{\frac{1}{2}}$$

If we define the integral operator

$$\tilde{K}f = \int_0^1 \sqrt{r(x)r(y)}\, K(x, y)\sqrt{r(y)}\, f(y)\, dy$$

we see readily that $\tilde{K}$ is self-adjoint and compact on $L_2[0, 1]$. Also

$$\|Kf\|_r = \|\tilde{K}f\|.$$

We can therefore select a subsequence from $\{f_n\}$ such that

$$\|Kf_n - Kf_m\|_r = \|\tilde{K}f_n - \tilde{K}f_m\| < \epsilon, \qquad n, m > N(\epsilon).$$

Accordingly K is compact and self-adjoint on $L_2([0, 1], r)$ and our general theory can be applied.

The same conclusions can be reached using more general boundary conditions than separated ones. As long as L satisfies the symmetry conditions

$$(Lu, v)_r = (u, Lv)_r$$

the resultant integral operator will be compact and self-adjoint.

EXAMPLE 7. We shall now consider an operator of the type (59), where some of the conditions imposed on $p(x)$, $q(x)$, $r(x)$ are relaxed. The previous conditions are sufficient, but not necessary to guarantee that the resultant integral operator is compact and self-adjoint.

The differential equation

$$u'' + \frac{1}{x} u' + \left(\lambda - \frac{v^2}{x^2}\right)u = 0 \tag{66}$$

is known as Bessel's equation. Its general solutions are known as Bessel functions and denoted by $J_v(\sqrt{\lambda}\, x)$, $N_v(\sqrt{\lambda}\, x)$. The second solution is sometimes known as a Neumann function. $J_v(x)$ is characterized by an infinite series expansion

$$J_v(x) = \left(\frac{x}{2}\right)^v \sum_{k=0}^{\infty} \frac{(i(x/2))^{2k}}{k!\, \Gamma(k + v + 1)}.$$

Equation (66) can be rewritten in the form

$$\frac{1}{x}\left[-(xu')' + \frac{v^2}{x} u\right] = \lambda u$$

which suggests that we study the operator

$$Lu = \frac{1}{x}\left[-(xu')' + \frac{v^2}{x} u\right].$$

For simplicity we will now consider the case $\nu = 1$ and focus our attention on

$$Lu = \frac{1}{x}\left[-(xu')' + \frac{1}{x}u \right],$$ (67)

and try to follow the steps outlined in Example 6. We see that

$$r(x) = x, \qquad p(x) = x, \qquad q(x) = \frac{1}{x}.$$

The space on which we will work is $L_2([0, 1], x)$. It consists of all functions for which $\int_0^1 x\,|f(x)|^2\,dx < \infty$. In Example 6 we demanded that $p(x)$ and $r(x)$ be positive and that $p(x)$, $r(x)$, and $q(x)$ be continuous. These conditions are obviously violated at $x = 0$. Nevertheless, as will be shown, the theory of compact, self-adjoint operators can be applied.

On the space $L_2([0, 1], x)$ we have

$$(f, g)_x = \int_0^1 xf(x)\overline{g(x)}\,dx$$

$$\|f\|_x = \left\{ \int_0^1 x\,|f(x)|^2\,dx \right\}^{1/2}.$$

Since $r(x) = x$ is no longer strictly positive, $\|f\|_x$ and $\|f\|$ will no longer be equivalent. In fact $L_2([0, 1], x)$ is a larger space than $L_2[0, 1]$, since a function like $f(x) = x^{-\alpha}$, $\frac{1}{2} \le \alpha < 1$ belongs to the former, but not the latter space.

In analogy to (64) we shall seek a Green's function $K(x, y)$ such that

$$-(xK_x(x, y))_x + \frac{1}{x} K(x, y) = \delta(x - y)$$ (68)

and we shall also impose the boundary condition

$$K(1, y) = 0.$$ (69)

A more general boundary condition could be used, but (69) will be used to keep matters simple. No boundary condition will be imposed at $x = 0$ for reasons to be explained shortly.

To solve for $K(x, y)$ we see, from (68), that

$$K(x, y) = C_1 x + \frac{C_2}{x}, \qquad x < y$$

$$= C_3 x + \frac{C_4}{x}, \qquad x > y$$

x belongs to $L_2([0, 1], x)$, but $1/x$ does not. For this reason we set $C_2 = 0$. The requirement that $K(x, y) \in L_2([0, 1], x)$ replaces a boundary condition at

$x = 0$. To satisfy (69) we let $C_4 = -C_3$. Finally to determine C_1 and C_3 we require, as before, that $K(x, y)$ be continuous at $x = y$ and that

$$\int_{y-\epsilon}^{y+\epsilon} \left[-(xK_x(x, y))_x + \frac{1}{x} K(x, y) \right] dx = 1.$$

Then one finds that

$$K(x, y) = \tfrac{1}{2} x_< \left(\frac{1}{x_>} - x_> \right). \tag{70}$$

The integral operator, inverse to L in analogy to (65), is

$$Kf = \int_0^1 yK(x, y)f(y) \, dy. \tag{71}$$

A calculation shows that

$$(Kf, g)_x = (f, Kg)_x$$

so that K is self-adjoint.

We shall now show that K is also a compact operator on $L_2([0, 1], x)$. Let $\{f_n\}$ be a uniformly bounded sequence, so that $\|f_n\|_x \leq M$. Evidently the sequence $\{\sqrt{y} f_n(y)\}$ is a uniformly bounded sequence on $L_2[0, 1]$. Now

$$\|Kf\|_x = \left\{ \int_0^1 x \left| \int_0^1 yK(x, y)f(y) \, dy \right|^2 dx \right\}^{\frac{1}{2}}$$

$$= \left\{ \int_0^1 \left| \int_0^1 \sqrt{xy} \, K(x, y)\sqrt{y} f(y) \, dy \right|^2 dx \right\}^{\frac{1}{2}}.$$

We now define the operator

$$\tilde{K}g = \int_0^1 \sqrt{xy} \, K(x, y)g(y) \, dy$$

on $L_2[0, 1]$. Using (70) it is easy to see that the kernel is an L_2 kernel, so that $\tilde{K}$ is a compact self-adjoint operator on $L_2[0, 1]$.

We have the obvious equality

$$\|Kf\|_x = \|\tilde{K}\sqrt{x} f(x)\|.$$

Now we can certainly find a subsequence of $\{f_n\}$ such that

$$\|Kf_n - Kf_m\|_x = \|\tilde{K}\sqrt{x} f_n(x) - \tilde{K}\sqrt{x} f_m(x)\| < \epsilon, \qquad n, m > N(\epsilon)$$

so that K is compact and self-adjoint on $L_2([0, 1], x)$. Its eigenvalues must be solutions of the transcendental equation

$$J_1 \left(\frac{1}{\sqrt{\mu_n}} \right) = 0$$

since this Bessel function satisfies the equation

$$LJ_1\left(\frac{x}{\sqrt{\mu_n}}\right) = \frac{1}{\mu_n} J_1\left(\frac{x}{\sqrt{\mu_n}}\right),$$

as well as the boundary condition.

We shall denote the eigenfunctions by $\{k_n J_1(\sqrt{\lambda_n}\, x)\}$, where $\lambda_n = (1/\mu_n)$ and k_n is so selected that

$$k_n^2 \int_0^1 x J_1^2(\sqrt{\lambda_n}\, x)\, dx = 1.$$

Finally using theorem 13 we conclude that for $f(x) \in L_2([0, 1], x)$ we have

$$f(x) = \sum_{n=1}^{\infty} \left(\int_0^1 y k_n J_1(\sqrt{\lambda_n}\, y) f(y)\, dy \right) k_n J_1(\sqrt{\lambda_n}\, x)$$

and convergence, of course, is in the mean on the space.

The differential operators treated so far, except the one in Example 7, are so-called regular operators. As a consequence the associated integral operators are compact and self-adjoint and a set of eigenvalues, eigenfunctions, and corresponding expansion theorems are assured.

Example 7 presented a singular operator, which nevertheless had many features in common with the regular operators. In general this need not be the case and there are no simple a priori methods for deciding such questions. In the next example a differential operator on an infinite interval will be discussed. Here again it will develop that the inverse integral operator is compact and self-adjoint, but in order to reach this decision some properties of the solutions of the differential equation will have to be analyzed.

EXAMPLE 8.   We consider the differential equation

$$u'' + (\lambda - x^2)u = 0 \tag{72}$$

defined on $L_2[-\infty, \infty]$. For a function $f(x)$ to be in this space we require that

$$\int_{-\infty}^{\infty} |f(x)|^2\, dx < \infty. \tag{73}$$

In general solutions of (72) will not belong to the space, but for special values of λ, namely eigenvalues this will be the case.

To analyze (72) we let $u = e^{-(x^2/2)} V$, and find that $V(x)$ satisfies

$$V'' - 2x V' + (\lambda - 1)V = 0. \tag{74}$$

We can find two independent solutions V_1 and V_2 that have power series expansions, using the standard method of Frobenius. Then

$$V_1(x) = \sum_{n=0}^{\infty} a_k x^{2k}, \qquad a_0 = 1$$

$$V_2(x) = \sum_{n=0}^{\infty} b_k x^{2k+1}, \qquad b_0 = 1.$$

The coefficients satisfy the recursion formulas

$$a_{k+1} = \frac{4k + 1 - \lambda}{(2k + 2)(2k + 1)} a_k$$

$$b_{k+1} = \frac{4k + 3 - \lambda}{(2k + 3)(2k + 2)} b_k.$$

(75)

If λ takes on any of the values $1, 3, 5, 7, \ldots$, we see that one of the above series will terminate. We have for

$$\lambda = 1 \qquad V_1(x) = 1$$
$$\lambda = 3 \qquad V_2(x) = x$$
$$\lambda = 5 \qquad V_1(x) = 1 - 2x^2$$
$$\lambda = 7 \qquad V_2(x) = x - \tfrac{2}{3}x^3$$
$$\vdots \qquad \vdots$$

In general if $\lambda_n = 2n + 1$ the associated polynomial solution of (74) is denoted by $H_n(x)$ and is known as an Hermite polynomial. In that case

$$u_n = e^{-(x^2/2)} H_n(x) \tag{76}$$

satisfies (72) and (73). A closer analysis of the recurrence formulas (75) allows one to show that in general $V_1(x)$ and $V_2(x)$ grow like e^{x^2} for large x so that the Hermite polynomials constitute the only possible solutions for which (76) can satisfy (73). We will show that the set $\{e^{-(x^2/2)} H_n(x)\}$ constitutes a complete orthogonal set of functions in $L_2[-\infty, \infty]$.

In order to accomplish the above goal we shall first analyze the case where $\lambda = 0$.

$$u'' - x^2 u = 0. \tag{77}$$

There are two independent solutions given by

$$u_e(x) = \sum_{n=0}^{\infty} c_n x^{4n}, \qquad c_0 = 1$$

$$u_0(x) = \sum_{n=0}^{n=\infty} d_n x^{4n+1}, \qquad d_0 = 1$$

where

$$c_{n+1} = \frac{c_n}{(4n + 4)(4n + 3)}$$

$$d_{n+1} = \frac{d_n}{(4n + 5)(4n + 4)}.$$

These solutions satisfy (77), but do not belong to $L_2[-\infty, \infty]$. Both u_e and u_0 have an infinite radius of convergence, and all coefficients are positive. Therefore both become infinitely large for large $|x|$.

We will show however that there is a solution that becomes small as $x \to \infty$. That solution will, of course, be a linear combination of u_e and u_0. To facilitate matters we let

$$u(x) = \frac{w(t)}{\sqrt{x}}, \qquad t = \frac{x^2}{2}$$

in (77) and work with $x > 0$. This leads to a new differential equation in w

$$w'' - w = -\frac{3}{16t^2} w. \tag{78}$$

We will show that for large t (78) has two solutions one behaving like e^t and the other like e^{-t}. Using the method of variation of parameters we can re-write (78) as a Volterra integral equation.

$$w(t) = e^{-t} + \frac{3}{16} \int_t^{\infty} \frac{\sinh(t - \tau)}{\tau^2} w(\tau) \, d\tau. \tag{79}$$

It is not clear whether the above integral exists. We will show however that the above integral equation has a solution for which the integral exists and for large t the resultant $w(t)$ behaves like e^{-t}.

To accomplish this we consider the iteration scheme

$$w_0(t) = 0$$

$$w_1(t) = e^{-t}$$

.

.

.

$$w_{n+1}(t) = e^{-t} + \frac{3}{16} \int_t^{\infty} \frac{\sinh(t - \tau)}{\tau^2} w_n(\tau) \, d\tau.$$

Clearly

$$|w_1 - w_0| \le e^{-t}$$

and

$$|w_2 - w_1| = \frac{3}{16} \int_t^\infty \frac{\sinh(\tau - t)}{\tau^2} e^{-\tau} \, d\tau = \tfrac{3}{16} e^{-t} \int_t^\infty \frac{1 - e^{-2(\tau - t)}}{2\tau^2} \, d\tau$$

$$\le \tfrac{3}{16} e^{-t} \int_t^\infty \frac{d\tau}{\tau^2} = \frac{3}{16t} e^{-t}.$$

A proof by induction now shows that

$$|w_{n+1} - w_n| \le \left(\frac{3}{16t}\right)^n e^{-t}.$$

Now

$$w_{n+1} = (w_{n+1} - w_n) + (w_n - w_{n-1}) + \cdots + (w_1 - w_0)$$

so that

$$|w_{n+1}| \le \sum_{k=0}^n |w_{k+1} - w_k| \le \frac{e^{-t}}{1 - (3/16t)}.$$

From the above it follows that $w(t) = \lim_{n \to \infty} w_n(t)$ exists and that it behaves like e^{-t} for large t so that the integral in (79) exists.

It follows that (77) has a solution such that

$$\lim_{x \to \infty} \sqrt{x} \, e^{x^2/2} u_s(x) = 1 \tag{80}$$

This solution must be of the form

$$u_s(x) = u_e(x) - \sigma u_0(x) \tag{81}$$

for a suitable positive constant σ. As $x \to -\infty$, $u_s(x)$ will become large, in view of the fact that $u_e(x)$ is even and $u_0(x)$ is odd so that

$$u_s(-x) = u_e(x) + \sigma u_0(x),$$

and as x becomes large so does $u_s(-x)$.

Equation (78) has a second solution that grows like e^t for large t. To see that, one sets up the Volterra integral equation

$$w(t) = e^t + \frac{3}{32} \int_t^\infty \frac{e^{t-\tau}}{\tau^2} w(\tau) \, d\tau + \frac{3}{32} \int_1^t \frac{e^{\tau-t}}{\tau^2} w(\tau) \, d\tau.$$

One can show that the solution of the equation satisfies

$$\lim_{t \to \infty} e^{-t} w(t) = 1$$

so that (77) has a second solution satisfying

$$\lim_{x \to \infty} \sqrt{x}\, e^{-(x^2/2)} u_t(x) = 1. \tag{82}$$

With these preliminaries we are prepared to discuss the differential operator

$$Lu = -u'' + x^2 u \tag{83}$$

acting on $L_2[-\infty, \infty]$. An inverse integral operator to L will be constructed. Its kernel will be obtained by solving the differential equation

$$-K_{xx}(x, y) + x^2 K(x, y) = \delta(x - y) \tag{84}$$

subject to the condition $\int_{-\infty}^{\infty} |K(x, y)|^2\, dx < \infty$. We have

$$K(x, y) = c_1(u_e(x) + \sigma u_0(x)), \qquad x < y$$
$$= c_2(u_e(x) - \sigma u_0(x)), \qquad x > y$$

$K(x, y)$ is so constructed that

$$\int_{-\infty}^{\infty} |K(x, y)|^2\, dx < \infty.$$

To determine c_1 and c_2 we require $K(x, y)$ to be continuous at $x = y$ and also use (84). Then

$$K(x, y) = \frac{(u_e(x_<) + \sigma u_0(x_<))(u_e(x_>) - \sigma u_0(x_>))}{2\sigma}$$
$$= \frac{u_s(-x_<)u_s(x_>)}{2\sigma}. \tag{85}$$

Now we have

$$Kf = \int_{-\infty}^{\infty} K(x, y) f(y)\, dy$$

and

$$LKf = f.$$

K is also self-adjoint and we shall show that $K(x, y)$ is an L_2 kernel so that K is compact. In this step the estimates (80) and (82) will be vital. Since $K(x, y) = K(y, x)$ we have

$$\int_{-\infty}^{\infty} \int_{-\infty}^{\infty} |K(x, y)|^2\, dx\, dy = 2 \int_{-\infty}^{\infty} \int_{x}^{\infty} |K(x, y)|^2\, dy\, dx$$
$$= \frac{1}{2\sigma^2} \int_{-\infty}^{0} u_s^2(-x) \int_{0}^{\infty} u_s^2(y)\, dy\, dx$$
$$+ \frac{1}{2\sigma^2} \int_{-\infty}^{0} u_s^2(-x) \int_{x}^{0} u_s^2(y)\, dy\, dx$$
$$+ \frac{1}{2\sigma^2} \int_{0}^{\infty} u_s^2(-x) \int_{x}^{\infty} u_s^2(y)\, dy\, dx.$$

It is easy to analyze the first of the three integrals on the right and to show that it is finite. The second and third present the same difficulties and we consider the third one only. Using (80) and (82) we find, for large x, that

$$u_s^2(-x) \int_x^\infty u_s^2(y)\, dy \to \frac{e^{x^2}}{x} \int_x^\infty \frac{e^{-y^2}}{y}\, dy.$$

Applying an integration by parts to the last term leads to

$$\left| \frac{e^{x^2}}{x} \left[\frac{e^{-x^2}}{2x^2} - \int_x^\infty \frac{e^{-y^2}}{y^3}\, dy \right] \right| = \left| \frac{1}{2x^3} - \frac{e^{x^2}}{x} \int_x^\infty \frac{e^{-y^2}}{y^3}\, dy \right|$$

$$\leq \frac{1}{2x^3} + \frac{1}{x} \int_x^\infty \frac{dy}{y^3} = \frac{1}{x^3},$$

which is integrable for large x. It follows that

$$\int_{-\infty}^\infty \int_{-\infty}^\infty |K(x, y)|^2\, dx\, dy < \infty$$

so that K is a compact, self-adjoint operator on $L_2[-\infty, \infty]$. We have seen earlier that the orthonormal eigenfunctions of K are given by $\phi_n(x) = k_n e^{-(x^2/2)} H_n(x)$, and k_n is so chosen that

$$\|\phi_n(x)\|^2 = \int_{-\infty}^\infty k_n^2 e^{-x^2} H_n^2(x)\, dx = 1.$$

Since K is compact these eigenfunctions form a complete orthonormal set on $L_2[-\infty, \infty]$.

5. POSITIVE OPERATORS

DEFINITION. A self-adjoint operator K acting on a Hilbert space H is said to be positive if $(Kf, f) \geq 0$ for all f in H.

If the space H is a complex space and K is bounded the requirement of self-adjointness in the above definition is superfluous. One can show that if (Kf, f) is real and K is bounded then K is self-adjoint. In that case the requirement $(Kf, f) \geq 0$ is enough to insure that K is self-adjoint if bounded.

If K is in addition compact, positivity is equivalent to saying that all eigenvalues are non-negative. To see this we use theorem 13. Then if $f \in H$ we have

$$f = \sum_i (f, \phi_i)\phi_i + f_0$$

$$Kf = \sum_i \mu_i(f, \phi_i)\phi_i.$$

In the above f_0 is in the null space of K and of course orthogonal to all ϕ_i. Then

$$(Kf, f) = \sum_i \mu_i |(f, \phi_i)|^2.$$

Clearly K is positive if and only if all $\mu_i \geq 0$.

More generally we speak also of semibounded operators.

DEFINITION.   A self-adjoint operator K acting on a Hilbert space H is said to be bounded below by λ^* if $(Kf, f) \geq \lambda^*(f, f)$ for all f in H. If $(Kf, f) \leq \lambda^*(f, f)$ for all f we say it is bounded above. Such operators are said to be semibounded.

Clearly K is positive if it is bounded below by 0. The operators treated in Examples 1, 5, and 8 in the last section are all positive since all their eigenvalues are non-negative.

EXAMPLE 1.   Consider

$$Lu = \frac{1}{x}\left[-(xu')' + \frac{1}{x}u\right], \qquad u(1) = 0. \tag{86}$$

This operator was treated in Example 7 of the last section. Its inverse integral operator is compact and self-adjoint, but the eigenvalues are solutions of the transcendental equation

$$J_1(\sqrt{\lambda}) = 0$$

and we have no other information about these values as yet.

We can show that the operator is positive. To do so we consider

$$(Lu, u)_x = \int_0^1 \left[-(xu')' + \frac{1}{x}u\right]\bar{u}\, dx$$

$$= -xu'\bar{u}\Big|_0^1 + \int_0^1 \left[x|u'|^2 + \frac{1}{x}|u|^2\right] dx. \tag{87}$$

The domain of the differential operator consists of those functions for which Lu belongs to $L_2([0, 1], x)$ and satisfies $u(1) = 0$. For those functions the integrated term vanishes and the integral in (87) exists so that

$$(Lu, u)_x = \int_0^1 \left[x|u'|^2 + \frac{1}{x}|u|^2\right] dx \geq 0.$$

The above integral can vanish only for $u = 0$ which is not an eigenfunction so that in fact the inequality is strict. One can conclude from that, that the lowest eigenvalue of L is positive.

In the above example it was shown that L is positive without an explicit knowledge of the eigenvalues. Somewhat later it will be shown that the regular second order differential operators discussed in Examples 3 and 6 in the last section are bounded below so that only a finite number of eigenvalues can be negative.

A simple way of generating positive operators is to consider an arbitrary operator K, its adjoint K^* and the product K^*K. Then

$$(K^*Kf, f) = (Kf, Kf) = \|Kf\|^2 \geq 0.$$

It follows that if K is self-adjoint then K^2 is positive.

In theorem 14 it was shown that, if $K(x, y)$ is an L_2 kernel, then

$$K(x, y) = \sum_{i=1}^{\infty} \mu_i \phi_i(x)\overline{\phi_i(y)}$$

and the above converges in the mean. This result can be strengthened considerably for positive operators with continuous kernels.

LEMMA. Let K be a positive operator on $L_2[0, 1]$, with a kernel $K(x, y)$ that is continuous. Then $K(x, x) \geq 0$ for $0 \leq x \leq 1$.

PROOF. We suppose that for some x_0, $K(x_0, x_0) < 0$. Since $K(x, y)$ is continuous $K(x, y) < 0$ in some neighborhood of the point (x_0, x_0). Then for some δ

$$K(x, y) < 0 \qquad \text{for} \quad |x - x_0| + |y - y_0| < \delta.$$

Now we consider a function $f_0(x)$ defined by

$$f_0(x) = 1, \qquad |x - x_0| \leq \tfrac{1}{2}\delta$$
$$= 0, \qquad |x - x_0| > \tfrac{1}{2}\delta.$$

Then

$$(Kf_0, f_0) = \int_0^1 \int_0^1 K(x, y)f_0(y)\,dy f_0(x)\,dx$$

$$= \int_{x_0-(\delta/2)}^{x_0+(\delta/2)} \int_{x_0-(\delta/2)}^{x_0+(\delta/2)} K(x, y)\,dy\,dx < 0$$

which contradicts the positivity of K so that $K(x, x) \geq 0$. ∎

The continuity of $K(x, y)$ is essential. For a general L_2 kernel $K(x, x)$ could be assigned arbitrarily without affecting the eigenvalues. However, we only needed the continuity of $K(x, y)$ near the line $y = x$. This leads us to the following corollary.

COROLLARY. Let K be a positive operator on $L_2[0, 1]$, with a kernel $K(x, y)$ that is continuous near $y = x$. Then $K(x, x) \geq 0$ for $0 \leq x \leq 1$.

PROOF. The proof of the previous lemma applies verbatim. ∎

We now come to one of the key theorems regarding positive operators.

THEOREM 17.   Mercer's Theorem.   Let $K(x, y)$ be the kernel of a positive operator on $L_2[0, 1]$ and suppose that $K(x, y)$ is continuous in both variables. Then

$$K(x, y) = \sum_{i=1}^{\infty} \mu_i \phi_i(x)\overline{\phi_i(y)} \tag{88}$$

where the series on the right converges uniformly and absolutely to $K(x, y)$.

PROOF.   We note that, since

$$\int_0^1 K(x, y)\phi_i(y) \, dy = \mu_i \phi_i(x)$$

and $K(x, y)$ is continuous, $\phi_i(x)$ is also continuous. Now consider the kernel

$$K^{(n)}(x, y) = K(x, y) - \sum_{i=1}^{n} \mu_i \phi_i(x)\overline{\phi_i(y)}.$$

Clearly the above kernel is continuous and the associated operator is positive. In fact its eigenvalues are $\mu_{n+1}, \mu_{n+2}, \ldots$. By the lemma

$$K^{(n)}(x, x) = K(x, x) - \sum_{i=1}^{n} \mu_i \, |\phi_i(x)|^2 \geq 0$$

so that

$$\sum_{i=1}^{n} \mu_i \, |\phi_i(x)|^2 \leq K(x, x) \qquad \text{for all } n. \tag{89}$$

It follows that the series on the left converges.

Now we consider

$$\left| \sum_{i=n}^{m} \mu_i \phi_i(x)\overline{\phi_i(y)} \right|^2 \leq \left[\sum_{i=n}^{m} \mu_i \, |\phi_i(x)\overline{\phi_i(y)}| \right]^2$$

$$\leq \sum_{i=n}^{m} \mu_i \, |\phi_i(x)|^2 \sum_{i=n}^{m} \mu_i \, |\phi_i(y)|^2. \tag{90}$$

Since $K(x, x) \leq M$ we have, using (89), $\sum_{i=1}^{n} \mu_i \, |\phi_i(x)|^2 \leq M$ independently of n and y. For a fixed y, say y_0, we have

$$\sum_{i=n}^{m} \mu_i \, |\phi_i(y_0)|^2 < \epsilon \qquad \text{for } n, m > N(\epsilon)$$

so that by (90) the sum in (88), for fixed y_0, converges uniformly and absolutely in x. It follows that the series on the right in (88) is a continuous function in x for every fixed y. Let

$$\tilde{K}(x, y_0) = K(x, y_0) - \sum_{i=1}^{\infty} \mu_i \phi_i(x)\overline{\phi_i(y_0)};$$

a calculation shows that

$$\int_0^1 |\tilde{K}(x, y_0)|^2 \, dx = \int_0^1 |K(x, y_0)|^2 \, dx - \sum_{i=1}^{\infty} \mu_i^2 \, |\phi_i(y_0)|^2 = 0$$

by Parseval's formula. Since $\tilde{K}(x, y_0)$ is also continuous in x we have $\tilde{K}(x, y_0) = 0$ so that

$$K(x, y_0) = \sum_{i=1}^{\infty} \mu_i \phi_i(x) \overline{\phi_i(y_0)}.$$

The above converges absolutely and uniformly for all $0 \leq x \leq 1$, and in particular for $x = y_0$ so that

$$K(y_0, y_0) = \sum_{i=1}^{\infty} \mu_i \, |\phi_i(y_0)|^2. \tag{91}$$

Equation (91) holds for arbitrary y_0. We now consider the sequence

$$k_n(y) = K(y, y) - \sum_{i=1}^{n-1} \mu_i \, |\phi_i(y)|^2.$$

$\{k_n(y)\}$ is a sequence of positive monotonically decreasing functions, and by virtue of (91) the sequence converges to 0 for every y. Therefore, for a given positive ϵ there exists an $N(\epsilon)$ such that

$$0 \leq k_n(y) \leq \epsilon, \qquad n > N(\epsilon).$$

The $N(\epsilon)$ is independent of y so that the convergence is uniform.

Finally, using the Cauchy-Schwarz inequality we have

$$\left| \sum_{i=n}^{\infty} \mu_i \phi_i(x) \overline{\phi_i(y)} \right|^2 \leq \left[\sum_{i=n}^{\infty} \mu_i \, |\phi_i(x) \overline{\phi_i(y)}| \right]^2 \leq k_n(x) k_n(y)$$

so that (88) converges absolutely and uniformly. ∎

There are two important corollaries that we can deduce immediately from Mercer's theorem.

COROLLARY 1. Let K be a compact self-adjoint operator that has at most a finite number of negative eigenvalues. Suppose the kernel $K(x, y)$ is continuous for $0 \leq x, y \leq 1$. Then

$$K(x, y) = \sum_{i=1}^{\infty} \mu_i \phi_i(x) \overline{\phi_i(y)} \tag{92}$$

where the series on the right converges absolutely and uniformly to $K(x, y)$.

PROOF. Suppose $\mu_1, \mu_2, \ldots, \mu_n$ are all the negative eigenvalues so that $\mu_{n+1}, \mu_{n+2}, \ldots$ are all positive. Consider the kernel

$$K^{(n)}(x, y) = K(x, y) - \sum_{i=1}^{n} \mu_i \phi_i(x) \overline{\phi_i(y)}. \tag{93}$$

The operator associated with $K^{(n)}(x, y)$ has no negative eigenvalues and hence is positive. Then

$$K^{(n)}(x, y) = \sum_{i=n+1}^{\infty} \mu_i \phi_i(x)\overline{\phi_i(y)}$$

converges absolutely and uniformly by Mercer's theorem. Combining this with (93) leads to the desired conclusion. ∎

COROLLARY 2.   Let K be as in Corollary 1. Then

$$\int_0^1 K(x, x)\,dx = \sum_{i=1}^{\infty} \mu_i. \tag{94}$$

PROOF.   Since the series in (92) converges uniformly term by term integration is permissible so that (94) follows. ∎

From theorem 13 we were able to conclude that $\sum \mu_i^2$ converges, but for positive operators the stronger conclusion $\sum_{i=1}^{\infty} \mu_i < \infty$ follows.

EXAMPLE 2.   Consider

$$Lu = -u''$$

$$u(0) = u(1) = 0. \tag{95}$$

This operator was considered in Example 1 of the previous section. There it was shown that

$$K(x, y) = x_<(1 - x_>)$$

$$\mu_n = \frac{1}{n^2\pi^2}.$$

Using (94) we have

$$\int_0^1 x(1 - x)\,dx = \frac{1}{\pi^2}\sum_{n=1}^{\infty}\frac{1}{n^2} = \frac{1}{6}. \tag{96}$$

EXAMPLE 3.   Consider

$$Lu = -u''$$

$$u(0) - u(1) = 0 \tag{97}$$

$$u'(0) - u'(1) = 0.$$

This operator was discussed in Example 5 of the previous section. It was shown that the nonzero eigenvalues of this operator are given by

$$\mu_n = \frac{1}{4n^2\pi^2}, \qquad n = 1, 2, \ldots.$$

Each of these is a double eigenvalue; that is, it has two independent eigenfunctions associated with it. The kernel of K is given by

$$K(x, y) = \tfrac{1}{12} - \tfrac{1}{2}(x_> - x_<) + \tfrac{1}{2}(x - y)^2$$

so that

$$\int_0^1 K(x, x)\, dx = \frac{2}{4\pi^2} \sum_{n=1}^\infty \frac{1}{n^2} = \frac{1}{12}. \tag{98}$$

The factor of 2 before the sum in (98) takes into account the fact that each eigenvalue must be counted twice in (94), since each is a double eigenvalue.

EXAMPLE 4. In Example 7 of the previous section we treated the operator

$$Lu = \frac{1}{x}\left[-(xu')' + \frac{1}{x}u \right] \tag{99}$$

$$u(1) = 0.$$

Its eigenvalues are solutions of $J_1(\sqrt{\lambda_n}) = 0$, and their explicit values are not known. We also found that

$$K(x, y) = \tfrac{1}{2}x_<\left(\frac{1}{x_>} - x_> \right)$$

and we worked in the space $L_2([0, 1], x)$. However it was shown that

$$\tilde{K}(x, y) = \sqrt{xy}\, K(x, y)$$

was an equivalent operator on $L_2[0, 1]$ so that

$$\int_0^1 \tfrac{1}{2}x^2\left(\frac{1}{x} - x \right) dx = \sum_{n=1}^\infty \mu_n = \tfrac{1}{8}. \tag{100}$$

The expansion (92), whether for continuous or general kernels is useful for the construction of the kernels of the iterated operators.

$$K_2(x, y) = \int_a^b K(x, z)K(z, y)\, dz$$

$$= \int_a^b \sum_{i=1}^\infty \mu_i \phi_i(x)\overline{\phi_i(z)} \sum_{i=1}^\infty \mu_i \phi_i(z)\overline{\phi_i(y)}\, dz$$

$$= \sum_{i=1}^\infty \mu_i^2 \phi_i(x)\overline{\phi_i(y)}.$$

More generally one finds that

$$K_n(x, y) = \sum_{i=1}^\infty \mu_i^{\,n} \phi_i(x)\overline{\phi_i(y)} \tag{101}$$

from which we have

$$\int_a^b K_n(x, x) \, dx = \sum_{i=1}^{\infty} \mu_i{}^n, \qquad n \geq 2. \tag{102}$$

In general (102) will converge only for $n \geq 2$, which follows from the corollary to theorem 14. If $K(x, y)$ is continuous on a bounded interval (102) converges for $n = 1$ as well, by Mercer's theorem.

We shall now discuss a number of results that enable us to compare operators in a certain sense. These results will be useful in showing that regular self-adjoint differential operators are lower semibounded.

LEMMA. Let K be a self-adjoint compact operator on a Hilbert space H. Suppose that H has a subspace M such that for some λ^*

$$(Kf, f) \leq \lambda^*(f, f) \qquad \text{for all } f \in M.$$

Let n be the number of eigenvalues of K that are less than or equal to λ^*. Then

$$n \geq \dim M.$$

PROOF.   Let $\mu_1, \mu_2, \ldots, \mu_n$ be the set of eigenvalues of K that are less than or equal to λ^*, and $\phi_1, \phi_2, \ldots, \phi_n$ the corresponding orthonormal eigenfunctions. Let H_0 be the space spanned by these eigenfunctions.

Let f be an element in H_0 so that

$$f = \sum_{i=1}^{n} (f, \phi_i)\phi_i$$

and

$$(Kf, f) = \sum_{i=1}^{n} \mu_i |(f, \phi_i)|^2 \leq \lambda^* \sum_{i=1}^{n} |(f, \phi_i)|^2 = \lambda^*(f, f). \tag{103}$$

Now let $H_0{}^\perp$ denote the complementary subspace of H_0 so that $H = H_0 \oplus H_0{}^\perp$.

Suppose that $m = \dim M > n$, and let $\psi_1, \psi_2, \ldots, \psi_m$ be an orthonormal set spanning M. Every ψ_i can be decomposed as

$$\psi_i = \psi_i^{(0)} + \psi_i^{(1)}, \qquad \psi_i^{(0)} \in H_0, \qquad \psi_i^{(1)} \in H_0{}^\perp.$$

The set $\{\psi_i^{(0)}\}$ must be linearly dependent in H_0 so that for a suitable set of scalars $\{\alpha_i\}$, not all of which vanish

$$\sum_{i=1}^{m} \alpha_i \psi_i^{(0)} = 0$$

and

$$\tilde{\psi} = \sum_{i=1}^{m} \alpha_i \psi_i = \sum_{i=1}^{m} \alpha_i \psi_i^{(1)} \in H_0{}^\perp \cap M.$$

Accordingly, since $\tilde{\psi} \in H_0^{\perp}$, $(K\tilde{\psi}, \tilde{\psi}) > \lambda^*(\tilde{\psi}, \tilde{\psi})$. But $\tilde{\psi} \in M$, so that $(K\tilde{\psi}, \tilde{\psi}) \leq \lambda^*(\tilde{\psi}, \tilde{\psi})$. This contradiction shows that dim $M \leq n$. ■

An inspection of the proof shows that the compactness of K was used in only one way. We used the fact that K has discrete eigenvalues, and that to each we could assign only one eigenfunction. The possibility of a finite number of eigenvalues having the same numerical value is not precluded. If K^{-1} exists it would have the same property. The differential operators treated in the previous section evidently had that property.

COROLLARY. Let L be a self-adjoint operator with discrete eigenvalues, such that all eigenvalues have finite multiplicity. Let M be a subspace of H such that for some λ^*

$$(Lf, f) \leq \lambda^*(f, f) \qquad \text{for all } f \in M.$$

Let n be the number of eigenvalues of L less than or equal to λ^*. Then

$$n \geq \dim M.$$

PROOF. The same proof as the one for the lemma can be used. ■

The above corollary enables us to prove an important comparison theorem.

THEOREM 18. Let K_1 and K_2 be two self-adjoint operators. Suppose both are of the type described in the above corollary; that is, they have discrete eigenvalues of finite multiplicity. Furthermore suppose

$$(K_1 f, f) \geq (K_2 f, f) \qquad \text{for all } f \in H.$$

Select any λ^* and let $\mu_1, \mu_2, \ldots, \mu_m$ be all eigenvalues of K_1 such that $\mu_i \leq \lambda^*$. Similarly let $\nu_1, \nu_2, \ldots, \nu_n$ be the eigenvalues of K_2 such that $\nu_i \leq \lambda^*$. Then

$$n \geq m.$$

N.B. Each eigenvalue is repeated in accordance with its multiplicity.

PROOF. Let M be the subspace spanned by $\phi_1, \phi_2, \ldots, \phi_m$, the ortho-normal eigenfunctions of K_1 corresponding to $\mu_1, \mu_2, \ldots, \mu_m$. We now have

$$(K_2 f, f) \leq (K_1 f, f) \leq \lambda^*(f, f) \qquad \text{for all } f \in M.$$

The above inequality is derived exactly as (103) was. Using the above inequality on K_2 combined with the corollary shows that

$$n \geq m. \quad ■$$

We shall now consider second order differential operators.

THEOREM 19.   Consider the differential operator

$$Lu = \frac{1}{r(x)} [-(p(x)u')' + q(x)u] \tag{104}$$

subject to the boundary conditions

$$a_0 u(0) + a_1 u'(0) + a_2 u(1) + a_3 u'(1) = 0$$
$$b_0 u(0) + b_1 u'(0) + b_2 u(1) + b_3 u'(1) = 0$$

$$\sum_{i=1}^{3} |a_i| \neq 0, \qquad \sum_{i=1}^{3} |b_i| \neq 0$$

and such that

$$(Lu, v)_r = (u, Lv)_r \tag{105}$$

for all u, v in $L_2([0, 1], r)$ for which Lu and Lv are also in $L_2([0, 1], r)$. The functions $r(x)$, $p(x)$ are to be positive and continuous on $[0, 1]$, and $p'(x)$, $q(x)$ are also continuous on $[0, 1]$. L is a semibounded operator that has only a finite number of negative eigenvalues.

PROOF.   If L does not have $\lambda = 0$ as an eigenvalue then an inverse integral operator K with kernel $K(x, y)$ exists. The kernel is an L_2 kernel so that K is compact and by virtue of (105) is also self-adjoint. $K(x, y)$ satisfies the differential equation.

$$-(p(x)K_x(x, y))_x + q(x)K(x, y) = \delta(x - y) \tag{106}$$

subject to the boundary condition. If $\lambda = 0$ were an eigenvalue (106) would have no solution. This case can be treated by simple modification of the type discussed in Example 5 of the previous section.

   Now we consider a second differential operator

$$\tilde{L} = \frac{1}{r(x)} [-(p(x)u')' + q(x)u] \tag{107}$$

defined on $L_2([0, 1], r)$ with the boundary conditions

$$u(0) = u(1) = 0.$$

With (107) we associate an inverse integral operator $\tilde{K}$ with a kernel $\tilde{K}(x, y)$. $\tilde{K}(x, y)$ satisfies

$$-(p(x)\tilde{K}_x(x, y))_x + q(x)\tilde{K}(x, y) = \delta(x - y),$$
$$\tilde{K}(0, y) = \tilde{K}(1, y) = 0. \tag{108}$$

$\tilde{K}$ is a compact self-adjoint operator. For the moment we shall assume that $q(x)$ is positive. This restriction will be lifted somewhat later.

We can now show that $\tilde{K}$ and $\tilde{L}$ are positive operators.

$$(\tilde{L}u, u)_r = \int_0^1 [-(p(x)u')' + q(x)u]\bar{u}\,dx$$
$$= -p(x)u'\bar{u}\big|_0^1 + \int_0^1 [p(x)\,|u'|^2 + q(x)\,|u|^2]\,dx.$$

The integrated term vanishes by use of the boundary conditions and the integral on the right is positive since $p(x)$ and $q(x)$ are positive. Then

$$(\tilde{L}u, u)_r \geq 0.$$

In fact $\lambda = 0$ is not an eigenvalue, because if $\tilde{L}u = 0$, then $u = 0$ but

$$(\tilde{L}u, u)_r > 0 \quad \text{if} \quad u \not\equiv 0.$$

Therefore $\tilde{K}$ is also positive with eigenvalues

$$\mu_1 \geq \mu_2 \geq \cdots > 0.$$

The eigenvalues of $\tilde{L}$ are $\lambda_1 = 1/\mu_1, \lambda_2 = 1/\mu_2 \cdots$ so that $0 < \lambda_1 \leq \lambda_2 \leq \cdots$ and they accumulate at infinity.

The possibility of having only a finite number of eigenvalues can be precluded as follows: Since the null space of $\tilde{K}$ is empty, the eigenfunctions corresponding to $\mu_1, \mu_2, \ldots$, form a complete set in $L_2([0, 1], r)$. The latter space is infinite dimensional so that a finite number of eigenfunctions cannot form a complete set.

We shall now compare K and $\tilde{K}$. We see that

$$\left[-\frac{\partial}{\partial x}\,p(x)\frac{\partial}{\partial x} + q(x)\right][K(x, y) - \tilde{K}(x, y)] = \delta(x - y) - \delta(x - y) = 0.$$

so that

$$K(x, y) - \tilde{K}(x, y) = a(y)u_1(x) + b(y)u_2(x). \tag{109}$$

Here $u_1(x)$ and $u_2(x)$ are two linearly independent solutions of

$$-(p(x)u')' + q(x)u = 0.$$

Since K is self-adjoint and $K(x, y)$ is real we have

$$K(x, y) = K(y, x)$$

and similarly for $\tilde{K}(x, y)$. Therefore,

$$a(y)u_1(x) + b(y)u_2(x) = a(x)u_1(y) + b(x)u_2(y)$$

from which we can deduce that

$$a(y) = \alpha u_1(y) + \beta u_2(y)$$
$$b(y) = \beta u_1(y) + \gamma u_2(y)$$

where α, β, and γ are suitable scalars. Thus $K - \tilde{K}$ is a self-adjoint degenerate operator, say D. Then

$$K - \tilde{K} = D$$

where

$$D(x, y) = \alpha u_1(x)u_1(y) + \beta(u_1(x)u_2(y) + u_2(x)u_1(y)) + \gamma u_2(x)u_2(y) \quad (110)$$

and

$$(Kf, f) = (\tilde{K}f, f) + (Df, f) \geq (Df, f) \quad (111)$$

since $\tilde{K}$ is positive.

D is a degenerate operator and therefore has at most a finite number of negative eigenvalues. In fact, by use of (110), it has at most two negative eigenvalues. It follows by theorem 18 that K has at most two negative eigenvalues. Since K is the operator inverse to L, L will have at most two negative eigenvalues as well. This completes the proof when $q(x)$ is positive.

$q(x)$ is continuous on $[0, 1]$, and therefore has a lower bound. In that case we can find a number τ such that $q(x) - \tau$ is positive. We now consider the operator

$$L_\tau u = \frac{1}{r(x)}\left[-(p(x)u')' + (q(x) - \tau)u\right]$$

subject to the same boundary conditions as L. We shall denote its eigenvalues by $\{\lambda_n\}$ and its eigenfunctions by $\{\phi_n(x)\}$. We have

$$\lambda_1 \leq \lambda_2 \leq \cdots$$

and the eigenvalues accumulate only at infinity, by the preceding discussion. We also know that from λ_3 on all are non-negative. Then

$$L_\tau \phi_n(x) = L\phi_n(x) - \tau\phi_n(x) = \lambda_n\phi_n(x)$$

so that

$$L\phi_n(x) = (\lambda_n + \tau)\phi_n(x).$$

It follows that L has the same eigenfunctions as L_τ and the eigenvalues $\{\lambda_n + \tau\}$. These are bounded below so that L is a semibounded operator. ■

EXAMPLE 5. Consider the operator

$$Lu = -u''$$
$$u(0) = 0, \qquad hu(1) + u'(1) = 0 \quad (112)$$

and the comparison operator

$$\tilde{L}u = -u''$$
$$u(0) = u(1) = 0. \quad (113)$$

For (113) we know that

$$\tilde{K}(x, y) = x_<(1 - x_>).$$

Similarly we can solve for $K(x, y)$ such that

$$LK(x, y) = \delta(x - y)$$
$$K(0, y) = 0, \qquad hK(1, y) + K_x(1, y) = 0.$$

A standard calculation shows that

$$K(x, y) = \frac{x_<(1 + h - hx_>)}{1 + h}$$

provided $1 + h \neq 0$.

If $h = -1$ we see that $\lambda = 0$ is an eigenvalue of L since $u = x$ satisfies $Lu = 0$ and also the boundary conditions.

For $1 + h \neq 0$ we now see that

$$K(x, y) - \tilde{K}(x, y) = D(x, y) = \frac{xy}{1 + h} \tag{114}$$

and clearly the operator D is a degenerate operator. To determine its eigenvalues and eigenfunctions we consider

$$\phi(x) = \frac{\lambda}{1 + h} x \int_0^1 y\phi(y)\, dy.$$

If $\lambda \neq 0$, $\phi(x)$ must be a multiple of x, say ax. Then

$$ax = \frac{\lambda}{1 + h} x \int_0^1 ay^2\, dy = \frac{a\lambda}{3(1 + h)} x$$

so that $\lambda = 3(1 + h)$ is the only nonzero eigenvalue. All functions orthogonal to x on $L_2[0, 1]$ will now satisfy

$$D\phi = \int_0^1 \frac{xy}{1 + h} \phi(y)\, dy = 0$$

so that D has the eigenvalue $\mu = 0$, which is of infinite multiplicity. It follows that if $h > -1$, D as given by (114) is a positive operator and since

$$(Kf, f) = (\tilde{K}f, f) + (Df, f) \geq (\tilde{K}f, f) > 0$$

K is also a positive operator. In other words L in (112) being K^{-1} is not only bounded below but positive.

For $h < -1$, D has one negative eigenvalue $\lambda = 3(1 + h)$. In this case L has at most one negative eigenvalue. In fact one can solve the differential equation

$$-u'' = \lambda u$$

subject to

$$u(0) = 0, \qquad hu(1) + u'(1) = 0.$$

Then

$$u = \sin \sqrt{\lambda}\, x$$

satisfies the equation and $u(0) = 0$. The second boundary condition leads to the transcendental equation

$$\tan \sqrt{\lambda} = -\frac{\sqrt{\lambda}}{h}$$

If λ is negative we let $\sqrt{\lambda} = i\rho$ and obtain

$$\tanh \rho = -\frac{\rho}{h}.$$

For $h < -1$, the above has precisely one positive solution so that L does have one negative eigenvalue.

In Example 5 the operator was a relatively simple one so that theorems 18 and 19 were useful, but not vital. A direct attack on the operator L would have yielded the same results. However in the next example we shall examine a case where more elementary tools no longer suffice.

EXAMPLE 6. Consider

$$Lu = \frac{1}{x}\left[-(xu')' + \frac{1}{x}u \right] \tag{115}$$

$$hu(1) + u'(1) = 0$$

and the comparison operator

$$\tilde{L}u = \frac{1}{x}\left[-(xu')' + \frac{1}{x}u \right] \tag{116}$$

$$u(1) = 0.$$

The kernels of the inverse integral operators are

$$K(x, y) = \frac{x_<[(1 - h)x_> + (1 + h)/x_>)]}{2(1 + h)}$$

$$\tilde{K}(x, y) = \tfrac{1}{2}x_<\left[\frac{1}{x_>} - x_> \right]$$

so that

$$D(x, y) = K(x, y) - \tilde{K}(x, y) = \frac{xy}{1 + h}.$$

The case $h = -1$ has to be treated separately since in that case L has an eigenvalue at 0. D has a single nonzero eigenvalue $1/[4(1 + h)]$. For $h > -1$, K and therefore L also will be positive. The analysis is identical to the one carried out in Example 5. If $h < -1$, L will have at most one negative eigenvalue.

Theorem 19 is not really restricted to second order differential operators, but can easily be extended to higher order operators. Let

$$Lu = \frac{1}{r(x)} \sum_{k=0}^{n} (-1)^k (p_k(x) u^{(k)})^{(k)}$$ (117)

be defined on those functions in $L_2([0, 1], r(x))$ for which $Lu \in L_2([0, 1], r(x))$ and assume that all $p_k(x)$ are sufficiently differentiable, that is, $p_k(x)$ has k continuous derivatives. We also assume that $p_n(x)$ and $r(x)$ are positive on $[0, 1]$. For boundary conditions we take $2n$ homogeneous conditions

$$\sum_{k=0}^{2n-1} a_{ik} u^{(k)}(0) + \sum_{k=0}^{2n-1} b_{ik} u^{(k)}(1) = 0, \qquad i = 1, 2, \ldots, 2n$$

$$\sum_{k=0}^{2n-1} |a_{ik}| + |b_{ik}| \neq 0.$$ (118)

We assume that the operator (117) under these boundary conditions is symmetric, so that

$$(Lu, v)_r = (u, Lv)_r.$$

In that case it is semibounded from below.

6. APPROXIMATION OF EIGENVALUES

The various expansions developed are often of great help in estimating eigenvalues of operators. If K is a compact self-adjoint operator with eigenfunctions $\{\phi_n\}$ and eigenvalues $\{\mu_n\}$ we have

$$Kf = \sum_{n=1}^{\infty} \mu_n (f, \phi_n) \phi_n.$$

It follows that

$$\frac{\|Kf\|}{\|f\|} = \left\{ \frac{\sum_{n=1}^{\infty} \mu_n^2 |(f, \phi_n)|^2}{\sum_{n=1}^{\infty} |(f, \phi_n)|^2} \right\}^{\frac{1}{2}}$$ (119)

From (119) it is obvious that, if μ_1 is the eigenvalue with largest magnitude,

$$|\mu_1| = \sup_{f \neq 0} \frac{\|Kf\|}{\|f\|} = \sup_{\|f\|=1} \|Kf\|.$$ (120)

For arbitrary f we have

$$\|Kf\| \leq |\mu_1| \|f\|$$ (121)

and one can quickly obtain lower bounds on $|\mu_1|$.

If K is an integral operator with kernel $K(x, y)$ and $K_2(x, y)$ is the iterated kernel we have

$$\int_a^b K_2(x, x)\, dx = \sum_{n=1}^\infty \mu_n^2$$

and it follows that

$$\mu_1^2 \le \int_a^b K_2(x, x)\, dx \tag{122}$$

yielding an upper estimate on μ_1^2.

EXAMPLE 1. The operator

$$\begin{aligned} Lu &= -u'' \\ u(0) &= u(1) = 0 \end{aligned} \tag{123}$$

has the inverse integral operator K with kernel $K(x, y) = x_<(1 - x_>)$.
We shall first use (121) with $f = 1$. Then $Kf = \frac{1}{2}x(1 - x)$ and

$$|\mu_1| \ge \|Kf\| = \left\{\int_0^1 \tfrac{1}{4}x^2(1 - x)^2\, dx\right\}^{\frac{1}{2}} = \frac{1}{\sqrt{120}}.$$

But we know that $\mu_1 = (1/\pi^2)$ so that the above is equivalent to

$$\pi \le (120)^{\frac{1}{4}} = 3.31 \cdots. \tag{124}$$

We can also use (94), namely

$$\int_0^1 K(x, x)\, dx = \sum_{n=1}^\infty \mu_n$$

so that

$$\frac{1}{\pi^2} = \mu_1 \le \int_0^1 K(x, x)\, dx = \frac{1}{6}$$

$$\pi \ge \sqrt{6} = 2.45 \cdots. \tag{125}$$

A better estimate is obtained using (122)

$$K_2(x, x) = \int_0^1 [K(x, y)]^2\, dy = \frac{x^2(1 - x)^2}{3}$$

so that

$$\frac{1}{\pi^4} = \mu_1^2 \le \int_0^1 K_2(x, x)\, dx = \frac{1}{90}$$

and

$$\pi \ge (90)^{\frac{1}{4}} = 3.08 \cdots. \tag{126}$$

Combining (124) and (126) we see that

$$3.08 \le \pi \le 3.31.$$

The estimate of (122) can be improved as follows

$$\int_a^b K_n(x, x)\, dx = \sum_{k=1}^{\infty} \mu_k^{\,n} = \mu_1^{\,n} \sum_{k=1}^{\infty} \left(\frac{\mu_k}{\mu_1}\right)^n.$$

From the above we see, if $|\mu_k/\mu_1| < 1$ for $k > 1$, that

$$\lim_{n \to \infty} \left[\int_a^b K_n(x, x)\, dx \right]^{1/n} = \mu_1. \tag{127}$$

Many physical problems are first formulated in terms of differential operators. In order to estimate the eigenvalues by the above techniques it is then necessary to construct the inverse integral operator. Depending on the nature of the differential operator L this may or may not be an easy task to carry out in practice. We shall now describe a technique that allows us to work directly with the differential operator.

Let L be a regular differential operator and let the set of orthonormal eigenfunctions $\{\phi_n\}$ form a complete set in H. Then for every $f \in H$

$$f = \sum_{n=1}^{\infty} (f, \phi_n)\phi_n.$$

If f is in the domain, that is, $Lf \in H$, we have

$$Lf = \sum_{n=1}^{\infty} \lambda_n(f, \phi_n)\phi_n.$$

In fact the domain can be characterized by requiring that

$$\sum_{n=1}^{\infty} |\lambda_n(f, \phi_n)|^2 < \infty. \tag{128}$$

Evidently if (128) is satisfied $Lf \in H$. We now have

$$\frac{\|Lf\|}{\|f\|} = \left\{ \frac{\sum_{n=1}^{\infty} \lambda_n^{\,2} |(f, \phi_n)|^2}{\sum_{n=1}^{\infty} |(f, \phi_n)|^2} \right\}^{\frac{1}{2}}. \tag{129}$$

If L is positive and λ_1 is its smallest eigenvalue we see from (129) that

$$\lambda_1 = \min_{f \neq 0} \frac{\|Lf\|}{\|f\|} = \min_{\|f\|=1} \|Lf\| \tag{130}$$

and more generally

$$\lambda_1 \|f\| \leq \|Lf\|. \tag{131}$$

In (130) and (131) we, of course, have to work with those f that are in the domain of L.

EXAMPLE 2. Let L be as in Example 1. A function f in the domain of L has to be such that $f'' \in L_2[0, 1]$ and $f(0) = f(1) = 0$. Such an f is given by

$$f(x) = x(1 - x)$$

$$\|f\| = \frac{1}{\sqrt{30}} \,.$$

$$Lf = -f''(x) = 2$$

$$\|Lf\| = 2$$

and as we know

$$\lambda_1 = \pi^2.$$

Using (131) we have

$$\pi^2 \leq 2\sqrt{30}$$

and

$$\pi \leq (120)^{\frac{1}{4}}$$

as in (124).

EXAMPLE 3. Let

$$Lu = \frac{1}{x}\left[-(xu')' + \frac{1}{x}u\right]$$ (132)

$$u(1) = 0.$$

In this case we work on the space $L_2([0, 1], x)$ and we had seen earlier that

$$K(x, y) = \tfrac{1}{2}x_<\left[\frac{1}{x_>} - x_>\right].$$

First we shall use (122).

$$K_2(x, x) = \int_0^1 yK^2(x, y)\, dy = \tfrac{1}{8}x^4 - \tfrac{1}{8}x^2 - \tfrac{1}{4}x^2 \ln x$$

so that

$$\mu_1 \leq \int_0^1 xK_2(x, x)\, dx = \frac{1}{192}$$

The reason for the x in the above integral has to do with the fact that we are working on $L_2([0, 1], x)$. We now recall that μ_1 satisfies

$$J_1\left(\frac{1}{\sqrt{\mu_1}}\right) = 0$$

and the first positive zero of $J_1(\tau)$ is known to be $\tau = 3.8317 \cdots$. From the above we obtain

$$\tau \geq (192)^{\frac{1}{4}} = 3.72 \cdots.$$ (133)

We can also use (131) with $f(x) = x(1 - x)$

$$\|f\| = \left[\int_0^1 x^3(1 - x)^2 \, dx \right]^{\frac{1}{2}} = \frac{1}{\sqrt{60}}$$

$$Lf(x) = 3$$

so that $\|Lf\| = \frac{3}{2}\sqrt{2}$ and if $\lambda_1 = \tau^2$

$$\tau \leq [\tfrac{3}{2}\sqrt{2}\sqrt{60}]^{\frac{1}{2}} = (270)^{\frac{1}{2}} = 4.05 \cdots. \tag{134}$$

Clearly (133) and (134) yield lower and upper estimates for the correct value. These can be improved by using better choices of $f(x)$ in (131) and using instead of (122) higher values of n to approximate μ as in (127).

7. FREDHOLM EQUATIONS WITH SELFADJOINT COMPACT OPERATORS

We shall consider the Fredholm equation of the second kind

$$\phi(x) - \lambda \int_a^b K(x, y)\phi(y) \, dy = f(x) \tag{135}$$

where $K(x, y)$ is an L_2 kernel and satisfies the symmetry property $K(x, y) = \overline{K(y, x)}$. In this case the integral operator in (135) is compact and self-adjoint. We suppose that $f(x) \in L_2[a, b]$.

Let $\{\phi_n(x)\}$ denote the orthonormal eigenfunctions of K corresponding to nonzero eigenvalues. Using theorem 13 we can represent $\phi(x)$ and $f(x)$ by

$$\phi = \sum_{n=1}^{\infty} (\phi, \phi_n)\phi_n + \phi_0 \tag{136}$$

$$f = \sum_{n=1}^{\infty} (f, \phi_n)\phi_n + f_0. \tag{137}$$

where f_0 and ϕ_0 are suitable elements in the null space of K. Insertion of these in (135) leads to

$$\sum_{n=1}^{\infty} (\phi, \phi_n)\phi_n + \phi_0 - \lambda \sum_{n=1}^{\infty} \mu_n(\phi, \phi_n)\phi_n = \sum_{n=1}^{\infty} (f, \phi_n)\phi_n + f_0. \tag{138}$$

Comparing corresponding terms leads to

$$\phi_0 = f_0$$

$$(\phi, \phi_n) - \lambda\mu_n(\phi, \phi_n) = (f, \phi_n)$$

so that

$$\phi = \sum_{n=1}^{\infty} \frac{(f, \phi_n)}{1 - \lambda\mu_n} \phi_n + f_0, \qquad |1 - \lambda\mu_n| \neq 0. \tag{139}$$

Insertion of (139) in (138) shows that ϕ is a solution. If $1 - \lambda\mu_n \neq 0$ for all n, (139) represents the unique solution of (135). Since the set of eigenvalues $\{\mu_n\}$ is either finite or accumulates at 0 we see that $|1 - \lambda\mu_n| \geq k > 0$ for all n and

$$\sum_{n=1}^{\infty} \left| \frac{(f, \phi_n)}{1 - \lambda\mu_n} \right|^2 \leq \frac{1}{k^2} \sum_{n=1}^{\infty} |(f, \phi_n)|^2 < \infty.$$

so that ϕ is indeed in $L_2[a, b]$.

If λ is such that $1 - \lambda\mu_n = 0$ for some set of μ_n (there may be several since μ_n appears as often as is warranted by the multiplicity) a solution will in general not exist. If however, the corresponding numerators (f, ϕ_n) also vanish a solution will exist, but it will no longer be unique. In such a case the coefficients of the corresponding ϕ_n can be assigned arbitrarily.

THEOREM 20. The equation

$$\phi - \lambda K\phi = f \tag{140}$$

where K is compact and self-adjoint and has eigenvalues $\{\mu_n\}$ and ortho-normal eigenfunctions $\{\phi_n\}$ has the solution

$$\phi = f + \lambda \sum_{n=1}^{\infty} \frac{\mu_n}{1 - \lambda\mu_n} (f, \phi_n)\phi_n \tag{141}$$

if $1 - \lambda\mu_n \neq 0$ for all n.

If $1 - \lambda\mu_n = 0$ a solution exists if and only if f is orthogonal to all solutions of

$$\phi - \frac{1}{\mu_n} K\phi = 0.$$

In that case

$$\phi = f + \lambda \sum_{\substack{n=1 \\ \lambda\mu_n \neq 1}}^{\infty} \frac{\mu_n}{1 - \lambda\mu_n} (f, \phi_n)\phi_n + \sum_{\lambda\mu_n=1} c_n\phi_n \tag{142}$$

where the c_n are arbitrary.

PROOF. For $1 - \lambda\mu_n \neq 0$, (139) has already been shown to satisfy (140). This solution can be rearranged so that

$$\phi = f_0 + \sum_{n=1}^{\infty} \frac{1 - \lambda\mu_n}{1 - \lambda\mu_n} (f, \phi_n)\phi_n + \lambda \sum_{n=1}^{\infty} \frac{\mu_n}{1 - \lambda\mu_n} (f, \phi_n)\phi_n$$

$$= f_0 + \sum_{n=1}^{\infty} (f, \phi_n)\phi_n + \lambda \sum_{n=1}^{\infty} \frac{\mu_n}{1 - \lambda\mu_n} (f, \phi_n)\phi_n$$

$$= f + \lambda \sum_{n=1}^{\infty} \frac{\mu_n}{1 - \lambda\mu_n} (f, \phi_n)\phi_n$$

thus proving (141).

For the case where $1 - \lambda\mu_n = 0$ for some n one can verify directly that (142) satisfies (140) for arbitrary n. ■

An inspection of this theorem shows that it is the Fredholm alternative for an equation of the second kind with a compact self-adjoint operator. We shall shortly extend this result to non-self-adjoint operators.

EXAMPLE 1. Consider

$$\phi(x) - \lambda \int_0^1 x_<(1 - x_>)\phi(y)\, dy = f(x). \tag{143}$$

The eigenfunctions associated with the above operator are $\{\sqrt{2} \sin n\pi x\}$ and the eigenvalues are $\{1/\pi^2 n^2\}$. Then

$$\phi(x) = f(x) + \lambda \sum_{n=1}^{\infty} \frac{1}{n^2\pi^2 - \lambda}\left(\int_0^1 \sqrt{2} \sin n\pi y f(y)\, dy\right)\sqrt{2} \sin n\pi x.$$

We shall now turn our attention to equations of the first kind. Let K be, as before, a compact self-adjoint operator on a Hilbert space H, and consider the equation

$$K\phi = f. \tag{144}$$

In order for (144) to have a solution f must necessarily belong to the range of the operator K.

The simplest procedure for the investigation of (144) is to use the expansions (136) and (137) in (144). This leads to the equation

$$\sum_{n=1}^{\infty} \mu_n(\phi, \phi_n)\phi_n = \sum_{n=1}^{\infty} (f, \phi_n)\phi_n + f_0.$$

In view of the fact that $K\phi_0 = 0$, this term does not appear. A comparison of corresponding terms shows that

$$f_0 = 0$$

$$(\phi, \phi_n) = \frac{1}{\mu_n}(f, \phi_n).$$

Evidently a necessary condition on f for (144) to have a solution is that $f_0 = 0$. Using the above we obtain the formal solution

$$\phi = \sum_{n=1}^{\infty} \frac{(f, \phi_n)}{\mu_n} \phi_n + \phi_0 \tag{145}$$

where ϕ_0 is an arbitrary element in the null space of K. Furthermore, in order for ϕ to be in H we require that

$$\sum_{n=1}^{\infty} \frac{|(f, \phi_n)|^2}{\mu_n^2} < \infty;$$

otherwise ϕ is not in the Hilbert space and (145) would make no sense.

THEOREM 21. Consider the Fredholm equation of the first kind

$$K\phi = f$$

where K is a compact self-adjoint operator, with eigenvalues $\{\mu_n\}$ and orthonormal eigenfunctions $\{\phi_n\}$. Necessary and sufficient for a unique solution to exist are the following conditions

1. the null space of K is empty.

2. $\sum_{n=1}^{\infty} \frac{|(f, \phi_n)|^2}{\mu_n^2} < \infty.$

In that case

$$\phi = \sum_{n=1}^{\infty} \frac{(f, \phi_n)}{\mu_n} \phi_n.$$

If condition 2 breaks down no solution can exist.

If condition 1 breaks down f must be orthogonal to the null space of K for a solution to exist. But as shown in (145) one can add to the solution an arbitrary element of the null space of K.

PROOF. Clearly the range of K consists of those f for which

$$\sum_{n=1}^{\infty} \frac{1}{\mu_n^2} |(f, \phi_n)|^2 < \infty.$$

If the null space of K is empty and f is in the range of K there is a unique ϕ satisfying (142). If the null space of K is not empty ϕ cannot be defined uniquely, even if a solution exists. A solution will still exist if f is in the range of K. Note that the range of K and the null space of K are orthogonal. ■

8. THE FREDHOLM ALTERNATIVE

THEOREM 22. Consider the Fredholm equation

$$\phi(x) - \lambda \int_a^b K(x, y)\phi(y)\, dy = f(x) \qquad (146)$$

where $K(x, y)$ is an L_2 kernel, so that K is a compact operator on $L_2[a, b]$. Equation (146) has a unique solution if and only if

$$\phi(x) - \lambda \int_a^b K(x, y)\phi(y) \, dy = 0 \qquad (147)$$

has only the trivial solution $\phi(x) = 0$.

Consider the adjoint equation to (147)

$$\psi(x) - \bar{\lambda} \int_a^b \overline{K(y, x)}\psi(y) \, dy = 0. \qquad (148)$$

Equations (147) and (148) have the same number of linearly independent solutions. If (147) has at least one nontrivial solution, (146) will have a solution only if

$$(f, \psi) = \int_a^b f(x)\overline{\psi(x)} \, dx = 0$$

for all $\psi(x)$ satisfying (148). In this case such a solution will not be unique.

Before we embark on a formal proof of this theorem we shall prove a lemma that is of great interest in its own right and vital for the proof of the theorem.

LEMMA. Let $K(x, y)$ be an L_2 kernel so that K is a compact operator on $L_2[a, b]$. It is possible to decompose $K(x, y)$ so that

$$K(x, y) = D(x, y) + S(x, y)$$

where $D(x, y)$ and $S(x, y)$ are again L_2 kernels. Furthermore, the corresponding operator D is degenerate and S is an operator such that $\|S\| < 1$.

PROOF. If the interval $[a, b]$ is finite the set

$$\{\sqrt{2/(b - a)} \sin n\pi(x - a)/(b - a)\}$$

is a complete orthonormal set on $L_2[a, b]$. On $L_2[-\infty, \infty]$ the set $\{k_n e^{-x^2/2} H_n(x)\}$ discussed in Section 5 is a complete orthonormal set. The $H_n(x)$ are the Hermite polynomials.

So far we have not yet discussed an orthonormal complete set on $L_2[0, \infty]$. We can show that $\{\sqrt{2} \, k_{2n} \, e^{-x^2/2} H_{2n}(x)\}$ is such a set on $L_2[0, \infty]$. Suppose that $f(x) \in L_2[0, \infty]$. Then we shall construct a function

$$\tilde{f}(x) = f(x), \qquad x \geq 0$$
$$= f(-x), \qquad x \leq 0$$

$\tilde{f}(x)$ is therefore an even function in $L_2[-\infty, \infty]$. We have

$$\tilde{f}(x) = \sum_{n=0}^{\infty} \int_{-\infty}^{\infty} \tilde{f}(y)\phi_n(y) \, dy\phi_n(x)$$

where $\phi_n(x) = k_n e^{-x^2/2} H_n(x)$. But for odd n, $\phi_n(x)$ is an odd function of x so that

$$\int_{-\infty}^{\infty} \tilde{f}(y)\phi_n(y)\, dy = 0, \qquad n \text{ odd}.$$

Also for n even

$$\int_{-\infty}^{\infty} \tilde{f}(y)\phi_n(y)\, dy = 2\int_0^{\infty} f(y)\phi_n(y)\, dy, \qquad n \text{ even}$$

so that

$$f(x) = \sum_{n=0}^{\infty} \left(\int_0^{\infty} f(y)\sqrt{2}\,\phi_{2n}(y)\, dy \right)\sqrt{2}\,\phi_{2n}(x), \qquad x \geq 0.$$

Since $K(x, y)$ is an L_2 kernel we can expand it in terms of an orthonormal set on that space. Let $\{\phi_n(x)\}$ be such a set. For fixed x we have

$$K(x, y) = \sum_{n=1}^{\infty} \int_a^b K(x, u)\overline{\phi_n(u)}\, du\,\phi_n(y)$$

since $\int_a^b |K(x, y)|^2\, dy$ exists for almost all x. Now we expand the coefficients so that

$$\int_a^b K(x, u)\overline{\phi_n(u)}\, du = \sum_{m=1}^{\infty} \int_a^b\int_a^b K(v, u)\overline{\phi_n(u)}\, du\,\overline{\phi_m(v)}\, dv\,\phi_m(x).$$

Note that $\int_a^b K(x, u)\overline{\phi_n(u)}\, du \in L_2[a, b]$. We now denote the coefficient in the above expansion by k_{nm} so that

$$K(x, y) = \sum_{n,m=1}^{\infty} k_{nm}\phi_m(x)\phi_n(y). \tag{149}$$

Equation (149) converges in the mean so that

$$\lim_{N \to \infty} \int_a^b\int_a^b \left| K(x, y) - \sum_{n,m \leq N} k_{nm}\phi_m(x)\phi_n(y) \right|^2 dx\, dy = 0 \tag{150}$$

We can now define

$$D(x, y) = \sum_{n,m \leq N} k_{n,m}\phi_m(x)\phi_n(y)$$

$$S(x, y) = \sum_{n,m > N} k_{n,m}\phi_m(x)\phi_n(y).$$

The operator associated with the kernel $D(x, y)$ is clearly degenerate. Furthermore, by theorem 9 of Chapter 1,

$$\|S\| \leq \left\{ \int_a^b\int_a^b |S(x, y)|^2\, dx\, dy \right\}^{\frac{1}{2}} = \left\{ \sum_{n,m > N} |k_{nm}|^2 \right\}^{\frac{1}{2}}$$

by use of (150). It follows that

$$\|S\| < 1$$

if N is sufficiently large. Finally

$$K(x, y) = D(x, y) + S(x, y). \quad \blacksquare$$

PROOF OF THEOREM 22.   First we consider the two equations (147) and (148). We let

$$\lambda K = D + S$$

so that (147) and (148) can be written as

$$\phi - D\phi - S\phi = 0$$
$$\psi - D^*\psi - S^*\psi = 0.$$

Since S is a contraction operator $I - S$ has an inverse (see (8) of Chapter 2)

$$(I - S)^{-1} = I + S + S^2 + \cdots \equiv T$$

so that the first equation for ϕ can be rewritten in the form

$$\phi - TD\phi = 0 \tag{151}$$

where TD is still a degenerate operator since

$$TD\phi = \sum_{n,m \leq N} k_{n,m}(T\phi_m(x)) \int_a^b \phi_n(y)\phi(y)\, dy.$$

In the second equation we let

$$(I - S^*)\psi = \tilde{\psi}$$

so that

$$\psi = T^*\tilde{\psi}$$

and $\tilde{\psi}$ satisfies

$$\tilde{\psi} - D^*T^*\tilde{\psi} = 0. \tag{152}$$

The operator D^*T^* is adjoint to TD and also degenerate since

$$D^*T^*\phi = \sum_{n,m \leq N} \overline{k_{n,m}}\, \overline{\phi_n(x)} \int_a^b \overline{\phi_m(y)}(T^*\phi(y))\, dy.$$

Now we can use the Fredholm alternative for degenerate operators (see Section 5 of Chapter 2). Comparing (151) and (152) we can conclude that both have the same number of independent solutions. The same conclusion holds therefore for (147) and (148).

Finally, we turn our attention to (146), and write it in the form

$$(I - S)\phi - D\phi = f.$$

We now apply $T = (I - S)^{-1}$ to both sides and obtain

$$\phi - TD\phi = Tf. \tag{153}$$

Again we can apply the Fredholm alternative for degenerate operators to (152) and (153). If (152) has only the trivial solution (153) has a unique solution.

If (152) has nontrivial solutions (153) will have a solution if and only if

$$(Tf, \tilde{\psi}) = 0$$

for all ψ satisfying (152). But

$$(Tf, \tilde{\psi}) = (Tf, T^{*-1}\psi) = (f, T^*T^{*-1}\psi) = (f, \psi) = \int_a^b f(x)\overline{\psi(x)}\, dx.$$

It follows that (146) has a solution if and only if

$$\int_a^b f(x)\overline{\psi(x)}\, dx = 0$$

for all $\psi(x)$ satisfying (146). ■

What makes the Fredholm alternative such a useful tool is that uniqueness results for one equation immediately yield existence results for another equation. Once we know that (147) has no solution, but the obvious and trivial one $\phi(x) = 0$, we are guaranteed that (146) has a solution and what is more it has a unique solution.

EXAMPLE 1. We wish to solve the differential equation

$$u'' + p(x)u' - q(x)u = f(x) \tag{154}$$

with $f(x) \in L_2[0, 1]$, subject to the boundary conditions

$$u(0) = u(1) = 0. \tag{155}$$

$p(x)$ and $q(x)$ are to be real and continuous on $[0, 1]$, and $q(x)$ is also positive. The latter condition will be seen to be a vital condition in the subsequent analysis.

We now define the functions

$$P(x) = \exp \int_0^x p(y)\, dy$$

$$R(x) = \int_0^x \frac{dy}{P(y)}$$

and multiply (154) by $P(x)$. This leads to

$$(P(x)u')' - P(x)q(x)u = P(x)f(x).$$

From the latter we obtain readily

$$u = c_1 R(x) + \int_0^x [R(x) - R(y)]P(y)[f(y) + q(y)u(y)]\, dy.$$

One can easily verify that the above satisfies (154) as well as the boundary condition $u(0) = 0$. To satisfy the second boundary condition we set

$$c_1 = -\int_0^1 \frac{R(1) - R(y)}{R(1)} P(y)[f(y) + q(y)u(y)]\, dy$$

and finally obtain

$$u - \int_0^1 \left[\frac{R(x_<)}{R(1)} - 1 \right] R(x_>)P(y)q(y)u(y)\, dy$$

$$= \int_0^1 \left[\frac{R(x_<)}{R(1)} - 1 \right] R(x_>)P(y)f(y)\, dy \quad (156)$$

Use of the properties of the functions $R(x)$, $P(x)$, and $q(x)$ allows one to show that if (156) has a solution in $L_2[0, 1]$, then it will have two derivatives and it will then also satisfy (154) almost everywhere.

Problem (154), (155) will have a solution if and only if (156) has a solution. Equation (156) will have a unique solution if and only if the corresponding homogeneous equation has none but the trivial solution. To investigate the latter we return to the homogeneous equation

$$-(P(x)u')' + P(x)q(x)u = 0.$$

Multiply the above by u and integrate. This leads to, after performing an integration by parts

$$-P(x)uu' \Big|_0^1 + \int_0^1 P(x)[u'^2 + q(x)u^2]\, dx = 0.$$

The first term vanishes by virtue of the boundary conditions. The second can vanish if and only if $u \equiv 0$, since both $P(x)$ and $q(x)$ are positive. It follows that the homogeneous problem has only the solution $u \equiv 0$ so that (156) has a unique solution for all $f(x) \in L_2[0, 1]$.

9. WEIGHTED INTEGRAL OPERATORS

In this section we shall discuss a broader class of operators. It will be seen that the preceding cases reduce to special cases of this new class of operators.

Let $\alpha(x)$ be a monotonically increasing real function, defined on the interval $[a, b]$, such that $\alpha(b) - \alpha(a)$ is finite. If $f(x)$ is a continuous function on that interval we can consider the following type of integral, known as a Stieltjes integral

$$\int_a^b f(x)\, d\alpha(x).$$

Such an integral can be defined in complete analogy to Riemann integrals. Let the interval $[a, b]$ be subdivided as follows

$$a = x_0 < x_1 < x_2 < \cdots < x_n = b$$

and consider the sum

$$\sum_{k=1}^n f(\xi_k)[\alpha(x_k) - \alpha(x_{k-1})]$$

where ξ_k is some point in the interval $[x_k, x_{k-1}]$. One can show that if a suitable limit is taken this sum converges. Let

$$\Delta = \max_{k=1,2,\ldots,n} |x_k - x_{k-1}|.$$

Then we can define

$$\int_a^b f(x)\, d\alpha(x) = \lim_{\Delta \to 0} \sum_{k=1}^n f(\xi_k)[\alpha(x_k) - \alpha(x_{k-1})]. \tag{157}$$

For $\alpha(x) = x$ the above reduces to the standard Riemann integral of course.

Stieltjes integrals have many properties in common with ordinary Riemann integrals. For example, if $f(x)$ and $g(x)$ are two continuous functions and c, d two complex scalars then

$$\int_a^b (cf(x) + dg(x))\, d\alpha(x) = c \int_a^b f(x)\, d\alpha(x) + d \int_a^b g(x)\, d\alpha(x).$$

If $\alpha_1(x)$ and $\alpha_2(x)$ are two monotonically increasing functions

$$\int_a^b f(x)\, d[\alpha_1(x) + \alpha_2(x)] = \int_a^b f(x)\, d\alpha_1(x) + \int_a^b f(x)\, d\alpha_2(x).$$

$$\int_a^b f(x)\, d\alpha(x) = \int_a^c f(x)\, d\alpha(x) + \int_c^b f(x)\, d\alpha(x), \qquad a < c < b$$

$$\left| \int_a^b f(x)\, d\alpha(x) \right| \le \int_a^b |f(x)|\, d\alpha(x)$$

$$\left| \int_a^b f(x)\, d\alpha(x) \right| \le [\alpha(b) - \alpha(a)] \max_{a \le x \le b} |f(x)|.$$

Many other properties can be derived involving, for example, integration by parts and laws of the mean under suitable hypotheses. The above will be the most important for our immediate purposes.

Now we consider the space $C[a, b]$ of continuous functions and assign to it the following inner product

$$(f, g) = \int_a^b f(x)\overline{g(x)}\, d\alpha(x). \tag{158}$$

It is easy to show that the above is indeed an inner product, and that

$$\|f\| = \left[\int_a^b |f(x)|^2 \, d\alpha(x) \right]^{\frac{1}{2}} \tag{159}$$

is indeed a norm. The fact that $\alpha(x)$ is monotonically increasing is crucial for this purpose.

The space $C[a, b]$ with the inner product (158) is now an inner product space. It is, however, not complete. As in the previous cases it is possible to form a completion by considering all possible Cauchy sequences and their limits. Let $\{f_n(x)\}$ be such a sequence. Then

$$\|f_n(x) - f_m(x)\|^2 = \int_a^b |f_n(x) - f_m(x)|^2 \, d\alpha(x) < \epsilon, \quad n, m > N(\epsilon).$$

Let (α, β) be any subinterval of $[a, b]$, and consider the sequence $\{\int_\alpha^\beta f_n(x) \, d\alpha(x)\}$. Using the Cauchy-Schwarz inequality we have

$$\left| \int_\alpha^\beta [f_n(x) - f_m(x)] \, d\alpha(x) \right| \leq \int_\alpha^\beta |f_n(x) - f_m(x)| \, d\alpha(x)$$

$$\leq \int_a^b |f_n(x) - f_m(x)| \, d\alpha(x) = (|f_n - f_m|, 1)$$

$$\leq \|f_n - f_m\| \, \|1\|$$

$$= \|f_n - f_m\| \, [\alpha(b) - \alpha(a)]^{\frac{1}{2}}.$$

It follows that the sequence $\{\int_\alpha^\beta f_n(x) \, d\alpha(x)\}$ is a Cauchy sequence of complex numbers which has a limit. While the sequence $\{f_n(x)\}$ has no limit in a point-wise sense, this procedure does assign a value to $\int_\alpha^\beta f(x) \, d\alpha(x)$ for arbitrary intervals (α, β) in $[a, b]$. We shall denote the resulting Hilbert space by $L_2([a, b], \alpha)$.

EXAMPLE 1. Let

$$\alpha(x) = x, \qquad 0 \leq x < \tfrac{1}{2}$$
$$= a + x, \qquad \tfrac{1}{2} \leq x \leq 1.$$

In the above a is a non-negative constant. For a continuous function $f(x)$ we have

$$\int_0^1 f(x) \, d\alpha(x) = \int_0^{\frac{1}{2}-\epsilon} f(x) \, dx + a \int_{\frac{1}{2}-\epsilon}^{\frac{1}{2}+\epsilon} f(x) \, d\alpha(x) + \int_{\frac{1}{2}+\epsilon}^1 f(x) \, dx$$

Now

$$\int_{\frac{1}{2}-\epsilon}^{\frac{1}{2}+\epsilon} f(x) \, d\alpha(x) = f(\xi)[\alpha(\tfrac{1}{2} + \epsilon) - \alpha(\tfrac{1}{2} - \epsilon)] = f(\xi)[a + 2\epsilon]$$

for a suitable mean value $\frac{1}{2} - \epsilon \leq \xi \leq \frac{1}{2} + \epsilon$. In the limit as $\epsilon \to 0$ we find

$$\int_0^1 f(x)\, d\alpha(x) = \int_0^1 f(x)\, dx + af(\tfrac{1}{2}).$$

We shall now examine the corresponding space $L_2([0, 1], \alpha)$. Let $\{f_n(x)\}$ be a Cauchy sequence of continuous functions. Then

$$\|f_n - f_m\|^2 = \int_0^1 |f_n(x) - f_m(x)|^2\, dx + a^2\, |f_n(\tfrac{1}{2}) - f_m(\tfrac{1}{2})|^2.$$

It follows that we can view the limit function $f(x)$ as a pair $\{f(x), f(\tfrac{1}{2})\}$ where $f(x) \in L_2[0, 1]$ and $f(\tfrac{1}{2})$ is a complex number.

Now that the space $L_2([a, b], \alpha)$ has been defined for finite intervals one can extend the definition to infinite intervals as before. We can also define integral operators by

$$Kf = \int_a^b K(x, y) f(y)\, d\alpha(y), \tag{160}$$

where $f(x) \in L_2([a, b], \alpha)$. Suppose $K(x, y)$ is continuous and bounded and such that $K(x, y) = \overline{K(y, x)}$. Then, as before,

$$(Kf, g) = (f, Kg) \tag{161}$$

and K is in fact a self-adjoint operator. In fact the following theorem can be stated.

THEOREM 23.   Let K be an integral operator on $L_2([a, b], \alpha)$

$$Kf = \int_a^b K(x, y) f(y)\, d\alpha(y)$$

where

$$\int_a^b \int_a^b |K(x, y)|^2\, d\alpha(x)\, d\alpha(y) < \infty. \tag{162}$$

Then K is a compact operator.

PROOF.   If the interval is finite and $K(x, y)$ is continuous the proof of theorem 6 can be applied almost verbatim. Suppose $\{f_n(x)\}$ is a uniformly bounded sequence in $L_2([a, b], \alpha)$, so that $\|f_n\| \leq M$ for all n. Now define $\{g_n(x)\}$ by

$$g_n(x) = \int_a^b K(x, y) f_n(y)\, d\alpha(y)$$

so that

$$|g_n(x)| \leq \left\{ \int_a^b |K(x, y)|^2\, d\alpha(y) \int_a^b |f_n(y)|^2\, d\alpha(y) \right\}^{1/2}$$

and

$$\|g_n\| \leq M \left\{ \int_a^b \int_a^b |K(x, y)|^2\, d\alpha(x)\, d\alpha(y) \right\}^{1/2}.$$

It follows that $\{g_n(x)\}$ is also uniformly bounded. Furthermore, this sequence is also equicontinuous since

$$|g_n(x_1) - g_n(x_2)| \leq M \left\{ \int_a^b |K(x_1, y) - K(x_2, y)|^2 \, d\alpha(y) \right\}^{1/2}.$$

By Arzelà's theorem it must be possible to extract a subsequence $\{g_{n'}(x)\}$ that converges uniformly to a continuous function. Accordingly, K is a compact operator.

If (162) holds, but $K(x, y)$ is not continuous the proof of theorem 7 can be applied. Finally, if the interval is infinite the proof of theorem 8 can be used. ∎

In a similar fashion the expansion theorem 14 for a compact self-adjoint operator applies here as well, and so does the corollary to theorem 14.

EXAMPLE 2. Let $K(x, y)$ be a continuous function for all $0 \leq x, y \leq 1$. Define $\alpha(x)$ by

$$
\begin{aligned}
\alpha(x) &= x, & 0 &\leq x < x_1 \\
&= x + a_1, & x_1 &< x < x_2 \\
&= x + a_1 + a_2, & x_2 &< x < x_3 \\
&= x + a_1 + \cdots + a_k, & x_k &< x < x_{k+1} \\
&= x + a_1 + \cdots + a_n, & x_n &< x \leq 1, \quad \text{all } a_k > 0.
\end{aligned}
$$

We can now define a compact operator acting on $L_2([0, 1], \alpha)$ by

$$
\begin{aligned}
Kf &= \int_0^1 K(x, y)f(y) \, d\alpha(y) \\
&= \int_0^1 K(x, y)f(y) \, dy + \sum_{i=1}^n a_i K(x, y_i)f(y_i).
\end{aligned}
$$

The above is a compact operator. If we assume that $K(x, y) = \overline{K(y, x)}$, K will also be self-adjoint. To verify this we see that

$$
\begin{aligned}
(Kf, g) &= \int_0^1 (Kf)\overline{g(x)} \, d\alpha(x) \\
&= \int_0^1 \int_0^1 K(x, y)f(y)\overline{g(x)} \, dy \, dx \\
&\quad + \sum_{i=1}^n a_i f(y_i) \int_0^1 K(x, y_i)\overline{g(x)} \, dx + \sum_{i=1}^n a_i \int_0^1 K(x_i, y)f(y) \, dy\overline{g(x_i)} \\
&\quad + \sum_{i,j=1}^n a_i a_j K(x_i, x_j)f(x_i)\overline{g(x_j)} = (f, Kg).
\end{aligned}
$$

Accordingly K has a countable set of nonzero eigenvalues $\{\mu_n\}$ and ortho-normal eigenfunctions $\{\phi_n(x)\}$. Any functions $f(x) \in L_2([0, 1], \alpha)$ can now be represented as

$$f(x) = \sum_1^\infty \alpha_k \phi_k(x) + f_0(x).$$

where

$$\alpha_k = (f, \phi_k) = \int_0^1 f(x)\overline{\phi_k(x)}\, dx + \sum_{i=1}^n a_i f(x_i)\phi_k(x_i)$$

and $f_0(x)$ is a suitable element in the null space of K.

EXAMPLE 3.   We consider the differential operator, defined in $L_2([0, 1], \alpha)$ with $\alpha(x)$ as defined in Example 1, with $a = 1$,

$$Lu = -u'', \qquad 0 \le x < \tfrac{1}{2}, \ \tfrac{1}{2} < x \le 1, \tag{163}$$

$$Lu = -[u'], \qquad x = \tfrac{1}{2}, \tag{164}$$

and the boundary conditions are given as

$$u(0) = u(1) = 0. \tag{165}$$

By $[u']$ we mean

$$[u'] = \lim_{\epsilon \to 0} u'(x + \epsilon) - u'(x - \epsilon).$$

If u' is continuous $[u'] = 0$.

Now we try to construct an integral operator that is inverse to L. If condition (164) were not part of the definition of L the appropriate inverse would be

$$Kf = \int_0^1 x_<(1 - x_>)f(y)\, dy.$$

The above is inverse to (163) with the boundary conditions (165). To include (164) the above operator must be so modified that a linear function is added to it, whose derivative has a suitable jump discontinuity at $x = \tfrac{1}{2}$. Consider

$$Kf = \int_0^1 x_<(1 - x_>)f(y)\, dy + \tfrac{1}{2}xf(\tfrac{1}{2}), \qquad x < \tfrac{1}{2}$$

$$= \int_0^1 x_<(1 - x_>)f(y)\, dy + \tfrac{1}{2}(1 - x)f(\tfrac{1}{2}), \qquad x > \tfrac{1}{2}. \tag{166}$$

For $x \ne \tfrac{1}{2}$ we have

$$-(Kf)'' = f$$

$$-Kf\big|_{x=0} = -Kf\big|_{x=1} = 0.$$

Also

$$(Kf)' = -\int_0^x yf(y)\,dy + \int_x^1 (1-y)f(y)\,dy + \tfrac{1}{2}f(\tfrac{1}{2}), \qquad x < \tfrac{1}{2}$$

$$= -\int_0^x yf(y)\,dy + \int_x^1 (1-y)f(y)\,dy - \tfrac{1}{2}f(\tfrac{1}{2}), \qquad x > \tfrac{1}{2}$$

so that

$$-[(Kf)'] = f(\tfrac{1}{2}), \qquad x = \tfrac{1}{2}.$$

Clearly K as defined in (166) is inverse to L. Equation (166) can be written more compactly as follows. Let $\alpha(x)$ be defined as in Example 1. Then

$$Kf = \int_0^1 x_<(1 - x_>)f(y)\,d\alpha(x). \tag{167}$$

By theorem (23) K as given in (167) is a compact, self-adjoint operator and all expansion theorems apply.

To obtain the eigenvalues and eigenfunctions we seek solutions of

$$Lu = \lambda u.$$

For $x \neq \tfrac{1}{2}$ we have

$$u'' + \lambda u = 0$$

$$u(0) = u(1) = 0$$

so that

$$u = \sin\sqrt{\lambda}\,x, \qquad x < \tfrac{1}{2}$$

$$= \sin\sqrt{\lambda}(1 - x), \qquad x > \tfrac{1}{2}.$$

To satisfy (164) we require that

$$-[u'] = \lambda u, \qquad x = \tfrac{1}{2}$$

so that

$$2\sqrt{\lambda}\cos\sqrt{\lambda}/2 = \lambda\sin\sqrt{\lambda}/2. \tag{168}$$

Equation (168) is a transcendental equation with an infinity of solutions. The resultant orthonormalized eigenfunctions are given by

$$\phi_n(x) = \left(\frac{2}{1 + \sin^2 \tfrac{1}{2}\sqrt{\lambda_n}}\right)^{1/2} \sin\sqrt{\lambda_n}\,x, \qquad x \leq \tfrac{1}{2}$$

$$= \left(\frac{2}{1 + \sin^2 \tfrac{1}{2}\sqrt{\lambda_n}}\right)^{1/2} \sin\sqrt{\lambda_n}(1 - x), \qquad x \geq \tfrac{1}{2}. \tag{169}$$

where, of course

$$\int_0^1 |\phi_n(x)|^2\,d\alpha(x) = 1.$$

Let $f(x)$ be any function in $L_2([0, 1], \alpha)$. One can easily show that the null space of K is empty. Then

$$f(x) = \sum_{n=1}^{\infty} a_n \phi_n(x) \tag{170}$$

where

$$a_n = \int_0^1 f(x)\phi_n(x)\, d\alpha(x)$$

$$= \int_0^1 f(x)\phi_n(x)\, dx + f(\tfrac{1}{2})\phi_n(\tfrac{1}{2}).$$

(170) of course converges in the mean.

We now take a function $f(x)$ defined as follows

$$f(x) = 0, \qquad x \neq \tfrac{1}{2}$$
$$= 1, \qquad x = \tfrac{1}{2}.$$

In $L_2[0, 1]$ such a function would be a null function. In $L_2([0, 1], \alpha) f(x)$ is not a null function. In this case

$$a_n = \int_0^1 f(x)\phi_n(x)\, d\alpha(x) = \phi_n(\tfrac{1}{2}).$$

Then

$$\lim_{n \to \infty} \int_0^1 \left| f(x) - \sum_{k=1}^n \phi_k(\tfrac{1}{2})\phi_k(x) \right|^2 d\alpha(x) = \lim_{n \to \infty} \left| 1 - \sum_{k=1}^n \phi_k^2(\tfrac{1}{2}) \right|^2 = 0$$

so that

$$\sum_{k=1}^{\infty} \phi_k^2(\tfrac{1}{2}) = 1. \tag{171}$$

EXERCISES

1. Show that for a degenerate operator $\mu = 0$ is an eigenvalue of infinite multiplicity.
2. Show that the Volterra operator

$$Kf = \int_0^x K(x, y)f(y)\, dy \quad \text{acting on } L_2[0, 1] \text{ is not self-adjoint.}$$

3. Verify the statement in (14) that $K_n f = Kf^{\perp}$.
4. Consider the differential operator

$$Lu = -u'', \qquad u(0) = u'(1) = 0.$$

Construct the integral operator inverse to L. Compute all eigenvalues and eigenfunctions, and expand the kernel $K(x, y)$ in terms of those eigenfunctions.

5. Do the same for
$$Lu = -u''$$
$$u(0) + u(1) = 0$$
$$u'(0) + u'(1) = 0.$$

6. Let $Kf = \int_0^1 K(x, y)f(y)\, dy$ and $K(x, y) = \overline{K(y, x)}$ and
$$\int_0^1 \int_0^1 |K(x, y)|^2\, dx\, dy < \infty.$$

Suppose K has only a finite number of nonzero eigenvalues. Show that K is a degenerate operator.

7. Show that the solution of (47) satisfies the differential equation
$$(-(p(x)K_x(x, y)))_x + q(x)K(x, y) = 0.$$

8. Prove theorem 16 in detail.

9. Show that the Green's function $K(x, y)$ derived in Example 5, given by (57) satisfies the differential equation (56) subject to the boundary conditions (51).

10. Let $r(x)$ be a continuous, positive function defined on $[0, 1]$. Show that the space of all continuous functions on $[0, 1]$ under the standard operations of addition and multiplication by scalars, and with the inner product
$$(f, g)_r = \int_0^1 r(x)f(x)\overline{g(x)}\, dx$$
forms an inner product space. Show that its completion under the norm
$$\|f\| = \left[\int_0^1 r(x)\, |f(x)|^2\, dx \right]^{\frac{1}{2}}$$
forms a Hilbert space.

11. Show that
$$J_\nu(x) = \left(\frac{x}{z}\right)^\nu \sum_{k=0}^\infty \frac{[i(x/2)]^{2k}}{k!\,\Gamma(k + \nu + 1)}$$
satisfies the differential equation
$$u'' + \frac{1}{x} u' + \left(1 - \frac{\nu^2}{x^2}\right) u = 0.$$

12. Solve Bessel's equation (Example 7, Section 4) for general $\nu > 1$. Show that the kernel is an L_2 kernel and $\tilde{K}$ is compact.

13. Show that the solutions $u_e(x)$ and $u_0(x)$ of $u'' - x^2 u = 0$ become large for large $|x|$. (See Example 8.)

14. Consider Eq. (79) and show that

$$|w_{n+1} - w_n| \leq \left(\frac{3}{16t}\right)^n e^{-t}$$

by induction.

15. Show that the solutions of the Volterra integral equations

$$w(t) = e^{-t} + \frac{3}{16} \int_t^\infty \frac{\sinh(t - \tau)}{\tau^2} w(\tau) \, d\tau$$

and

$$w(t) = e^t + \frac{3}{32} \int_t^\infty \frac{e^{t-\tau}}{\tau^2} w(\tau) \, d\tau + \frac{3}{32} \int_1^t \frac{e^{t-\tau}}{\tau^2} w(\tau) \, d\tau$$

satisfy the differential equation

$$w'' - w = \frac{-3}{16t^2} w.$$

16. Prove the corollary on p. 122.

17. Consider the operator

$$Lu = \frac{1}{x}\left[-(xu')' + \frac{4}{x}u\right]$$

$$u(1) = 0.$$

Construct the integral operator inverse to L, and work in the Hilbert space $L_2([0, 1], x)$.

18. Find the eigenvalues and eigenfunctions of the differential operator defined by

$$Lu = -u'', \qquad 0 \leq x < 1$$
$$Lu = u'(1), \qquad x = 1$$
$$u(0) = 0.$$

Investigate the associated expansion theorems. *Hint:* Use an integral operator inverse to L of the form:

$$Kf = \int_0^1 K(x, y)f(y) \, d\alpha(y)$$

where

$$\alpha(y) = y, \qquad 0 \leq y < 1$$
$$\alpha(1) = 2.$$

19. Find as many estimates of the eigenvalues as you conveniently can for problem 17. Do the same for problems 4 and 5.

20. Find as many estimates of the eigenvalues as you conveniently can for Example 3, p. 104.

21. For the case that $1 - \lambda\mu_n = 0$ for some n, verify that (142) satisfies (140).

4

APPLICATIONS TO
PARTIAL DIFFERENTIAL
EQUATIONS

SUMMARY

An abstract technique is developed for studying boundary value problems associated with certain ordinary and partial differential equations. The results show that the inverses of these operators are compact and self-adjoint so that the general theorems proved in Chapter 3 can be used. Some applications to heat flow and wave propagation are then discussed.

1. INTRODUCTION

The primary focus of this chapter will be a discussion of a partial differential operator of the type

$$Lu = \frac{1}{r(x, y)}\left[-\frac{\partial}{\partial x} p_1(x, y) \frac{\partial u}{\partial x} - \frac{\partial}{\partial y} p_2(x, y) \frac{\partial u}{\partial y} + q(x, y)u \right] \tag{1}$$

where p_1 and p_2 are differentiable functions on some domain D in the x, y plane. r, p_1, p_2, and q will also be assumed to be positive in D. D is to be a bounded region with a smooth boundary denoted by ∂D. By a smooth boundary we mean one with a continuously turning tangent. Furthermore, the functions u that constitute the domain of L will satisfy the boundary condition.

$$u = 0 \quad \text{on} \quad \partial D. \tag{2}$$

In order to analyze the above operator we shall introduce the Hilbert space $L_2[D]$. That space can be viewed as the completion of the space of

123

continuous functions defined on D, with the inner product

$$(u, v) = \iint_D r(x, y)u(x, y)\overline{v(x, y)} \, dx \, dy. \tag{3}$$

The norm is then given by

$$\|u\| = \left\{ \iint_D r(x, y) |u(x, y)|^2 \, dx \, dy \right\}^{\frac{1}{2}} \tag{4}$$

The operator L is defined on those functions u in $L_2[D]$, that are sufficiently differentiable so that Lu exists and $\|Lu\| < \infty$.

What we would like to do is to show that L has an inverse integral operator K as was done in the case of ordinary differential operators in Section 4 of Chapter 3. An explicit construction of the type developed for those cases is no longer possible. Nevertheless we will develop some abstract techniques that will enable us to show that L has an inverse operator K that is also compact. We will then be able to obtain eigenfunction expansions for the partial differential operator (1).

2. LINEAR FUNCTIONALS

In our subsequent program the concept of a linear functional will play a fundamental role.

DEFINITION.   Let H be a Hilbert space. A bounded linear functional is a mapping from the space H into the field of complex numbers with the following properties. Let f, g be elements in H and α, β scalars in C. Then let $l(f)$ denote such a functional. We require that

1. $l(\alpha f + \beta g) = \alpha l(f) + \beta l(g)$
2. there exists a constant k such that

$$|l(f)| \leq k \|f\| \qquad \text{for all} \quad f \in H.$$

Condition 1 describes the linearity of l and 2 its boundedness.

EXAMPLE 1.   Let H be the space $L_2[0, 1]$, and define

$$l(f) = \int_0^1 f(x) \, dx.$$

Clearly l is linear and

$$|l(f)| \leq \int_0^1 |f(x)| \, dx \leq \left\{ \int_0^1 dx \int_0^1 |f(x)|^2 \, dx \right\}^{\frac{1}{2}} = \|f\|$$

so that l is bounded.

EXAMPLE 2. Let H be a Hilbert space and g a given element in H. Define

$$l(f) = (f, g).$$

l is linear and also bounded since

$$|l(f)| \le \|g\| \, \|f\|.$$

EXAMPLE 3. Let H be $L_2[0, 1]$ and x_0 a given point in $[0, 1]$. Let

$$l(f) = f(x_0).$$

One can show that l is linear, but unbounded.

Example 2 is a representative case in the following sense.

THEOREM 1: Riesz Representation Theorem. Let l be a bounded linear functional on H. There exists a unique $g \in H$ such that

$$l(f) = (f, g) \tag{5}$$

for all $f \in H$.

PROOF. This theorem shows that every bounded linear functional can be represented in terms of an inner product.

Let M denote the null space of l so that

$$l(f) = 0, \qquad f \in M.$$

If M is the whole space, l is the zero functional and $g = 0$ satisfies (5). Suppose now that M is not the whole space. Let $M^\perp$ denote the orthogonal complement of M so that

$$H = M \oplus M^\perp$$

We can show that $M^\perp$ is in fact one dimensional. Suppose f_1 and $f_2 \in M^\perp$ so that

$$l(f_1) = a, \qquad l(f_2) = b.$$

It follows that

$$l(bf_1 - af_2) = 0.$$

Therefore $bf_1 - af_2$ belongs to both M and $M^\perp$. But the only element they have in common is 0. Hence f_1 and f_2 are linearly dependent so that $M^\perp$ has at most one independent element and thus is one-dimensional.

Now let $h \in M^\perp$ and such that $\|h\| = 1$. A general element in H can be written as

$$f = (f, h)h + f_0$$

where $f_0 \in M$. Then

$$l(f) = (f, h)l(h),$$

since $l(f_0) = 0$. We now define $g = \overline{l(h)}\, h$ and the above is equivalent to

$$l(f) = (f, g) \qquad \text{for all} \quad f \in H.$$

To show that g is unique, suppose

$$l(f) = (f, g_1) = (f, g_2) \qquad \text{for all} \quad f \in H.$$

Then $(f, g_1 - g_2) = 0$ for all f, in which case $g_1 - g_2 = 0$ so that g is indeed unique. ∎

DEFINITION. Let $(f \mid g)$ be a mapping that assigns to every pair f, g in H a scalar in C. $(f \mid g)$ will be said to be a bounded sesquilinear functional if

1. $(\alpha f_1 + \beta f_2 \mid g) = \alpha(f_1 \mid g) + \beta(f_2 \mid g)$ for all f_1, f_2, g in H and α, β in C.
2. $(f \mid g) = \overline{(g \mid f)}$.
3. There exists a constant k such that for all f, g in H

$$|(f \mid g)| \leq k \, \|f\| \, \|g\|.$$

EXAMPLE 4. Let $K(x, y)$ be an L_2 kernel on $L_2[a, b]$. A typical sesquilinear functional is given by

$$(f \mid g) = \int_a^b \int_a^b K(x, y) f(y) \overline{g(x)} \, dy \, dx.$$

EXAMPLE 5. Let H be a Hilbert space. The inner product (f, g) is a bounded sesquilinear functional.

Theorem 1 shows that every bounded linear functional can be represented in terms of an inner product. We shall now develop a representation theorem for sesquilinear functionals.

THEOREM 2. Let $(f \mid g)$ denote a bounded sesquilinear functional on a Hilbert space H. There exists a unique bounded self-adjoint operator A, such that

$$(f \mid g) = (f, Ag) = (Af, g). \tag{6}$$

PROOF. For a fixed g $(f \mid g)$ is in fact a bounded linear functional, and theorem 1 can be applied to it. There exists an element $h(g)$ such that

$$(f \mid g) = (f, h(g)). \tag{7}$$

The symbolism $h(g)$ indicates that (7) holds for all g, but that $h(g)$ will vary with g.

Our next task is to explore the relationship between g and $h(g)$. From (7), and the definition of $f(g)$ we see that

$$(f \mid \alpha g_1 + \beta g_2) = (f, h(\alpha g_1 + \beta g_2))$$
$$(f \mid \alpha g_1 + \beta g_2) = \bar{\alpha}(f \mid g_1) + \bar{\beta}(f \mid g_2)$$
$$= \bar{\alpha}(f, h(g_1)) + \bar{\beta}(f, h(g_2))$$
$$= (f, \alpha h(g_1) + \beta h(g_2))$$

so that

$$h(\alpha g_1 + \beta g_2) = \alpha h(g_1) + \beta h(g_2).$$

The above is nothing but the definition of a linear operator so that

$$h(g) = Ag. \tag{8}$$

We have

$$(f \mid g) = (f, Ag),$$

and we also have

$$(f \mid g) = \overline{(g \mid f)} = \overline{(g, Af)} = (Af, g)$$

so that

$$(f, Ag) = (Af, g). \tag{9}$$

Finally we have to show that A is bounded. From the definition we know that

$$|(f \mid g)| \leq k \, \|f\| \, \|g\|$$

so that

$$|(f, Ag)| \leq k \, \|f\| \, \|g\|.$$

In particular for $f = Ag$ we see that

$$\|Ag\|^2 \leq k \, \|Ag\| \, \|g\|$$

and, provided $Ag \neq 0$,

$$\|Ag\| \leq k \, \|g\|. \tag{10}$$

For $Ag = 0$ (10) is trivial.

By (8), (9), and (10) we see that A is a linear bounded self-adjoint operator. Finally we still need to show that A is unique. Suppose

$$(f \mid g) = (f, A_1 g) = (f, A_2 g) \qquad \text{for all } f, g \in H$$

Then

$$(f, (A_1 - A_2)g) = 0 \qquad \text{for all } f, g \in H$$

So that

$$(A_1 - A_2)g = 0 \qquad \text{for all } g.$$

Then $A_1 = A_2$. ∎

It is precisely this theorem that will enable us to construct the inverses to differential operators via certain sesquilinear functionals.

3. ORDINARY DIFFERENTIAL OPERATORS

In this section we shall examine the operator

$$Lu = \frac{1}{r(x)} \left[-(p(x)u')' + q(x)u \right] \tag{11}$$

acting on suitable functions in $L_2([0, 1], r)$. We shall assume that $p'(x)$ is continuous and $r(x)$, $p(x)$, $q(x)$ are all positive on $[0, 1]$. This operator was studied in Chapter 3, but we shall now see how we can reach similar conclusions without explicitly constructing the kernel $K(x, y)$ and the integral operator K. In the next section the same technique will be applied to partial differential operators.

The Hilbert space on which we are working will be as before $L_2([0, 1], r)$ with the inner product

$$(f, g)_r = \int_0^1 r(x)f(x)\overline{g(x)}\, dx. \tag{12}$$

Let $C_0'[0, 1]$ denote that subspace of $L_2([0, 1], r)$ on which all elements have continuous first derivatives and satisfy the boundary conditions $u(0) = u(1) = 0$. Then, for functions having two continuous derivatives

$$
\begin{aligned}
(Lu, v)_r &= \int_0^1 [-(p(x)u')' + q(x)u]\bar{v}\, dx \\
&= -p(x)u'\bar{v}\Big|_0^1 + \int_0^1 [p(x)u'\bar{v}' + q(x)u\bar{v}]\, dx \\
&= \int_0^1 [p(x)u'\bar{v}' + q(x)u\bar{v}]\, dx.
\end{aligned}
$$

We shall now define a new inner product on $C_0'[0, 1]$ by

$$(u, v)_L = \int_0^1 [p(x)u'\bar{v}' + q(x)u\bar{v}]\, dx \tag{13}$$

and as a norm we take

$$\|u\|_L = \left\{ \int_0^1 [p(x)\,|u'|^2 + q(x)\,|u|^2]\, dx \right\}^{1/2}. \tag{14}$$

It is easy to verify that (13) and (14) have all the requisite properties to qualify as inner product and norm on $C_0'[0, 1]$. The resultant space is an inner product space. We can form its completion with respect to (13) and (14) and thus obtain a new Hilbert space to be denoted by H. In the subsequent development it will be shown that H is a subspace of $L_2([0, 1], r)$. Since $p(x)$,

$r(x)$, and $q(x)$ are bounded and positive on $[0, 1]$ we have for suitable constants

$$0 < p_0 \leq p(x) \leq p_1$$
$$0 < q_0 \leq q(x) \leq q_1$$
$$0 < r_0 \leq r(x) \leq r_1.$$

It follows that

$$\|u\|_L{}^2 \geq \int_0^1 q(x)\,|u|^2\,dx \geq \frac{q_0}{r_1} \int_0^1 r(x)\,|u|^2\,dx = \frac{q_0}{r_1}\,\|u\|_r{}^2. \tag{15}$$

From (15) we see that if $\{f_n\}$ is a Cauchy sequence in H it will also be a Cauchy sequence in $L_2([0, 1], r)$ and accordingly a subsequence will have a limit in both spaces. Accordingly H is a subspace of $L_2([0, 1], r)$.

We shall now consider the inner product $(u, v)_r$, but restrict u, v to lie in H. Then, in view of the fact that,

$$|(u, v)_r| \leq \|u\|_r \|v\|_r \leq \frac{r_1}{q_0} \|u\|_L \|v\|_L \tag{16}$$

where $(u, v)_r$ is a sesquilinear functional on H and by (16) it is bounded. We can now conclude, by theorem 2, that there exists a self-adjoint bounded operator K on H such that

$$(u, v)_r = (u, Kv)_L = (Ku, v)_L. \tag{17}$$

Insertion of (17) in (16) and letting $u = Kv$ shows that

$$\|Kv\|_L \leq \frac{r_1}{q_0} \|v\|_L$$

so that K is bounded and $\|K\| \leq (r_1/q_0)$. This operator will be seen to be the inverse of L and furthermore will be shown to be compact. Letting $u = v$ in (17) shows that K is also a positive operator.

Let $\{f_n\}$ be a bounded set in H, so that

$$\|f_n\|_L{}^2 = \int_0^1 [p(x)\,|f_n'(x)|^2 + q(x)\,|f_n(x)|^2]\,dx \leq M^2.$$

Since H is the completion of $C_0'[0, 1]$ $f_n(x)$ need not be differentiable. But $f_n(x)$ will be the limit of a sequence of differentiable functions $\{f_{n,k}(x)\}$ in $C_0'[0, 1]$, and $f_n'(x)$ will be the limit of $\{f_{n,k}'(x)\}$. From the above inequality it follows that

$$\int_0^1 |f_n'(x)|^2\,dx \leq N^2$$

for some N. We now consider the functions

$$f_n(x) = \int_0^x f_n'(y)\,dy$$

as elements in $L_2([0, 1], r)$. The above holds, of course, almost everywhere. Clearly

$$|f_n(x)| \leq \left\{ \int_0^x 1 \, dy \int_0^x |f_n'(y)|^2 \, dy \right\}^{1/2} \leq N \qquad (18)$$

so that

$$\|f\|_r \leq N \left\{ \int_0^1 r(x) \, dx \right\}^{1/2}.$$

We also have

$$|f_n(x_1) - f_n(x_2)| = \left| \int_{x_1}^{x_2} f_n'(y) \, dy \right| \leq \sqrt{|x_1 - x_2|} \, N. \qquad (19)$$

By use of (18) and (19) it can be shown that we can select a subsequence $\{f_{n'}(x)\}$ that is a Cauchy sequence in $L_2([0, 1], r)$. The argument will be a repetition of one used in the proof of theorem 6 in Chapter 3. This result is a consequence of Arzelà's theorem. Accordingly the sequence $\{f_{n'}(x)\}$ converges to an element $f(x)$ in $L_2([0, 1], r)$.

For convenience we shall now reindex the sequence and denote it by $\{f_n(x)\}$. We let $u = v = f_n - f_m$ in (17) so that

$$(K(f_n - f_m), f_n - f_m)_L = \|f_n - f_m\|_r^2 < \epsilon, \quad n, m > N(\epsilon). \qquad (20)$$

At this point we shall use a result regarding bounded, positive operators. Every such operator K has a square root; that is, there exists a bounded positive operator T such that $T^2 = K$. We shall defer the proof of this result until later so as not to interrupt the present argument. Then (20) becomes

$$(T^2(f_n - f_m), f_n - f_m)_L = \|T(f_n - f_m)\|_L^2 < \epsilon, \quad n, \quad m > N(\epsilon)$$

so that $\{Tf_n\}$ is a Cauchy sequence in H. It follows that T is compact and then $T^2 = K$ is also compact.

Using our general theory of self-adjoint compact operators we know that K has an infinite number of positive eigenvalues $\{\mu_n\}$ and orthonormal functions $\{\phi_n\}$. Since the null space of K is empty (see (17) with $u = v$) we have

$$f = \sum_{i=1}^{\infty} (f, \phi_i)_L \phi_i$$

$$Kf = \sum_{i=1}^{\infty} \mu_i (f, \phi_i)_L \phi_i$$

for all f in H. The set $\{\phi_n\}$ is an orthogonal set not only in H, but also in $L_2([0, 1], r)$. To verify this we use (17) and then

$$(\phi_n, \phi_m)_r = (\phi_n, K\phi_m)_L = \mu_m(\phi_n, \phi_m)_L = 0, \qquad n \neq m.$$

Using the same result we see that

$$\|\phi_n\|_r = \sqrt{\mu_n}\,\|\phi_n\|_L = \sqrt{\mu_n}.$$

If we let $\psi_n = (1/\sqrt{\mu_n})\phi_n$, we see that $\{\psi_n\}$ is an orthonormal set in $L_2([0, 1], r)$. In fact each ψ_n is an eigenfunction of K with respect to $L_2([0, 1], r)$ also. Up to now K was an operator defined on H. But since the set $\{\psi_n\}$ lies in $L_2([0, 1], r)$ as well as H, K can be defined on $L_2([0, 1], r)$ as well. From (15) we have

$$\|K\psi_n - \mu_n\psi_n\|_r \leq \left(\frac{r_1}{q_0}\right)^{\!1/2}\|K\psi_n - \mu_n\psi_n\|_L = 0$$

so that $K\psi_n = \mu_n\psi_n$ in $L_2([0, 1], r)$. It follows that $\{\phi_n\}$ is a set of orthonormal eigenfunctions of K with respect to H, and $\{\psi_n\}$ is a set of orthonormal eigenfunctions with respect to $L_2([0, 1], r)$. The former set is a complete set in H. We shall now show that $\{\psi_n\}$ is a complete set in $L_2([0, 1], r)$.

$L_2([0, 1], r)$ is the completion of the space of continuous functions with respect to its norm. Therefore, if $f(x)$ is a function in $L_2([0, 1], r)$ there exists a continuous function $f_\epsilon(x)$ such that

$$\|f - f_\epsilon\|_r < \epsilon.$$

Such a function must exist for every positive ϵ. The space H is the completion of $C_0'[0, 1]$, the space of functions having a continuous derivative and vanishing at 0 and 1, with respect to its norm. If $g(x)$ is a given continuous function there must exist a $g_\epsilon(x) \in C_0'[0, 1]$, such that

$$\|g - g_\epsilon\|_L < \epsilon.$$

A simple way of showing this is to construct a trigonometric polynomial approximating $g(x)$. We can certainly find such a polynomial so that

$$\int_0^1 r(x)\left|g(x) - \sum_{k=1}^n \alpha_k\sqrt{2}\,\sin k\pi x\right|^2 dx$$

$$\leq r_1 \int_0^1 \left|g(x) - \sum_{k=1}^n \alpha_k\sqrt{2}\,\sin k\pi x\right|^2 dx < \epsilon$$

and such a polynomial is certainly in $C_0'[0, 1]$.

Finally we see that for every $f(x)$ in $L_2([0, 1], r)$ we can find an $h_\epsilon(x)$ in H such that

$$\|f - h_\epsilon\|_r < \epsilon. \tag{20}$$

To construct such an $h_\epsilon(x)$ we first approximate $f(x)$ by a continuous function $f_\epsilon(x)$ in $L_2([0, 1], r)$. In turn such an $f_\epsilon(x)$ can be approximated by a trigonometric polynomial that is in H. This polynomial can be defined to be $h_\epsilon(x)$.

Now we can find a set of coefficients β_k such that

$$\left\|h_\epsilon(x) - \sum_{k=1}^{n} \beta_k \phi_k(x)\right\|_r \leq \left(\frac{r_1}{q_0}\right)^{1/2} \left\|h_\epsilon(x) - \sum_{k=1}^{n} \beta_k \phi_k(x)\right\|_L < \epsilon.$$

Then

$$\left\|f(x) - \sum_{k=1}^{n} \beta_k \sqrt{\mu_k}\, \psi_k(x)\right\|_r \leq \|f(x) - h_\epsilon(x)\|_r + \left\|h_\epsilon(x) - \sum_{k=1}^{n} \beta_k \phi_k(x)\right\|_r < 2\epsilon.$$

And it follows that the set $\{\psi_n(x)\}$ is a complete orthonormal set in $L_2([0, 1], r)$.

We still have to show that the compact operator is the inverse of the operator L. We recall that

$$(f, g)_r = (f, Kg)_L = (f, LKg)_r$$

for all $f \in H$ and all g for which LKg exists and is in $L_2([0, 1], r)$ so that

$$(f, (I - LK)g))_r = 0. \tag{21}$$

Equation (21) certainly holds for all $f \in H$, but it follows that it holds for all $f \in L_2([0, 1], r)$ since by (20) every element in $L_2([0, 1], r)$ can be approximated by elements in H. Then (21) can vanish only if

$$(I - LK)g = 0$$

so that

$$LKg = g. \tag{22}$$

We now have

$$LK\phi_n = \phi_n$$

Up to now it was not clear that $L\phi_n$ exists, but from the above we see that $LK\phi_n$ exists and lies in $L_2([0, 1], r)$. Since $K\phi_n = \mu_n\phi_n$ it follows that

$$L\phi_n = \frac{1}{\mu_n}\, \phi_n. \tag{23}$$

We now consider the expression

$$K(x, y) = \sum_{n=1}^{\infty} \mu_n \psi_n(x)\overline{\psi_n(y)}. \tag{24}$$

So far there is no way of telling whether this series converges in any sense whatsoever. In a formal sense we have

$$Kf = \int_0^1 r(y)K(x, y)f(y)\, dy = \sum_{n=1}^{\infty} \mu_n \left(\int_0^1 r(y)f(y)\overline{\psi_n(y)}\, dy\right)\psi_n(x). \tag{25}$$

(25) does converge to Kf in $L_2([0, 1], r)$.

It is possible to obtain an upper bound on μ_1. From (15) with $u = \psi_1$ we have

$$\frac{1}{\mu_1} = (L\psi_1, \psi_1)_r = \|\psi_1\|_L^2 \geq \frac{q_0}{r_1} \|\psi_1\|_r^2 = \frac{q_0}{r_1} \qquad (26)$$

so that $\mu_1 \leq (r_1/q_0)$.

To complete the preceding discussion we still have to prove that the operator K has a square root.

THEOREM 3. Let K be a bounded, positive (and therefore self-adjoint) operator on a Hilbert space H. There exists a bounded, positive (self-adjoint) operator T such that $T^2 = K$. T is said to be the square root of K.

PROOF. Consider the sequence of operators defined by

$$T_{n+1} = T_n + \frac{1}{2\alpha}(K - T_n^2), \qquad \alpha = \|K\|^{1/2} \qquad (27)$$

For $K = 0$ the result is trivial with $T = 0$. Without loss of generality we can assume therefore that $K \neq 0$ so that $\alpha > 0$. We shall show that T_n converges to an operator such that $T^2 = K$. (27) can be rewritten as

$$\alpha I - T_{n+1} = \frac{1}{2\alpha}(\alpha I - T_n)^2 + \frac{1}{2\alpha}(\alpha^2 I - K). \qquad (28)$$

From the above we see that $\alpha I - T_n$ is a positive operator for all n. Furthermore we see that

$$(\alpha I - T_n) - (\alpha I - T_{n+1}) = \frac{1}{2\alpha}[(\alpha I - T_{n-1})^2 - (\alpha I - T_n)^2]$$

and it follows that

$$T_{n+1} - T_n = \frac{1}{2\alpha}[2\alpha I - T_n - T_{n-1}](T_n - T_{n-1}).$$

An inspection of the above shows that the bracketed term on the right is a positive operator. $T_1 - T_0 = \frac{1}{2}K$ is also a positive operator and it follows that $T_{n+1} - T_n$ is positive for all n. Since $T_0 = 0$, T_1 is positive and so are all subsequent T_n. More generally, we see that $T_n - T_m$ is positive for $n \geq m$ and so are $T_n^2 - T_m T_n$ and $T_n T_m - T_m^2$.

We utilize these operators as follows. Since they are positive we obtain the following inequalities

$$\|T_n f\|^2 = (T_n^2 f, f) \geq (T_m T_n f, f) \geq (T_m^2 f, f) = \|T_m f\|^2. \qquad (29)$$

This tells us that $\|T_n f\|$ is a monotonically increasing sequence. But from (28)

$$\|T_n f\| \leq \alpha \|f\|$$

so that the sequence is bounded above and converges. From (29) we see that

$$\lim_{n \to \infty} \|T_n f\|^2 = \lim_{m,n \to \infty} (T_m T_n f, f).$$

Finally

$$\lim_{n \to m} \|T_n f - T_m f\|^2 = \lim_{n,m \to \infty} \{\|T_n f\|^2 - 2(T_n T_m f, f) + \|T_m f\|^2\} = 0$$

shows that $\{T_n f\}$ is a Cauchy sequence. That limit is defined to be Tf. From (27) we see that

$$\lim_{n \to \infty} T_n^2 f = Kf.$$

It follows that $\lim_{n \to \infty} T_n f = Tf$ exists for all f and $T^2 f = Kf$, so that T is a square root of K.

Since $\{T_n\}$ is a sequence of positive, bounded operators that converges to T, T is also positive and bounded. ■

In fact one can show that the positive square root of a positive operator is unique.

We shall now reexamine (24) to see whether more can be said regarding its convergence. In fact, we will show that $|n^2 \mu_n| < M$ for some M and all n. This will be accomplished by a comparison technique of the type discussed in theorem 18 of Chapter 3.

The following estimate is immediate.

$$\frac{(Lu, u)_r}{(u, u)_r} \geq \frac{\int_0^1 [p_0 |u'| + q_0 |u|^2] \, dx}{r_1 \int_0^1 |u|^2 \, dx}. \tag{30}$$

We now define the operator τ on $L_2[0, 1]$ by

$$\tau u = -\frac{p_0}{r_1} u'' + \frac{q_0}{r_1} u$$
$$u(0) = 0, \qquad u(1) = 0 \tag{31}$$

acting in $L_2[0, 1]$. We now see that

$$(\tau u, u) = \int_0^1 \left[\frac{p_0}{r_1} |u'|^2 + \frac{q_0}{r_1} |u|^2 \right] dx$$

by an integration by parts so that (30) takes the form

$$\frac{(Lu, u)}{(u, u)_r} \geq \frac{(\tau u, u)}{(u, u)} \tag{32}$$

The left side of (32) is defined on $L_2([0, 1], r)$ and the right side on $L_2[0, 1]$. But we have the obvious inequalities

$$r_0^{1/2} \|u\|^2 \leq \|u\|_r^2 \leq r_1^{1/2} \|u\|^2 \leq \left(\frac{r_1}{r_0}\right)^{1/2} \|u\|_r^2$$

which show that both spaces contain the same functions, but under different norms.

L is a positive operator and we suppose that its eigenvalues are arranged so that $0 < \lambda_1 \leq \lambda_2 \leq \lambda_3 \leq \cdots$, with corresponding eigenfunctions $\phi_1(x)$, $\phi_2(x)$, ... Now let M be that subspace of $L_2([0, 1], r)$ that is spanned by the first n eigenfunctions $\phi_1(x)$, $\phi_2(x)$, ..., $\phi_n(x)$. M is clearly finite dimensional. For any u in M we have

$$u = \sum_1^n \alpha_k \phi_k$$

$$Lu = \sum_1^n \lambda_k \alpha_k \phi_k$$

so that

$$(Lu, u)_r = \sum_1^n \lambda_k |\alpha_k|^2 \leq \lambda_n \sum_1^n |\alpha_k|^2 = \lambda_n(u, u)_r.$$

Using this result we now have

$$\lambda_n \geq \frac{(Lu, u)_r}{(u, u)_r} \geq \frac{(\tau u, u)}{(u, u)} \quad \text{on} \quad M. \tag{33}$$

Let $\{\tilde{\lambda}_n\}$ denote the eigenvalues of τ. By the lemma preceding theorem 18 of Chapter 3, (33) tells us that τ has at least n eigenvalues less than or equal to λ_n, so that

$$\tilde{\lambda}_n \leq \lambda_n. \tag{34}$$

Obviously (34) holds for all n.

The eigenvalues of τ can be obtained trivially by solving the differential equation

$$p_0 u'' + (\lambda r_1 - q_0)u = 0$$
$$u(0) = u(1) = 0$$

so that

$$\tilde{\lambda}_n = \frac{n^2 \pi^2 p_0 + q_0}{r_1}.$$

Then (34) reduces to

$$\frac{1}{\mu_n} = \lambda_n \geq \frac{n^2 \pi^2 p_0 + q_0}{r_1}$$

so that

$$\mu_n \leq \frac{r_1}{n^2 \pi^2 p_0 + q_0} \tag{35}$$

Returning to (24) we see that both

$$\int_0^1 \int_0^1 |K(x, y)|^2 \, dx \, dy = \sum_{n=1}^{\infty} \mu_n^{\,2} \tag{36}$$

$$\int_0^1 K(x, x) \, dx = \sum_{n=1}^{\infty} \mu_n. \tag{37}$$

exist and converge.

4. PARTIAL DIFFERENTIAL OPERATORS

Let D be a bounded domain in the x, y plane with a smooth boundary. Its boundary ∂D will be taken to be a simple rectifiable curve. The Hilbert space we will work with is $L_2(D, r)$, which consists of those functions $f(x, y)$ for which

$$\|f\|_r^{\,2} = \iint_D r(x, y) \, |f(x, y)|^2 \, dx \, dy < \infty. \tag{38}$$

$r(x, y)$ will be a continuous, positive function on $\bar{D}$, the closure of D. $L_2(D, r)$ can be viewed as the completion of the space of continuous functions defined on D under the above norm. As inner product we have

$$(f, g)_r = \iint_D r(x, y) f(x, y) \overline{g(x, y)} \, dx \, dy. \tag{39}$$

We can also construct the space $L_2(D)$ defined by the inner product and norm

$$(f, g) = \iint_D f(x, y) \overline{g(x, y)} \, dx \, dy$$

$$\|f\|^2 = \iint_D |f(x, y)|^2 \, dx \, dy$$

and if $0 < r_0 \leq r(x, y) \leq r_1$ we have as before

$$r_0^{\frac{1}{2}} \|f\|^2 \leq \|f\|_r^{\,2} \leq r_1^{\frac{1}{2}} \|f\|^2 \leq \left(\frac{r_1}{r_0}\right)^{\frac{1}{2}} \|f\|_r^{\,2}.$$

We now consider the following operator

$$Lu = \frac{1}{r(x, y)} \left[-\frac{\partial}{\partial x} p_1(x, y) \frac{\partial}{\partial x} u - \frac{\partial}{\partial y} p_2(x, y) \frac{\partial}{\partial y} u + q(x, y)u \right] \tag{40}$$

$$u = 0, \quad (x, y) \quad \text{on} \quad \partial D.$$

p_1, p_2, q, and r are continuous positive functions on $\bar{D}$ and p_1, p_2 have continuous first derivatives. Then

$$0 < p_0 \leq p_i(x, y) \leq p_1, \qquad i = 1, 2,$$
$$0 < q_0 \leq q(x, y) \leq q_1,$$
$$0 < r_0 \leq r(x, y) \leq r_1.$$

The subsequent analysis is completely analogous to the one in the previous section. By integration by parts we obtain

$$(Lu, v)_r = \iint\limits_{D} \left[-\frac{\partial}{\partial x} p_1(x, y) \frac{\partial}{\partial x} u - \frac{\partial}{\partial y} p_2(x, y) \frac{\partial}{\partial y} u + q(x, y)u \right] \bar{v} \, dx \, dy$$

$$= \left(-p_1(x, y) \frac{\partial}{\partial x} u \right) \bar{v} - \left(p_2(x, y) \frac{\partial}{\partial y} \right) \bar{v} \bigg|_{\partial D}$$

$$+ \iint\limits_{D} \left[p_1(x, y) \frac{\partial u}{\partial x} \frac{\partial \bar{v}}{\partial x} + p_2(x, y) \frac{\partial u}{\partial y} \frac{\partial \bar{v}}{\partial y} + q(x, y)u\bar{v} \right] dx \, dy$$

$$= \iint\limits_{D} \left[p_1(x, y) \frac{\partial u}{\partial x} \frac{\partial \bar{v}}{\partial x} + p_2(x, y) \frac{\partial u}{\partial y} \frac{\partial \bar{v}}{\partial y} + q(x, y)u\bar{v} \right] dx \, dy. \qquad (41)$$

We now consider the following subspace of $L_2(D, r)$. Let $C_0'(D, r)$ denote the space of continuous functions on $\bar{D}$ that vanish on ∂D and that have continuous first derivatives in $\bar{D}$. We shall define as an inner product on $C_0'(D, r)$

$$(u, v)_L = (Lu, v)_r, \qquad u, v \in C_0'(D, r) \qquad (42)$$

as given in (41). We now form the completion of $C_0'(D, r)$ with respect to $\|u\|_L$ and denote the resultant Hilbert space by H. As in (15) we have the inequality

$$\|u\|_L^2 \geq \iint\limits_{D} q(x, y) |u|^2 \, dx \, dy \geq \frac{q_0}{r_1} \|u\|_r^2$$

which shows that H is a subspace of $L_2([0, 1], r)$.

The inner product $(u, v)_r$ when restricted to elements in H is a sesquilinear functional. Also

$$|(u, v)_r| \leq \|u\|_r \|v\|_r \leq \frac{r_1}{q_0} \|u\|_L \|v\|_L$$

in analogy to (16). The functional is therefore bounded so that there exists a self-adjoint bounded operator K on H, such that

$$(u, v)_r = (u, Kv)_L = (Ku, v)_L. \qquad (43)$$

This operator, as will be shown, is the inverse of L and is also compact. It is clearly bounded since

$$|(u, Kv)_L| \leq \frac{r_1}{q_0} \|u\|_L \|v\|_L$$

and if $u = Kv$

$$\|Kv\|_L \leq \frac{r_1}{q_0} \|v\|_L.$$

From (43) we see that K is positive and we know that such an operator has a square root, say T, so that $T^2 = K$.

Let $\{f_n\}$ be a bounded set in H.

$$\|f_n\|_L{}^2 = \iint\limits_D \left[p_1(x, y) \left| \frac{\partial}{\partial x} f_n \right|^2 + p_2(x, y) \left| \frac{\partial}{\partial y} f_n \right|^2 + q(x, y) |f_n|^2 \right] dx\, dy \leq M^2$$

so that

$$\iint\limits_D \left| \frac{\partial}{\partial x} f_n \right|^2 dx\, dy \leq N^2, \qquad \iint\limits_D \left| \frac{\partial}{\partial y} f_n \right|^2 dx\, dy \leq N^2.$$

for a suitable N. We now examine the function

$$F_n(x, y) = \int_{x_0}^{x} \int_{y_0}^{y} \frac{\partial}{\partial \eta} f_n(\xi, \eta)\, d\eta\, d\xi$$

where (x, y) is any point in D and (x_0, y_0) is an arbitrary point on ∂D. The following estimates are immediate consequences of the bounds on f_n and the Cauchy–Schwarz inequality.

$$|F_n(x, y)| \leq \iint\limits_D \left| \frac{\partial}{\partial \eta} f_n \right|^2 d\eta\, d\xi \iint\limits_D d\eta\, d\xi \leq N^2 A \tag{44}$$

where A denotes the area of D.

$$|F_n(x_1, y_1) - F_n(x_2, y_2)| = \left| \int_{x_0}^{x_2} \int_{y_1}^{y_2} + \int_{x_1}^{x_2} \int_{y_0}^{y_1} \frac{\partial}{\partial \eta} f_n(\xi, \eta)\, d\eta\, d\xi \right|$$

It could happen that the rectangles over which we are integrating partially fall outside of D. In that case we define f_n to vanish identically outside. Let d denote the diameter of D; that is the largest distance between boundary points of D. Then

$$\left| \int_{x_0}^{x_2} \int_{y_1}^{y_2} \frac{\partial}{\partial \eta} f_n(\xi, \eta)\, d\eta\, d\xi \right|^2 \leq N^2 \int_{x_0}^{x_2} \int_{y_1}^{y_2} d\xi\, d\eta = N^2 d\, |y_2 - y_1|$$

so that

$$|F_n(x_1, y_1) - F_n(x_2, y_2)| \leq \{N^2 d[|y_2 - y_1| + |x_2 - x_1|]\}^{1/2}. \tag{45}$$

As in the preceding section, one can show that it is possible to select a subsequence from $\{F_n\}$ that converges uniformly to a continuous function in $\bar{D}$. We shall again, for convenience, denote that sequence by $\{F_n\}$. We shall now suppose that $y_0 = \min y$ over all $y \in \partial D$. Clearly such a y_0 exists. Then $f_n(x, y_0) = 0$ for all x, whether (x, y_0) is outside D or on ∂D. Now

$$\iint_D \left[\frac{\partial}{\partial x}(f_n(x, y) - f_m(x, y))\right][\overline{F_n(x, y)} - \overline{F_m(x, y)}] \, dx \, dy$$

$$= -\iint_D (f_n(x, y) - f_m(x, y)) \frac{\partial}{\partial x}(\overline{F_n(x, y)} - \overline{F_m(x, y)})) \, dx \, dy$$

$$= -\iint_D |f_n(x, y) - f_m(x, y)|^2 \, dx \, dy. \tag{46}$$

In the above, an integration by parts was performed, where the integrated terms vanish since all $f_n(x, y)$ vanish on ∂D.

Use of (46) now shows that

$$\|f_n - f_m\|_r^2 \le r_1 \|f_n - f_m\|^2 \le r_1 \left\|\frac{\partial}{\partial x}(f_n - f_m)\right\|^2 \|F_n - F_m\|^2.$$

But

$$\left\|\frac{\partial}{\partial x}(f_n - f_m)\right\|^2 \le \frac{1}{p_0} \|f_n - f_m\|_L^2 \le \frac{4}{p_0} M^2$$

and

$$\lim_{n, m \to \infty} \|F_n - F_m\| = 0.$$

It follows that $\{f_n\}$ is a Cauchy sequence in $L_2([0, 1], r)$. Combining this with (43) shows that

$$\|f_n - f_m\|_r^2 = (K(f_n - f_m), f_n - f_m)_L = \|T(f_n - f_m)\|_L^2$$

so that $\{Tf_n\}$ is a Cauchy sequence in H. It follows that T, and also $T^2 = K$, are compact operators on H.

Accordingly K has an infinite set of eigenvalues $\{\mu_n\}$ and orthonormal eigenfunctions $\{\phi_n\}$ in H. These eigenfunctions form a complete set. The functions $\{\psi_n\}$, where $\psi_n = (1/\sqrt{\mu_n})\phi_n$, are eigenfunctions of K in $L_2([0, 1], r)$ and are also orthonormal. They form a complete set in $L_2([0, 1], r)$. To see this we observe that if $f(x) \in L_2([0, 1], r)$, and ϵ is any positive number we can find an element $f_\epsilon(x) \in H$ such that

$$\|f - f_\epsilon\|_r < \epsilon.$$

Then there exists a set of scalars $\{\beta_k\}$ such that

$$\left\| f_\epsilon(x) - \sum_{k=1}^{n} \beta_k \phi_k(x) \right\|_r \leq \left(\frac{r_1}{q_0} \right)^{1/2} \| f_\epsilon(x) - \sum \beta_k \phi_k(x) \|_L < \epsilon$$

and

$$\left\| f(x) - \sum_{k=1}^{n} \beta_k \sqrt{\mu_k} \, \psi_k(x) \right\|_r < \epsilon$$

so that $\{\psi_n\}$ is complete in $L_2([0, 1], r)$.

As in (21) we find that

$$(f, (I - LK)g)_r = 0 \qquad \text{for all } f \in L_2([0, 1], r)$$

so that

$$LKg = g.$$

Thus K is the inverse of f. We obtain the formal expansion

$$K(x, y; \xi, \eta) = \sum_{n=1}^{\infty} \mu_n \psi_n(x, y) \overline{\psi_n(\xi, \eta)} \tag{47}$$

where $K\psi_n = \mu_n \psi_n$ and $L\psi_n = (1/\mu_n)\psi_n$. Using (47) we find

$$Kf = \iint_D r(\xi, \eta) K(x, y; \xi, \eta) f(\xi, \eta) \, d\xi \, d\eta$$

$$= \sum_{n=1}^{\infty} \mu_n (f, \psi_n)_r \psi_n(x, y)$$

which does converge to Kf in $L_2([0, 1], r)$. We find that

$$\frac{1}{\mu_1} = (L\psi_1, \psi_1) = \| \psi_1 \|_L^2 \geq \frac{q_0}{r_1} \| \psi_1 \|_r^2 = \frac{q_0}{r_1}$$

so that

$$\mu_1 \leq \frac{r_1}{q_0}. \tag{48}$$

It can be shown that $n\mu_n$ is bounded, which is similar to (35) the corresponding result for ordinary differential operators. But the analysis is considerably more difficult. In that case we find as in (36) that

$$\iint_D \iint_D |K(x, y; \xi, \eta)|^2 \, dx \, dy \, d\xi \, d\eta = \sum_{n=1}^{\infty} \mu_n^2. \tag{49}$$

In both the ordinary and partial differential operators (11) and (40) it was assumed that q was positive. This assumption is in fact not vital. If q were not strictly positive we could find a constant τ such that $q + \tau$ is positive. Then the analysis would apply to the operator $L + \tau$. Then L and $L + \tau$ would have the same eigenfunctions. The eigenvalues would be uniformly shifted by τ.

5. PARTIAL DIFFERENTIAL EQUATIONS

In various problems of mathematical physics one encounters the following type of differential equation

$$\frac{\partial u}{\partial t} = \frac{1}{r(x, y)}\left[\frac{\partial}{\partial x}\, p_1(x, y)\frac{\partial u}{\partial x} + \frac{\partial}{\partial y}\, p_2(x, y)\frac{\partial u}{\partial y} - q(x, y)u\right], \qquad (50)$$

where p_1 and p_2 have continuous first partial derivatives in a bounded domain D with boundary ∂D. Therefore p_1, p_2, r, and q are continuous in the closure $\bar{D}$ and p_1, p_2, and r are also positive. The solutions of (50) should satisfy the boundary condition

$$u = 0, \qquad x, y \text{ on } \partial D \qquad (51)$$

and the initial condition

$$u = f(x, y) \qquad \text{at} \quad t = 0. \qquad (52)$$

Using the notation of (40) we can rewrite the problem in the form

$$\frac{\partial u}{\partial t} = -Lu \qquad (53)$$

$$u = f(x, y) \qquad \text{at} \quad t = 0,\, u = 0 \text{ on } \partial D.$$

By use of the eigenfunctions of L we can expand $f(x, y)$ (we are of course assuming that $f(x, y) \in L_2(D, r)$) so that

$$f(x, y) = \sum_{n=1}^{\infty} f_n \psi_n(x, y). \qquad (54)$$

We shall seek a solution of (53) in the form

$$u = \sum_{n=1}^{\infty} a_n(t)\psi_n(x, y). \qquad (55)$$

Then we insert the above in (53).

$$\sum_{n=1}^{\infty} a'_n(t)\psi_n(x, y) = -\sum_{n=1}^{\infty} \lambda_n a_n(t)\psi_n(x, y)$$

where $\lambda_n = (1/\mu_n)$. The functions $a_n(t)$ must satisfy

$$a'_n(t) = -\lambda_n a_n(t)$$

so that

$$a_n(t) = c_n e^{-\lambda_n t}$$

If we select $c_n = f_n$ we obtain

$$u = \sum_{n=1}^{\infty} f_n e^{-\lambda_n t} \psi_n(x, y). \tag{56}$$

The above clearly satisfies the differential equation, the boundary conditions, and the initial condition.

Equations of the type (50) arise in heat flow problems, and more generally in diffusion processes. In wave propagation problems a typical problem is

$$\frac{\partial^2 u}{\partial t^2} = -Lu \tag{57}$$

$$u = f(x, y), \qquad \frac{\partial u}{\partial t} = g(x, y) \qquad \text{at} \quad t = 0.$$

L is as in (50), and the boundary condition $u = 0$ on ∂D is implicit in L. We again seek a solution in the form (55), and we suppose that

$$f(x, y) = \sum_{n=1}^{\infty} f_n \psi_n(x, y)$$

$$g(x, y) = \sum_{n=1}^{\infty} g_n \psi_n(x, y).$$

The function $a_n(t)$ must satisfy

$$a_n''(t) = -\lambda_n a_n(t)$$

so that

$$a_n(t) = c_n \frac{\sin \sqrt{\lambda_n} t}{\sqrt{\lambda_n}} + d_n \cos \sqrt{\lambda_n} t.$$

If we set $c_n = g_n$, $d_n = f_n$ we obtain as a solution

$$u = \sum_{n=1}^{\infty} \left(f_n \cos \sqrt{\lambda_n} t + g_n \frac{\sin \sqrt{\lambda_n} t}{\sqrt{\lambda_n}} \right) \psi_n(x, y). \tag{58}$$

A simple check shows that (58) satisfies the equation and the boundary and initial conditions.

EXERCISES

1. A linear functional is said to be continuous if for every $\epsilon > 0$ there exists a $\delta(\epsilon)$ such that $|l(f_1 - f_2)| < \epsilon$ for $\|f_1 - f_2\| < \delta(\epsilon)$. Show that a necessary and sufficient condition for l to be bounded is that it is continuous.

2. Consider the Hilbert space $L_2[0, 1]$, and the functional $l(f) = f(x_0)$ where $x_0 \in [0, 1]$. Show that l is linear but unbounded.

3. Let $K(x, y)$ be an L_2 kernel on $L_2[a, b]$. Show that

$$(f \mid g) = \int_a^b \int_a^b K(x, y)f(y)\overline{g(x)} \, dy \, dx$$

is a sesquilinear functional.

4. Let H be a Hilbert space. Show that the inner product (f, g) is a bounded sesquilinear functional.

5. Show that $(u, v)_L$ and $\|u\|_L$ as defined in (13) and (14) have all the properties of an inner product and norm respectively.

6. Let $C[a, b]$ be the space of continuous functions on $[a, b]$. Let $f, g \in C[a, b]$ with norm

$$\|f\| = \sup_{t \in [a,b]} |f(t)|.$$

Let f be fixed. Define $l_f(g)$ for any g as follows

$$l_f(g) = \int_a^b f(t)g(t) \, dt.$$

Show that $l_f(g)$ is a bounded linear functional.

7. Apply the techniques of Section 3 to

$$Lu = \frac{1}{x}\left[-(xu')' + \frac{1}{x}u \right]$$

$$u'(1) = 0.$$

8. Apply the techniques of Section 3 to

$$Lu = \frac{1}{x}\left[-\left(\frac{1}{x}u'\right)' + xu \right]$$

$$u(1) = 0.$$

9. Apply the techniques of Section 3 to

$$Lu = \frac{1}{x}\left[-(xu')' + \frac{v^2}{x}u \right]$$

$$u(1) = 0, \quad \text{for} \quad v = 3.$$

10. Apply the techniques of Section 4 to

$$Lu = \frac{1}{xy}\left[-\frac{\partial}{\partial x}(xy)\frac{\partial}{\partial x}u - \frac{\partial}{\partial y}(xy)\frac{\partial}{\partial y}u + \frac{1}{xy}u \right].$$

11. Referring to (47) and (48). Show that $n\mu_n$ is bounded. Prove (49).

5

FOURIER TRANSFORMS

SUMMARY

The Fourier transform is shown to be a unitary, isometric integral operator on $L_2[-\infty, \infty]$ with a particular kernel. Properties of Laplace, Hankel, and Mellin transforms can then be derived by relating them to Fourier transforms. Some of their applications to problems in mathematical physics are discussed.

The Hilbert transform too can be discussed by means of Fourier transforms. By using this tool, the projection method can be developed and applied to Wiener–Hopf equations. A complete theory for these equations in $L_2[-\infty, \infty]$ is developed for all cases where the index may be positive, negative, or zero. It is then shown that a number of dual integral equations can also be solved by use of the projection method.

1. FOURIER TRANSFORMS

This chapter will be devoted to a special integral operator and some of its applications. We define the Fourier transform of a function $f(x)$ by

$$F(f) = \frac{1}{\sqrt{2\pi}} \int_{-\infty}^{\infty} e^{isx} f(x)\, dx. \tag{1}$$

The kernel of the above operator is $(1/\sqrt{2\pi})e^{isx}$ and the operator acts on the space $L_2[-\infty, \infty]$. It is not clear, as yet, that the above integral exists for all functions in that space. But we will now show that (1) defines a bounded, and more specifically a unitary operator on $L_2[-\infty, \infty]$.

DEFINITION. An operator U acting on a Hilbert space H is said to be unitary if it is

(a) isometric; that is

$$\|Uf\| = \|f\| \quad \text{for all } f \text{ in } H,$$

(b) U has an inverse on all of H given by U^*, the adjoint of U.

144

In Example 8, Section 4, Chapter 3, we constructed a complete set of functions $\{\phi_n(x)\}$ on $L_2[-\infty, \infty]$. These were of the form $k_n e^{-(x^2/2)} H_n(x)$, where the $H_n(x)$ are the Hermite polynomials; they are of precise degree n.

THEOREM 1. The Hermite polynomials can be represented in the form

$$H_n(x) = c_n e^{x^2} \left(\frac{d}{dx}\right)^n e^{-x^2} \tag{2}$$

where c_n is a constant that depends on the particular normalization being used. Equation (2) is generally known as Rodriguez' formula.

PROOF. The functions $u_n = e^{-(x^2/2)} H_n(x)$ satisfy

$$u_n'' + (\lambda_n - x^2)u_n = 0$$
$$\lambda_n = 2n + 1, \qquad n = 0, 1, 2, \ldots.$$

Let $u_n = e^{x^2/2} V_n$, so that $V_n = e^{-x^2} H_n(x)$ and

$$V_n'' + 2xV_n' + (2n + 2)V_n = 0. \tag{3}$$

Consider

$$V_0'' + 2xV_0' + 2V_0 = 0.$$

A solution of the above is $V_0 = e^{-x^2}$. If we differentiate the equation for V_0 we obtain

$$V_0''' + 2xV_0'' + 4V_0' = 0.$$

Successive differentiations finally yield

$$V_0^{(n+2)} + 2xV_0^{(n+1)} + (2n + 2)V_0^{(n)} = 0.$$

We now identify $V_0^{(n)}$ with V_n and compare the above to (3). Evidently

$$V_n = \left(\frac{d}{dx}\right)^n V_0 = \left(\frac{d}{dx}\right)^n e^{-x^2}$$

so that

$$H_n(x) = e^{x^2} V_n = e^{x^2} \left(\frac{d}{dx}\right)^n e^{-x^2}.$$

If necessary the above has to be multiplied by a suitable constant, depending on the normalization desired. ■

THEOREM 2. The set

$$\left\{ \frac{1}{\sqrt{2^n n! \sqrt{\pi}}} e^{x^2/2} \left(\frac{d}{dx}\right)^n e^{-x^2} \right\}$$

is a complete orthonormal set on $L_2[-\infty, \infty]$.

PROOF. That the set $\phi_n(x) = k_n e^{(-x^2/2)} H_n(x)$ is a complete orthogonal set on $L_2[-\infty, \infty]$ was shown in Chapter 3. Use of theorem 1 shows that

$$\phi_n(x) = c_n k_n e^{x^2/2} \left(\frac{d}{dx}\right)^n e^{-x^2}$$

so that we will establish the theorem if we evaluate $c_n k_n$.

$$\|\phi_n(x)\|^2 = c_n^2 k_n^2 \int_{-\infty}^{\infty} e^{x^2} \left[\left(\frac{d}{dx}\right)^n e^{-x^2}\right]^2 dx.$$

If we integrate the above integral by parts n times we obtain

$$\|\phi_n(x)\|^2 = c_n^2 k_n^2 (-1)^n \int_{-\infty}^{\infty} e^{-x^2} \left(\frac{d}{dx}\right)^n e^{x^2} \left(\frac{d}{dx}\right)^n e^{-x^2} dx.$$

In theorem 1 we saw that $e^{x^2}(d/dx)^n e^{-x^2}$ is a polynomial of degree n. A straightforward differentiation shows that

$$e^{x^2} \left(\frac{d}{dx}\right)^n e^{-x^2} = (-2x)^n + \cdots.$$

It follows therefore that

$$\left(\frac{d}{dx}\right)^n e^{x^2} \left(\frac{d}{dx}\right)^n e^{-x^2} = (-2)^n n!$$

so that

$$1 = \|\phi_n(x)\|^2 = c_n^2 k_n^2 (2^n n!) \int_{-\infty}^{\infty} e^{-x^2} dx = c_n^2 k_n^2 (2^n n!) \sqrt{\pi}.$$

This yields the value of $c_n k_n$ and thereby establishes the theorem. ■

By means of Rodriguez' formula it is a relatively simple matter to determine the Fourier transform of $\phi_n(x)$.

LEMMA. Let

$$\phi_n(x) = \frac{1}{\sqrt{2^n n! \sqrt{\pi}}} e^{x^2/2} \left(\frac{d}{dx}\right)^n e^{-x^2}.$$

Then

$$F(\phi_n) = \frac{1}{\sqrt{2\pi}} \int_{-\infty}^{\infty} e^{isx} \phi_n(x) \, dx = i^n \phi_n(s). \tag{4}$$

PROOF. We first introduce a new variable of integration $y = x + is$ so that

$$F(\phi_n) = \frac{1}{\sqrt{2\pi}} \frac{1}{\sqrt{2^n n! \sqrt{\pi}}} e^{s^2/2} \int_{-\infty}^{\infty} e^{y^2/2} \left(\frac{d}{dy}\right)^n e^{-(y-is)^2} \, dy.$$

We also have the identity

$$\frac{d}{dy} e^{-(y-is)^2} = i \frac{d}{ds} e^{-(y-is)^2}$$

so that

$$F(\phi_n) = \frac{1}{\sqrt{2\pi}} \frac{i^n}{\sqrt{2^n n! \sqrt{\pi}}} e^{s^2/2} \left(\frac{d}{ds}\right)^n \int_{-\infty}^{\infty} e^{y^2/2} e^{-(y-is)^2} dy.$$

The above integral reduces to

$$e^{-s^2} \int_{-\infty}^{\infty} e^{-[(y/\sqrt{2})-i\sqrt{2}\,s]^2} dy = \sqrt{2\pi}\, e^{-s^2}$$

so that finally

$$F(\phi_n) = i^n \frac{1}{\sqrt{2^n n! \sqrt{\pi}}} e^{s^2/2} \left(\frac{d}{ds}\right)^n e^{-s^2} = i^n \phi_n(s). \quad \blacksquare$$

By means of the above lemma we can show that the operator F defined on all $\phi_n(x)$ is in fact a continuous operator defined on all of $L_2[-\infty, \infty]$. To see this we consider a function $f(x) \in L_2[-\infty, \infty]$ and represent it by

$$f(x) = \sum_{k=0}^{\infty} f_k \phi_k(x).$$

The sequence $\{p_n(x)\}$ defined by

$$p_n(x) = \sum_{k=0}^{n} f_k \phi_k(x)$$

is now a Cauchy sequence in $L_2[-\infty, \infty]$ whose limit is $f(x)$ and evidently defines a second Cauchy sequence in $L_2[-\infty, \infty]$, since

$$\|F(p_n) - F(p_m)\|^2 = \sum_{k=m+1}^{n} |f_k|^2 = \|p_n(x) - p_m(x)\|^2.$$

Now

$$\lim_{n \to \infty} F(p_n) = \sum_{k=0}^{\infty} i^k f_k \phi_k(s) = F(f). \tag{5}$$

From (5) we can conclude that

$$\int_{-\infty}^{\infty} |F(f)|^2 ds = \sum_{n=0}^{\infty} |f_n|^2 = \int_{-\infty}^{\infty} |f(x)|^2 dx \tag{6}$$

and (6) shows that F is an isometric operator.

More generally we see that if

$$f(x) = \sum_{n=0}^{\infty} f_n \phi_n(x), \qquad g(x) = \sum_{n=0}^{\infty} g_n \phi_n(x)$$

then

$$F(f) = \sum_{n=0}^{\infty} f_n i^n \phi_n(s)$$

$$F(g) = \sum_{n=0}^{\infty} g_n i^n \phi_n(s)$$

and

$$\int_{-\infty}^{\infty} f\bar{g} \, dx = \int_{-\infty}^{\infty} F(f)\overline{F(g)} \, ds. \tag{7}$$

We shall now examine the adjoint operator of F.

$$F^*(f) = \frac{1}{\sqrt{2\pi}} \int_{-\infty}^{\infty} e^{-isx} f(x) \, dx. \tag{8}$$

As in (4) we find that

$$F^*(\phi_n(x)) = (-i)^n \phi_n(s) \tag{9}$$

so that

$$F(F^*(\phi_n(x))) = F^*(F(\phi_n(x))) = \phi_n(x), \tag{10}$$

and it follows that the inverse of F is F^*.

We can also investigate the eigenvalues of F by the use of (4). Suppose

$$F(f) = \sum_{n=0}^{\infty} f_n i^n \phi_n(s) = \lambda \sum_{n=0}^{\infty} f_n \phi_n(s)$$

for some $f(x)$ and some λ. Then

$$(\lambda - i^n) f_n = 0$$

for all n. Evidently λ can have only four eigenvalues, $1, i, -1, -i$. If $\lambda = 1$ we find f_n is arbitrary if $n = 4k$, and $f_n = 0$ otherwise.

$$F(f) = f \qquad \text{if} \quad f = \sum_{k=0}^{\infty} f_{4k} \phi_{4k}(x).$$

Similarly,

$$F(f) = if \qquad \text{if} \quad f = \sum_{k=0}^{\infty} f_{4k+1} \phi_{4k+1}(x)$$

$$F(f) = -f \qquad \text{if} \quad f = \sum_{k=0}^{\infty} f_{4k+2} \phi_{4k+2}(x)$$

$$F(f) = -if \qquad \text{if} \quad f = \sum_{k=0}^{\infty} f_{4k+3} \phi_{4k+3}(x).$$

Each of these eigenvalues is of infinite multiplicity. This result makes it clear that F is a bounded, but definitely not compact operator.

We shall formally collect these results in the form of a theorem.

THEOREM 3. The integral operator

$$F(f) = \frac{1}{\sqrt{2\pi}} \int_{-\infty}^{\infty} e^{isx} f(x)\, dx$$

maps the space $L_2[-\infty, \infty]$ into itself. The operator is unitary so that

$$\int_{-\infty}^{\infty} |F(f)|^2\, ds = \int_{-\infty}^{\infty} |f|^2\, dx$$

(the above is known as Parseval's formula) and

$$F^*(f) = \frac{1}{\sqrt{2\pi}} \int_{-\infty}^{\infty} e^{-isx} f(x)\, dx$$

the adjoint of F, is also its inverse. More generally we have

$$\int_{-\infty}^{\infty} F(f)\overline{F(g)}\, ds = \int_{-\infty}^{\infty} f\bar{g}\, dx$$

if f, g are any elements in $L_2[-\infty, \infty]$.

The operator F has precisely four eigenvalues $1, i, -1, -i$, each of infinite multiplicity.

Another result of considerable importance is the following theorem.

THEOREM 4: Convolution Theorem. Let $f(x)$ and $g(x)$ be two functions in $L_2[-\infty, \infty]$. We define their convolution to be

$$(f * g)(x) = \int_{-\infty}^{\infty} f(y)g(x - y)\, dy. \tag{11}$$

The above is another function of x and its Fourier transform is given by

$$F(f * g) = \sqrt{2\pi}\, F(f)F(g). \tag{12}$$

PROOF. $f * g$ in general will not be a function in $L_2[-\infty, \infty]$. By the Cauchy–Schwarz inequality

$$|f * g| = \left| \int_{-\infty}^{\infty} f(y)g(x - y)\, dy \right|$$

$$\leq \left\{ \int_{-\infty}^{\infty} |f(y)|^2\, dy \int_{-\infty}^{\infty} |g(x - y)|^2\, dy \right\}^{1/2} = \|f\|\, \|g\|$$

so that $f * g$ is bounded. We shall proceed formally for the moment and find

$$F(f * g) = \frac{1}{\sqrt{2\pi}} \int_{-\infty}^{\infty} e^{isx} dx \int_{-\infty}^{\infty} f(y)g(x - y)\, dy$$

$$= \frac{1}{\sqrt{2\pi}} \int_{-\infty}^{\infty} f(y)\, dy \int_{-\infty}^{\infty} e^{isx} g(x - y)\, dx$$

$$= \int_{-\infty}^{\infty} f(y)F(g(x - y))\, dy.$$

But

$$F(g(x - y)) = \frac{1}{\sqrt{2\pi}} \int_{-\infty}^{\infty} e^{isx} g(x - y)\, dx = \frac{e^{isy}}{\sqrt{2\pi}} \int_{-\infty}^{\infty} e^{isu} g(u)\, du = e^{isy} F(g)$$

where the substitution $x = y + u$ was used. Now

$$F(f * g) = F(g) \int_{-\infty}^{\infty} e^{isy} f(y)\, dy = \sqrt{2\pi}\, F(f)F(g).$$

This establishes (12) formally.

The function $F(f)F(g)$ is a product of two functions in $L_2[-\infty, \infty]$ and is therefore integrable. It follows that

$$F^*(\sqrt{2\pi}\, F(f)F(g)) = \int_{-\infty}^{\infty} e^{-isx} F(f)F(g)\, ds$$

exists, since

$$\left| \int_{-\infty}^{\infty} e^{-isx} F(f)F(g)\, ds \right| \leq \|F(f)\|\, \|F(g)\|.$$

Now

$$F^*(\sqrt{2\pi}\, F(f)F(g)) = \frac{1}{\sqrt{2\pi}} \int_{-\infty}^{\infty} f(y)\, dy \int_{-\infty}^{\infty} e^{-is(x-y)} F(g)\, ds$$

$$= \int_{-\infty}^{\infty} f(y)g(x - y)\, dy \tag{13}$$

and we see that $F^*F(f * g) = f * g$.

One can show that the Fourier transform of an integrable function is integrable so that $f * g$ is not only bounded but also integrable. ∎

There are two notable transforms associated with the Fourier transforms, namely the Fourier cosine and the Fourier sine transforms.

We now consider $L_2[0, \infty]$ and take $f(x)$ to be a function in it. We can extend $f(x)$ as an even function to $L_2[-\infty, \infty]$ by letting

$$f(-x) = f(x) \qquad \text{all} \quad x.$$

Then

$$F(f) = \frac{1}{\sqrt{2\pi}} \int_{-\infty}^{\infty} e^{isx} f(x)\, dx = \frac{1}{\sqrt{2\pi}} \int_{-\infty}^{\infty} \cos sx f(x)\, dx$$

$$= \sqrt{\frac{2}{\pi}} \int_{0}^{\infty} \cos sx f(x)\, dx.$$

by the symmetry of the integrand. Note that $F(f)$ as a function of s is again an even function. By use of the inverse transform we have

$$f(x) = F^*(F(f)) = \frac{1}{\sqrt{2\pi}} \int_{-\infty}^{\infty} e^{-isx} F(f)\, dx = \sqrt{\frac{2}{\pi}} \int_{0}^{\infty} \cos sx F(f)\, ds \quad (14)$$

and we also find that

$$\int_{-\infty}^{\infty} |f|^2\, dx = 2 \int_{0}^{\infty} |f|^2\, dx = \int_{-\infty}^{\infty} |F(f)|^2\, dx = 2 \int_{0}^{\infty} |F(f)|^2\, ds.$$

We thus find that the operator

$$F_c(f) = \sqrt{\frac{2}{\pi}} \int_{0}^{\infty} \cos sx f(x)\, dx \qquad (15)$$

is a unitary operator on $L_2[0, \infty]$. Inspection of (14) and (15) shows that the operator F_c^* defined by

$$F_c^*(F_c(f)) = \sqrt{\frac{2}{\pi}} \int_{0}^{\infty} \cos sx F_c(f)\, ds = f$$

in inverse to F_c and equal to F_c. In other words F_c is its own inverse. We also have

$$\int_{0}^{\infty} F_c(f)\overline{F_c(g)}\, ds = \int_{0}^{\infty} f\bar{g}\, dx \qquad (16)$$

for all f and g in $L_2[0, \infty]$. One can easily check that

$$F_c(\phi_{4n}(x)) = \phi_{4n}(s) \qquad (17)$$

$$F_c(\phi_{4n+2}(x)) = -\phi_{4n+2}(s) \qquad (18)$$

so that F_c has the eigenvalues ± 1, where F_c is known as the Fourier cosine transform.

In a similar fashion $f(x) \in L_2[0, \infty]$ can be extended to $L_2[-\infty, \infty]$ as an odd function. This leads to the Fourier sine transform.

$$F_s(f) = \sqrt{\frac{2}{\pi}} \int_0^\infty \sin sx f(x) \, dx \tag{19}$$

$$F_s^*(F_s(f)) = \sqrt{\frac{2}{\pi}} \int_0^\infty \sin sx F_s(f) \, ds \tag{20}$$

$$\int_0^\infty F_s(f)\overline{F_s(g)} \, ds = \int_0^\infty f\bar{g} \, dx \tag{21}$$

F, has the eigenvalues $+1$ and -1 corresponding to the eigenfunctions $\phi_{4n+1}(x)$ and $\phi_{4n+3}(x)$ respectively.

2. APPLICATIONS OF FOURIER TRANSFORMS

In numerous applications integral equations of the type

$$\int_{-\infty}^\infty K(x - y)\phi(y) \, dy = f(x) \tag{22}$$

are encountered. The integral on the left is a convolution. If the functions $K(x), f(x) \in L_2[-\infty, \infty]$ one can take the Fourier transform of both sides. Then we obtain

$$\sqrt{2\pi} \, F(K)F(\phi) = F(f)$$

and

$$F(\phi) = \frac{F(f)}{\sqrt{2\pi} \, F(K)}, \tag{23}$$

provided $F(K)$ does not vanish.

If the right side, as a function of s, is in $L_2[-\infty, \infty]$ we finally obtain

$$\phi = \frac{1}{\sqrt{2\pi}} F^*\left(\frac{F(f)}{F(K)}\right). \tag{24}$$

EXAMPLE 1. Consider

$$\phi(x) - \lambda \int_{-\infty}^\infty e^{-|x-y|}\phi(y) \, dy = f(x), \qquad f(x) \in L_2[-\infty, \infty]. \tag{25}$$

By a direct integration

$$F(e^{-|x|}) = \frac{\sqrt{2/\pi}}{1 + s^2}$$

so that

$$F(\phi) - \lambda\sqrt{2\pi}\,\frac{\sqrt{2/\pi}}{1 + s^2}\,F(\phi) = F(f)$$

and

$$F(\phi) = \frac{1 + s^2}{1 + s^2 - 2\lambda}\,F(f). \tag{26}$$

If $\lambda \geq \frac{1}{2}$ the numerator in (26) will vanish for some real value of s. For all other values of λ, whether real or complex, $F(\phi) \in L_2[-\infty, \infty]$ if $F(f)$ does also. Then

$$\phi = F^*\left(\frac{1 + s^2}{1 + s^2 - 2\lambda}\,F(f)\right). \tag{27}$$

As a special case, suppose

$$f(x) = e^{-|x|}.$$

Then

$$\phi = F^*\left(\frac{\sqrt{2/\pi}}{1 - 2\lambda + s^2}\right) = \frac{1}{\pi}\int_{-\infty}^{\infty}\frac{e^{-isx}}{1 - 2\lambda + s^2}\,ds$$

and for $\lambda < \frac{1}{2}$

$$\phi = \frac{e^{-\sqrt{1-2\lambda}\,|x|}}{\sqrt{1 - 2\lambda}}.$$

It is not evident that the integral operator in (25) is a bounded operator. But by means of the Fourier transform, one can show not only that it is bounded, but in fact that its norm is bounded by 2. Let

$$K\phi = \int_{-\infty}^{\infty} e^{-|x-y|}\phi(y)\,dy$$

so that

$$F(K\phi) = \frac{2}{1 + s^2}\,F(\phi).$$

Since F is unitary

$$\|F(K\phi)\| = \|K\phi\| \leq 2\,\|\phi\|$$

so that

$$\|K\| \leq 2.$$

EXAMPLE 2.   Consider the differential equation

$$\frac{\partial u}{\partial t} = \frac{\partial^2 u}{\partial x^2} \tag{28}$$

$$u(x, 0) = g(x)$$

where $g(x) \in L_2[-\infty, \infty]$. The techniques of the previous chapter cannot be applied here. There we dealt with finite intervals in the x domain. Here we have an infinite interval. We shall suppose that u, u_x, and u_{xx} are all in $L_2[-\infty, \infty]$. Then repeated integrations by parts show that

$$F(u_{xx}) = \frac{1}{\sqrt{2\pi}} \int_{-\infty}^{\infty} u_{xx} e^{isx} \, dx = -s^2 F(u).$$

Applying the Fourier transform to (28) yields

$$\frac{\partial F(u)}{\partial t} = -s^2 F(u)$$

$$F(u)\big|_{t=0} = F(g)$$

By solving this ordinary differential equation we find

$$F(u) = e^{-s^2 t} F(g).$$

Use of the inversion formula (13) shows that

$$u = F^*(e^{-s^2 t} F(g)) = \frac{1}{\sqrt{2\pi}} \int_{-\infty}^{\infty} F^*(e^{-s^2 t}) g(x-y) \, dy.$$

From (4) it follows that

$$F(e^{-(x^2/2)}) = e^{-(s^2/2)}$$

so that by a change of variable

$$F\left(\frac{1}{\sqrt{2t}} e^{-(x^2/4t)}\right) = e^{-s^2 t}.$$

Finally,

$$u = \int_{-\infty}^{\infty} \frac{e^{-(y^2/4t)}}{2\sqrt{\pi t}} g(x-y) \, dy \tag{29}$$

satisfies (28). One can now verify that u, u_x, and u_{xx} are all in $L_2[-\infty, \infty]$ as was assumed initially.

EXAMPLE 3. Solve the wave equation

$$\frac{\partial^2 u}{\partial t^2} = \frac{\partial^2 u}{\partial x^2}$$

$$u(x, 0) = f(x), \qquad \frac{\partial u}{\partial t}(x, 0) = 0, \tag{30}$$

where $f(x) \in L_2[-\infty, \infty]$. Proceeding as in Example 2, Eq. (30) is reduced to

$$\frac{\partial^2 F(u)}{\partial t^2} = -s^2 F(u)$$

$$F(u)\big|_{t=0} = F(f), \qquad \frac{\partial F(u)}{\partial t}\bigg|_{t=0} \doteq 0.$$

It follows that

$$F(u) = F(f)\cos st$$

and

$$u = \frac{1}{\sqrt{2\pi}} \int_{-\infty}^{\infty} e^{-isx} \cos st F(f)\, ds$$

$$= \frac{1}{2\sqrt{2\pi}} \int_{-\infty}^{\infty} [e^{-is(x-t)} + e^{-is(x+t)}] F(f)\, ds.$$

The above is a sum of two inverse Fourier transforms where x was replaced by $x - t$ and $x + t$ respectively. Then

$$u = \tfrac{1}{2}[f(x+t) + f(x-t)] \tag{31}$$

EXAMPLE 4.   Solve the wave equation

$$\frac{\partial^2 u}{\partial t^2} = \frac{\partial^2 u}{\partial x^2}$$

$$u(x, 0) = 0, \qquad \frac{\partial u}{\partial t}(x, 0) = g(x). \tag{32}$$

As in the previous example, (32) can be reduced to

$$\frac{\partial^2 F(u)}{\partial t^2} = -s^2 F(u)$$

$$F(u)\big|_{t=0} = 0, \qquad \frac{\partial F(u)}{\partial t}\bigg|_{t=0} = F(g)$$

so that

$$F(u) = \frac{1}{s} F(g) \sin st.$$

First we note that by applying an integration by parts

$$F\left(\int_0^x g(y)\, dy\right) = \frac{i}{s} F(g).$$

It is assumed that g is such that the contributions from the integrated terms vanish. Then

$$u = \frac{1}{\sqrt{2\pi}} \int_{-\infty}^{\infty} e^{-isx} \frac{1}{s} F(g) \sin st \, ds$$

$$= \frac{-1}{2\sqrt{2\pi}} \int_{-\infty}^{\infty} [e^{-is(x-t)} - e^{-is(x+t)}] F\left(\int_0^x g(y) \, dy\right) ds$$

$$= -\frac{1}{2}\left[\int_0^{x-t} g(y) \, dy - \int_0^{x+t} g(y) \, dy\right] = \frac{1}{2}\int_{x-t}^{x+t} g(y) \, dy \qquad (33)$$

satisfies (32).

EXAMPLE 5. Solve

$$\int_{-\infty}^{\infty} K(x + y)\phi(y) \, dy = f(x), \qquad f(x) \in L_2[-\infty, \infty]. \qquad (34)$$

Although (34) does not involve a convolution type integral we can apply Fourier transforms. Then, in view of the fact that

$$\frac{1}{\sqrt{2\pi}} \int_{-\infty}^{\infty} K(x + y)e^{isx} \, dx = e^{-isy}F(K),$$

(34) becomes

$$\sqrt{2\pi}\, F(K)F^*(\phi) = F(f).$$

Finally, provided $F(K) \neq 0$,

$$\phi = F\left(\frac{F(f)}{\sqrt{2\pi}\, F(K)}\right) \qquad (35)$$

assuming that our formal calculations may be carried out.

EXAMPLE 6. Solve

$$\frac{1}{\pi} \int_0^{\infty} \frac{\phi(y)}{x + y} \, dy = f(x), \qquad f(x) \in L_2[0, \infty]. \qquad (36)$$

At first glance the method of Fourier transforms does not seem to apply to (36). However, we can change the variables via the following substitutions,

$$x = e^{2\xi}, \qquad y = e^{2\eta}, \qquad \phi(e^{2\eta})e^{\eta} = \psi(\eta), \qquad f(e^{2\xi})e^{\xi} = g(\xi).$$

Then (36) becomes

$$\frac{1}{\pi} \int_{-\infty}^{\infty} \frac{\psi(\eta)}{\cosh(\xi - \eta)} \, d\eta = g(\xi), \qquad (37)$$

which is of the form (22).

Applying Fourier transforms to (37) we obtain

$$F(\psi) = \sqrt{\frac{\pi}{2}} \frac{F(g)}{F(1/\cosh \xi)}.$$

Now

$$F\left(\frac{1}{\cosh \xi}\right) = \frac{1}{\sqrt{2\pi}} \int_{-\infty}^{\infty} \frac{e^{is\xi}}{\cosh \xi} \, d\xi.$$

In order to evaluate the above integral we shall use a residue integration. We see that

$$\lim_{R \to \infty} \int_{-R}^{R} \frac{e^{is\xi}}{\cosh \xi} \, d\xi + \int_{0}^{\pi} \frac{e^{isz}}{\cosh z} \, iRe^{i\theta} \, d\theta \bigg|_{z=Re^{i\theta}} = \int_{-\infty}^{\infty} \frac{e^{is\xi}}{\cosh \xi} \, d\xi$$

since the contribution of the integration over the semicircle $|z| = R$ vanishes in the limit if $s > 0$. The integrand has poles at all zeros of $\cosh \xi$, namely $\xi = (n + \frac{1}{2})\pi i$, $n = 0, 1, 2, \ldots$. The residues at these points are given by

$$\lim_{\xi \to (n+\frac{1}{2})\pi i} \frac{[\xi - (n + \frac{1}{2})\pi i] e^{is\xi}}{\cosh \xi} = -i(-1)^n e^{-s(2n+1)(\pi/2)} e^{-s(2n+1)(\pi/2)}$$

so that

$$F\left(\frac{1}{\cosh \xi}\right) = \sqrt{2\pi} \sum_{n=0}^{\infty} (-1)^n e^{-s(2n+1)(\pi/2)} = \frac{\sqrt{\pi/2}}{\cosh (\pi/2)s}$$

For $s < 0$ one can close the contour over the lower half plane and the same answer as above is obtained. Finally

$$\phi(e^{2\xi})e^{\xi} = \psi(\xi) = \frac{1}{\sqrt{2\pi}} \int_{-\infty}^{\infty} e^{-is\xi} \cosh \frac{\pi}{2} sF(g) \, ds \tag{38}$$

provided the integral on the right exists.

Similarly we could analyze the equation

$$\phi(x) - \frac{\lambda}{\pi} \int_{0}^{\infty} \frac{\phi(y)}{x + y} \, dy = f(x).$$

One then obtains

$$\phi(e^{2\xi})e^{\xi} = \frac{1}{\sqrt{2\pi}} \int_{-\infty}^{\infty} \frac{e^{-is\xi} \cosh (\pi/2)sF(g)}{\cosh (\pi/2)s - \lambda} \, ds.$$

For $\lambda < 1$ the above integral exists and it follows that the integral operator in (36) is bounded. In fact

$$\left\| \frac{1}{\pi} \int_{0}^{\infty} \frac{\phi(y)}{x + y} \, dy \right\| \leq \|\phi\|. \tag{39}$$

EXAMPLE 7. We shall consider the equation

$$\frac{1}{\pi} \int_{-\infty}^{\infty *} \frac{\phi(y)}{x - y} \, dy = f(x), \qquad f(x) \in L_2[-\infty, \infty]. \tag{40}$$

The operator on the left is known as the Hilbert transform. The * above the integral indicates that the integral is a so-called principal value integral. That is,

$$\int_{-\infty}^{\infty *} \frac{\phi(y)}{x - y} \, dy = \lim_{\epsilon \to 0} \int_{-\infty}^{x-\epsilon} + \int_{x+\epsilon}^{\infty} \frac{\phi(y)}{x - y} \, dy$$

We then find that

$$\frac{1}{\sqrt{2\pi}} \int_{-\infty}^{\infty} e^{isx} \int_{-\infty}^{\infty *} \frac{\phi(y)}{x - y} \, dy \, dx = F(\phi) \int_{-\infty}^{\infty *} \frac{e^{isy}}{y} \, dy. \tag{41}$$

But

$$\int_{-\infty}^{\infty *} \frac{e^{isy}}{y} \, dy = \int_{-\infty}^{\infty *} \frac{\cos sy}{y} \, dy + i \int_{-\infty}^{\infty *} \frac{\sin sy}{y} \, dy$$

and the first term vanishes since the integrand is an odd function of y. Accordingly

$$\int_{-\infty}^{\infty *} \frac{e^{isy}}{y} \, dy = 2i \int_{0}^{\infty} \frac{\sin sy}{y} \, dy = \pi i \, \text{sgn} \, s. \tag{42}$$

where $\text{sgn} \, s = \pm 1$ depending on whether s is positive or negative.

Applying Fourier transforms to (40) we obtain

$$i \, \text{sgn} \, s F(\phi) = F(f)$$

Using (41), (42) now shows that

$$\phi(x) = -\frac{1}{\pi} \int_{-\infty}^{\infty *} \frac{f(y)}{x - y} \, dy. \tag{43}$$

The integral operator in (40) is customarily denoted by the symbol H, in honor of Hilbert. Then (40) and (43) can be rewritten in the form

$$H(\phi) = f$$

$$\phi = -Hf$$

and it follows that

$$H(H(\phi)) = -\phi. \tag{44}$$

From the above we can also observe that

$$F(H\phi) = -i \, \text{sgn} \, s F(\phi)$$

which indicates that $H(\phi) \in L_2[-\infty, \infty]$. This operator is a very important one and will play a significant role in later sections of this chapter.

EXAMPLE 8. Solve

$$\frac{1}{\pi} \int_0^{\infty *} \frac{\phi(y)}{x - y}\, dy = f(x), \qquad f(x) \in L_2[0, \infty]. \tag{45}$$

Using the same substitutions as in Example 6, the above is reduced to

$$\frac{1}{\pi} \int_{-\infty}^{\infty *} \frac{\psi(\eta)}{\sinh(\xi - \eta)}\, d\eta = g(\xi). \tag{46}$$

In order to apply the Fourier transform technique we require the integral

$$\int_{-\infty}^{\infty *} \frac{e^{is\eta}}{\sinh \eta}\, d\eta = 2\pi i \left[\tfrac{1}{2} + \sum_{n=1}^{\infty} (-1)^n e^{-ns\pi} \right] = \frac{\pi i \sinh \pi s/2}{\cosh \pi s/2}$$

The evaluation is made using a residue integration. The $\frac{1}{2}$ in the bracket above is the contribution from $\eta = 0$, where as a result of the principal value evaluation only half the residue is taken. Now (46) reduces to

$$F(\psi) = \frac{1}{i} \frac{\cosh \pi s/2}{\sinh \pi s/2} F(g)$$

and

$$\psi(\xi) = C + \frac{1}{2\pi i} \int_{-\infty}^{\infty} g(\xi - \eta)\, d\eta \int_{-\infty}^{\infty *} \frac{e^{-is\eta} \cosh \pi s/2}{\sinh \pi s/2}\, ds \tag{47}$$

where C is an arbitrary constant. Clearly, since

$$\int_{-\infty}^{\infty *} \frac{C}{\sinh(\xi - \eta)}\, d\eta = 0$$

the homogeneous form of (45) has nontrivial solutions which are constants.

The inner integral in (47) can be evaluated by a residue integration, as in Example 6. Then

$$\int_{-\infty}^{\infty *} \frac{e^{-is\eta} \cosh \pi s/2}{\sinh \pi s/2}\, ds = -2\pi i \left[\frac{1}{\pi} + \frac{2}{\pi} \sum_{n=1}^{\infty} e^{-2n\eta} \right] = -2i \frac{\cosh \eta}{\sinh \eta}$$

and

$$\psi(\xi) = C - \frac{1}{\pi} \int_{-\infty}^{\infty *} \frac{\cosh(\xi - \eta)}{\sinh(\xi - \eta)} g(\eta)\, d\eta.$$

One can now revert to the original variables to obtain

$$\phi(x) = \frac{C}{\sqrt{x}} - \frac{1}{2\pi} \int_0^{\infty^*} \frac{x+y}{x-y} \cdot \frac{f(y)}{\sqrt{yx}} \, dy \tag{48}$$

As in (39) one can show that

$$\left\| \frac{1}{\pi} \int_0^{\infty^*} \frac{\phi(y)}{x-y} \, dy \right\| \leq \|\phi\| \tag{49}$$

so that the operator is bounded.

EXAMPLE 9. In this example we shall consider an operator that can be analyzed by means of Fourier series. It will be shown, however, that the results will enable us to provide a new analysis of the operator investigated in Example 8. Consider

$$K\phi = \frac{1}{\pi} \int_0^{\pi^*} \frac{\sin y}{\cos x - \cos y} \phi(y) \, dy = f(x), \qquad f(x) \in L_2[0, \pi]. \tag{50}$$

Evidently

$$K^*\phi = -\frac{1}{\pi} \int_0^{\pi^*} \frac{\sin x}{\cos x - \cos y} \phi(y) \, dy. \tag{51}$$

We shall first determine the effect that K has on all $\sqrt{2/\pi} \sin nx$. In view of the fact that this set is a complete orthonormal set on $L_2[0, \pi]$ this computation will show how K acts on this space.

$$\begin{aligned}
K \sin nx &= \frac{1}{\pi} \int_0^{*\pi} \frac{\sin y \sin ny}{\cos x - \cos y} \, dy \\
&= \frac{1}{2\pi} \int_0^{*\pi} \frac{\cos (n-1)y - \cos (n+1)y}{\cos x - \cos y} \, dy \\
&= \frac{1}{4\pi} \int_{-\pi}^{*\pi} \frac{\cos (n-1)y - \cos (n+1)y}{\cos x - \cos y} \, dy \\
&= \frac{1}{4\pi} \int_{-\pi}^{*\pi} \frac{e^{i(n-1)y} - e^{i(n+1)y}}{\cos x - \cos y} \, dy
\end{aligned}$$

In the last two steps the symmetry of the integrand was taken into account. Now the new variable of integration $e^{iy} = z$ is introduced.

$$K \sin nx = \frac{1}{2\pi i} \int_C \frac{z^{n-1} - z^{n+1}}{(z - e^{ix})(z - e^{-ix})} \, dz$$

Here C is the unit circle. The poles of the integrand lie on the circle so that, if residue integration is used, half the residues must be taken. Then

$$K \sin nx = \frac{-[\sin (n-1)x - \sin (n+1)x]}{2 \sin x} = \cos nx, \qquad n = 1, 2, \ldots. \tag{52}$$

The same calculation shows that

$$K^* \cos nx = \sin nx, \qquad n = 1, 2, \ldots. \tag{53}$$

$$K^*1 = 0$$

Combining (52) and (53) we observe that

$$K^*K \sin nx = \sin nx \tag{54}$$

so that $K^*K = I$. Similarly

$$KK^* \cos nx = \cos nx, \qquad n \neq 0, \tag{55}$$

$$KK^*1 = 0.$$

It follows that KK^* is not the identity since $KK^*1 = 0$.

We now return to (50). $\phi(x)$ can be expressed in the form

$$\phi(x) = \sum_{n=1}^{\infty} a_n \sqrt{\frac{2}{\pi}} \sin nx$$

so that

$$K\phi = \sum_{n=1}^{\infty} a_n \sqrt{\frac{2}{\pi}} \cos nx = f(x). \tag{56}$$

Clearly (50) or equivalently (56) can have a solution only if $\int_0^\pi f(x)\, dx = 0$. In that case

$$a_n = \sqrt{\frac{2}{\pi}} \int_0^\pi f(y) \cos ny\, dy$$

so that

$$\phi(x) = \sum_{n=1}^{\infty} \left(\sqrt{\frac{2}{\pi}} \int_0^\pi f(y) \cos ny\, dy \right) \sqrt{\frac{2}{\pi}} \sin nx \tag{57}$$

is the solution of (50). It can also be expressed in the form

$$\phi(x) = K^*f = -\frac{1}{\pi} \int_0^{\pi^*} \frac{\sin x}{\cos x - \cos y} f(y)\, dy. \tag{58}$$

EXAMPLE 10. In this example we consider the equation

$$K^*\phi = -\frac{1}{\pi} \int_0^{\pi^*} \frac{\sin x}{\cos x - \cos y} \phi(y)\, dy = f(x), \qquad f(x) \in L_2[0, \pi]. \tag{59}$$

We now expand

$$\phi(x) = \frac{1}{\sqrt{\pi}} a_0 + \sum_{n=1}^{\infty} a_n \sqrt{\frac{2}{\pi}} \cos nx$$

so that

$$K^*\phi = \sum_{n=1}^{\infty} a_n \sqrt{\frac{2}{\pi}} \sin nx = f(x)$$

and

$$a_n = \sqrt{\frac{2}{\pi}} \int_0^{\pi} f(y) \sin ny \, dy.$$

Unlike (50) no condition need be imposed on $f(x)$, since solutions exist for all $f(x) \in L_2[0, \pi]$. In fact all a_n, $n \geq 1$, are determined by $f(x)$. a_0, however can be assigned arbitrarily. Finally,

$$\phi(x) = \frac{1}{\sqrt{\pi}} a_0 + \sum_{n=1}^{\infty} \sqrt{\frac{2}{\pi}} \int_0^{\pi} f(y) \sin ny \, dy \sqrt{\frac{2}{\pi}} \cos nx$$

$$= \frac{1}{\sqrt{\pi}} a_0 + \frac{1}{\pi} \int_0^{\pi}{}^* \frac{\sin y}{\cos x - \cos y} f(y) \, dy. \tag{60}$$

In this case the solution (60) is no longer unique since a_0 can be assigned arbitrarily.

EXAMPLE 11.   We return to Eq. 45

$$\frac{1}{\pi} \int_0^{\pi}{}^* \frac{\phi(y)}{x - y} \, dy = f(x), \qquad f(x) \in L_2[0, \infty], \tag{61}$$

and use the new variables

$$y = \frac{1}{1 + \cos \eta} - \frac{1}{2}, \qquad x = \frac{1}{1 + \cos \xi} - \frac{1}{2}$$

$$\psi(\eta) = \frac{\phi(y) \sin \eta}{1 + \cos \eta}, \qquad g(\xi) = \frac{f(x) \sin \xi}{1 + \cos \xi}.$$

Substituting these in (61) leads to

$$K^*\psi = -\frac{1}{\pi} \int_0^{\pi}{}^* \frac{\sin \xi}{\cos \xi - \cos \eta} \psi(\eta) \, d\eta = g(\xi)$$

and use of (60) shows that

$$\psi(\xi) = C + \frac{1}{\pi} \int_0^{\pi}{}^* \frac{\sin \eta}{\cos \xi - \cos \eta} g(\eta) \, d\eta$$

where C is arbitrary.

Now we return to the original variables to obtain

$$\phi(x) = \frac{C}{\sqrt{x}} - \frac{1}{\pi} \int_0^{\infty *} \frac{(x + \frac{1}{2})/(y + \frac{1}{2})}{x - y} \sqrt{\frac{y}{x}} f(y)\, dy \qquad (62)$$

as a solution of (61). At first glance (48) and (62) are not identical. However,

$$\int_0^{\infty *} \frac{(x + \frac{1}{2})/(y + \frac{1}{2})}{x - y} \sqrt{\frac{y}{x}} f(y)\, dy - \frac{1}{2} \int_0^{\infty *} \frac{x + y}{x - y} \frac{f(y)}{\sqrt{yx}}\, dy$$

$$= \frac{1}{2\sqrt{x}} \int_0^{\infty} \frac{y - \frac{1}{2}}{y + \frac{1}{2}} \frac{f(y)}{\sqrt{y}}\, dy$$

so that the difference between (48) and (62) can be absorbed in the term $C/\sqrt{x}$.

There is an alternative procedure. We could have introduced, instead of $\psi(\eta)$ and $g(\xi)$, the variables $\tilde{\psi}(\eta)$ and $\tilde{g}(\xi)$ defined by

$$\tilde{\psi}(\eta) = \frac{\phi(y)}{1 + \cos \eta}, \qquad \tilde{g}(\xi) = \frac{f(x)}{1 + \cos \xi}$$

so that (61) becomes

$$K\tilde{\psi} = \frac{1}{\pi} \int_0^{\pi *} \frac{\sin \xi}{\cos \xi - \cos \eta} \tilde{\psi}(\eta)\, d\eta = -\tilde{g}(\xi).$$

Use of (58) now shows that

$$\tilde{\psi}(\xi) = -K^* \tilde{g}(\xi) = \frac{1}{\pi} \int_0^{\pi *} \frac{\sin \xi}{\cos \xi - \cos \eta} \tilde{g}(\eta)\, d\eta$$

provided that the compatibility condition

$$\int_0^{\pi} \tilde{g}(\eta)\, d\eta = 0$$

is satisfied.

Returning to the original variables we find

$$\phi(x) = -\frac{1}{\pi} \int_0^{\infty *} \frac{(y + \frac{1}{2}/x + \frac{1}{2})\sqrt{x/y}\, f(y)\, dy}{x - y} \qquad (63)$$

subject to the compatibility condition

$$\int_0^{\infty} \frac{f(y)}{\sqrt{y}}\, dy = 0 \qquad (64)$$

The difference between solutions (62) and (63) subject to condition (64) is of the form $(C/\sqrt{x})$.

EXAMPLE 12. Solve

$$\frac{1}{\pi} \int_0^{\infty *} \left(\frac{1}{x+y} - \frac{1}{x-y} \right) \phi(y)\, dy = f(x), \qquad f(x) \in L_2[0,\, \infty]. \qquad (65)$$

Using the same change of variables as in Example 6, and then applying Fourier transforms (65) is reduced to

$$\frac{1}{\pi} \int_{-\infty}^{\infty *} \left[\frac{1}{\cosh\,(\xi - \eta)} - \frac{1}{\sinh\,(\xi - \eta)} \right] \psi(\eta)\, d\eta = g(\xi)$$

so that

$$F(\psi) = \frac{\cosh\,\pi s/2}{1 - i\,\sinh\,\pi s/2}\, F(g)$$

$$\psi(\xi) = \frac{1}{2\pi} \int_{-\infty}^{\infty} g(\xi - \eta)\, d\eta \int_{-\infty}^{\infty} \frac{e^{-is\eta}\,\cosh\,\pi s/2}{1 - i\,\sinh\,\pi s/2}\, ds.$$

The inner integral can be evaluated by a residue integration so that

$$\psi(\xi) = \frac{2}{\pi} \int_{-\infty}^{\infty *} \frac{g(\xi - \eta)e^{\eta}}{\sinh\,2\eta}\, d\eta.$$

Returning to the original variables we have

$$\phi(x) = \frac{1}{\pi} \int_0^{\infty *} \left(\frac{1}{x+y} + \frac{1}{x-y} \right) f(y)\, dy. \qquad (66)$$

One can show as before that

$$\left\| \frac{1}{\pi} \int_0^{\infty *} \left(\frac{1}{x+y} - \frac{1}{x-y} \right) \phi(y)\, dy \right\| \leq \|\phi\|. \qquad (67)$$

There is an alternate procedure for finding the solution (66). Equation (65) is defined on $L_2[0,\, \infty]$, but can be extended to $L_2[-\infty,\, \infty]$ by continuing $\phi(x)$ from $[0,\, \infty)$ to $(-\infty,\, 0]$ as an odd function. If $f(x)$ is continued as an even function (65) can be rewritten as

$$\frac{1}{2\pi} \int_{-\infty}^{\infty *} \left(\frac{1}{x+y} - \frac{1}{x-y} \right) \phi(y)\, dy = f(x).$$

Use of the operator H [defined in (40)] allows us to rewrite the above equation in the form

$$\tfrac{1}{2}[H\phi(-x) - H\phi(x)] = f(x).$$

In (44) we saw that $H^2 = -I$ so that

$$-\tfrac{1}{2}[\phi(-x) - \phi(x)] = Hf(x).$$

Finally, use of the symmetry properties of $\phi(x)$ and $f(x)$ shows that

$$\phi(x) = \frac{1}{\pi} \int_0^\infty \left(\frac{1}{x+y} + \frac{1}{x-y} \right) f(y) \, dy.$$

EXAMPLE 13. Another equation that can be treated by this method is the so-called airfoil equation

$$\frac{1}{\pi} \int_{-1}^{*1} \frac{\phi(y)}{x-y} \, dy = f(x), \qquad f(x) \in L_2[-1, 1]. \tag{68}$$

We now introduce the new variables

$$x = \cos \xi, \qquad y = \cos \eta$$

$$g(\xi) = f(\cos \xi) \sin \xi, \qquad \psi(\eta) = \phi(\cos \eta) \sin \eta$$

so that (68) becomes

$$\frac{1}{\pi} \int_0^{*\pi} \frac{\sin \xi}{\cos \xi - \cos \eta} \, \psi(\eta) \, d\eta = g(\xi). \tag{69}$$

Using Example 10, in particular (60), and reverting to the original variables one obtains

$$\phi(x) = \frac{C}{\sqrt{1-x^2}} - \frac{1}{\pi} \int_{-1}^{*1} \sqrt{\frac{1-y^2}{1-x^2}} \frac{f(y)}{x-y} \, dy, \tag{70}$$

where C is an arbitrary constant.

Had we used the variables

$$\tilde{g}(\xi) = f(\cos \xi), \qquad \tilde{\psi}(\eta) = \phi(\cos \eta)$$

(69) would have been replaced by

$$\frac{1}{\pi} \int_0^{*\pi} \frac{\sin \eta}{\cos \xi - \cos \eta} \, \tilde{\psi}(\eta) \, d\eta = \tilde{g}(\xi)$$

whose solution, by (58), would lead to

$$\phi(x) = -\frac{1}{\pi} \int_{-1}^{*1} \sqrt{\frac{1-x^2}{1-y^2}} \frac{f(y)}{x-y} \, dy \tag{71}$$

subject to the compatibility condition

$$\int_{-1}^{1} \frac{f(y)}{\sqrt{1 - y^2}} \, dy = 0 \tag{72}$$

A comparison of (70) and (71) shows that the difference of the two solutions, providing that (72) holds, is of the form $C/\sqrt{1 - x^2}$.

3. LAPLACE TRANSFORMS

A transform closely related to the Fourier transform is the Laplace transform. It is defined by

$$\mathscr{F}(\sigma) = \int_{0}^{\infty} e^{-\sigma x} f(x) \, dx, \tag{73}$$

provided that the integral exists for suitable (possibly complex) values of σ, and functions $f(x)$.

In order to investigate the above integral, it will be convenient first to assume that $f(x) \in L_2[0, \infty]$. We can now extend $f(x)$ by defining

$$f(x) = 0, \qquad x < 0$$

so that the resultant function belongs to $L_2[-\infty, \infty]$. In that case

$$F(f) = \frac{1}{\sqrt{2\pi}} \int_{-\infty}^{\infty} e^{isx} f(x) \, dx = \frac{1}{\sqrt{2\pi}} \int_{0}^{\infty} e^{isx} f(x) \, dx, \tag{74}$$

and

$$f = \frac{1}{\sqrt{2\pi}} \int_{-\infty}^{\infty} e^{-isx} F(f) \, ds \tag{75}$$

by use of the Fourier inversion formula. If we now replace s by $i\sigma$, (74) and (75) become

$$F(f) = \frac{1}{\sqrt{2\pi}} \int_{0}^{\infty} e^{-\sigma x} f(x) \, dx$$

$$f = \frac{1}{i\sqrt{2\pi}} \int_{-i\infty}^{i\infty} e^{\sigma x} F(f) \, d\sigma.$$

Finally a comparison with (73) shows that

$$\mathscr{F}(\sigma) = \int_{0}^{\infty} e^{-\sigma x} f(x) \, dx \tag{76}$$

$$f(x) = \frac{1}{2\pi i} \int_{-i\infty}^{i\infty} e^{\sigma x} \mathscr{F}(\sigma) \, d\sigma. \tag{77}$$

Equations (76) and (77) are the Laplace and inverse Laplace transforms. So far they are defined only for functions $f(x) \in L_2[0, \infty]$. In order to extend (76) to a broader class of functions we will consider functions of exponential growth. That is, we suppose that there exists a real number p such that $e^{-px}f(x) \in L_2[0, \infty]$. Then

$$\mathscr{F}(\sigma) = \int_0^\infty e^{-(\sigma+p)x}f(x)\,dx \tag{78}$$

$$e^{-px}f(x) = \frac{1}{2\pi i}\int_{-i\infty}^{i\infty} e^{\sigma x}\mathscr{F}(\sigma)\,d\sigma. \tag{79}$$

In view of the fact that (74) was defined for real s, (76) is defined for σ on the imaginary axis. In fact (76) will be defined for all σ with non-negative real parts, since the integral will still exist, if the real part of σ is allowed to grow. Finally we shall replace $\sigma + p$ by σ. Then (78) and (79) become

$$\mathscr{F}(\sigma) = \int_0^\infty e^{-\sigma x}f(x)\,dx \tag{80}$$

$$f(x) = \frac{1}{2\pi i}\int_{-i\infty+p}^{i\infty+p} e^{\sigma x}\mathscr{F}(\sigma)\,d\sigma. \tag{81}$$

Formally (80) and (81) are very similar to (76) and (77). In (80) however, we require that Re $\sigma \geq p$ and in (81) the integration is carried out along the line Re $\sigma = p$.

We can now apply (6) to (76) and (77). We have immediately

$$\int_0^\infty |f|^2\,dx = \frac{1}{2\pi i}\int_{-i\infty}^{i\infty} |\mathscr{F}(\sigma)|^2\,d\sigma, \tag{82}$$

and if

$$\mathscr{G}(\sigma) = \int_0^\infty e^{-\sigma x}g(x)\,dx$$

in analogy to (7)

$$\int_0^\infty f\bar{g}\,dx = \frac{1}{2\pi i}\int_{-i\infty}^{i\infty} \mathscr{F}(\sigma)\overline{\mathscr{G}(\sigma)}\,d\sigma. \tag{83}$$

In (82) and (83) f and $g \in L_2[0, \infty]$, of course. In order to apply (82) and (83) to functions $f(x)$ and $g(x)$ that are of exponential growth, we rewrite (81) as

$$e^{-px}f(x) = \frac{1}{2\pi i}\int_{-i\infty}^{i\infty} e^{\sigma x}\mathscr{F}(\sigma + p)\,d\sigma. \tag{84}$$

(84) is obtained by multiplying (81) by e^{-px} and replacing the variable σ by $\sigma + p$. Then (83) yields

$$\int_0^\infty e^{-2px}f(x)\overline{g(x)}\,dx = \frac{1}{2\pi i}\int_{-i\infty}^{i\infty}\mathscr{F}(\sigma+p)\overline{\mathscr{G}(\sigma+p)}\,d\sigma. \qquad (85)$$

The above results are in complete analogy to those stated in theorem 3. We can also define a convolution operation, whose properties are discussed in theorem 4. We define

$$(f*g)(x) = \int_0^x f(y)g(x-y)\,dy. \qquad (86)$$

One can now apply theorem 4 to (86). It follows that

$$\int_0^\infty e^{-\sigma x}(f*g)(x)\,dx = \mathscr{F}(\sigma)\mathscr{G}(\sigma) \qquad (87)$$

and using (81)

$$(f*g)(x) = \frac{1}{2\pi i}\int_{-i\infty+p}^{i\infty+p} e^{\sigma x}\mathscr{F}(\sigma)\mathscr{G}(\sigma)\,d\sigma. \qquad (88)$$

4. APPLICATIONS OF LAPLACE TRANSFORMS

Laplace transforms are useful in solving initial value problems, and also Volterra integral equations with kernels of the type $K(x-y)$. Several examples will illustrate some of these applications.

EXAMPLE 13.   We consider the heat flow equation

$$\frac{\partial u}{\partial t} = \frac{\partial^2 u}{\partial x^2} \qquad (89)$$

with the boundary conditions

$$u(0, t) = u(L, t) = 0 \qquad (90)$$

and the initial condition

$$u(x, 0) = f(x). \qquad (91)$$

In order to apply Laplace transforms we let

$$U(x, \sigma) = \int_0^\infty e^{-\sigma t}u(x, t)\,dt.$$

In this case we take transforms with respect to the time vari it follows that

$$\frac{\partial^2}{\partial x^2}U(x, \sigma) = \int_0^\infty e^{-\sigma t}\frac{\partial^2}{\partial x^2}u(x, t)\,dt.$$

For the moment our procedures are formal, since we are not certain that $u(x, t)$ has a transform. The fact that it does will be established later. We also need to calculate the transform of $\partial u / \partial t$. Then

$$\int_0^\infty e^{-\sigma t} \frac{\partial u(x, t)}{\partial t}\, dt = e^{-\sigma t} u(x, t) \Big|_0^\infty + \sigma \int_0^\infty e^{-\sigma t} u(x, t)\, dt$$

$$= \sigma U(x, \sigma) - f(x)$$

by an integration by parts and use of the initial condition (91). It follows that (89) can now be rewritten in the form

$$\int_0^\infty e^{-\sigma t} \left(\frac{\partial u}{\partial t} - \frac{\partial^2 u}{\partial x^2} \right) dt = \sigma U(x, \sigma) - f(x) - \frac{\partial^2}{\partial x^2} U(x, \sigma) = 0$$

so that we obtain

$$\frac{\partial^2 U}{\partial x^2} - \sigma U = -f(x)$$

$$U(0, \sigma) = U(L, \sigma) = 0$$

In other words the partial differential equation has been reduced to the above ordinary differential equation. By the methods of Chapter 3 we find that

$$U(x, \sigma) = \int_0^L \frac{\sinh \sqrt{\sigma}\, x_< \sinh \sqrt{\sigma}(L - x_>)}{\sqrt{\sigma} \sinh \sqrt{\sigma}\, L} f(y)\, dy.$$

Although the kernel in the above integral appears to be a function of $\sqrt{\sigma}$, it is in fact a ratio of entire functions of σ. To show this we note that

$$\sinh \sqrt{\sigma}\, a = \sqrt{\sigma}\, a \sum_{n=0}^\infty \frac{a^{2n} \sigma^n}{(2n + 1)!}.$$

If the above is substituted in the integral all $\sqrt{\sigma}$ terms cancel and only integral powers of σ are left.

Finally, to find $u(x, t)$ we apply the inverse Laplace transform to $U(x, \sigma)$.

$$u(x, t) = \frac{1}{2\pi i} \int_{-i\infty}^{i\infty} e^{\sigma t} \int_0^L \frac{\sinh \sqrt{\sigma}\, x_< \sinh \sqrt{\sigma}(L - x_>)}{\sqrt{\sigma} \sinh \sqrt{\sigma}\, L} f(y)\, dy\, d\sigma$$

$$= \int_0^L f(y) \left\{ \frac{1}{2\pi i} \int_{-i\infty}^{i\infty} \frac{e^{\sigma t} \sinh \sqrt{\sigma}\, x_< \sinh \sqrt{\sigma}(L - x_>)}{\sqrt{\sigma} \sinh \sqrt{\sigma}\, L}\, d\sigma \right\} dy.$$

To evaluate the inner integral we note that for $t > 0$ the line integral can be completed to a contour integral that encloses the left half plane. The term $e^{\sigma t}$ dominates the integrand and becomes exponentially small as $\sigma \to -\infty$ in that half plane. The denominator vanishes at all points $\sqrt{\sigma}\, L = in\pi$ or equivalently $\sigma = -(n^2\pi^2/L^2)$. By a residue integration the inner integral then reduces to

$$\frac{2}{L} \sum_{n=1}^{\infty} e^{-(n^2\pi^2 t/L^2)} \sin \frac{n\pi x}{L} \sin \frac{n\pi y}{L}$$

so that

$$u(x, t) = \sum_{n=1}^{\infty} \left(\frac{2}{L} \int_0^L \sin \frac{n\pi y}{L} f(y)\, dy \right) \sin \frac{n\pi x}{L}\, e^{-(n^2\pi^2 t/L^2)} \tag{93}$$

Equation (93) is the solution to the original problem. One can now show that all the Laplace transform operations performed are indeed legitimate, by checking that $u(x, t)$, as given in (93) is such that all previously used operations are permissible.

EXAMPLE 14.   We shall now consider the wave equation

$$\frac{\partial^2 u}{\partial t^2} = \frac{\partial^2 u}{\partial x^2} \tag{94}$$

subject to the conditions

$$u(0, t) = u(L, t) = 0$$

$$u(x, 0) = f(x), \qquad \frac{\partial u}{\partial t}(x, 0) = 0. \tag{95}$$

Again, we let

$$U(x, \sigma) = \int_0^{\infty} e^{-\sigma t} u(x, t)\, dt$$

and by a double integration by parts, combined with the initial conditions find that

$$\int_0^{\infty} e^{-\sigma t} \frac{\partial^2 u}{\partial t^2}\, dt = \sigma^2 U(x, \sigma) - \sigma f(x).$$

Equation (94) can now be rewritten in the form

$$\int_0^{\infty} e^{-\sigma t} \left(\frac{\partial^2 u}{\partial t^2} - \frac{\partial^2 u}{\partial x^2} \right) dt = \sigma^2 U(x, \sigma) - \sigma f(x) - \frac{\partial^2}{\partial x^2} U(x, \sigma) = 0$$

so that we obtain the system

$$\frac{\partial^2}{\partial x^2} U - \sigma^2 U = -\sigma f(x)$$

$$U(0, \sigma) = U(L, \sigma) = 0.$$

The solution to the above is given by

$$U(x, \sigma) = \int_0^L \frac{f(x) \sinh \sigma x_< \sinh \sigma(L - x_>)}{\sinh \sigma L} dy. \tag{96}$$

In order to obtain the solution $u(x, t)$ of (94) we have to evaluate the integral

$$\frac{1}{2\pi i} \int_{-i\infty}^{i\infty} \frac{e^{\sigma t} \sinh \sigma x_< \sinh \sigma(L - x_>)}{\sinh \sigma L} d\sigma. \tag{97}$$

Once this is found the inverse Laplace transform of $U(x, \sigma)$ can be constructed. For σ in the left half plane the integrand of the above integral has the following asymptotic form,

$$e^{\sigma(t + x_> - x_<)}$$

and the latter vanishes as $\sigma \to -\infty$ in that half plane and $t > 0$. This enables us to close the contour in (97) and to perform a residue integration. (97) then reduces to

$$\sum_{n=-\infty}^{\infty} \frac{e^{in\pi t/L} \sin (n\pi/L)x_< \sin (n\pi/L)(L - x_>)}{L \cos n\pi}$$

$$= \frac{2}{L} \sum_{n=1}^{\infty} \cos \frac{n\pi t}{L} \sin \frac{n\pi x}{L} . \sin \frac{n\pi y}{L}$$

since the residues are found at $\sigma = in\pi/L$. Insertion of the latter in (96) leads to

$$u(x, t) = \sum_{n=1}^{\infty} \left(\frac{2}{L} \int_0^L f(y) \sin \frac{n\pi y}{L} dy \right) \cos \frac{n\pi t}{L} \sin \frac{n\pi x}{L} .$$

EXAMPLE 15. Solve

$$\phi(x) - \lambda \int_0^x K(x - y)\phi(y) dy = f(x). \tag{98}$$

Let

$$\Phi(\sigma) = \int_0^\infty e^{-\sigma x} \phi(x) dx$$

$$\mathscr{F}(\sigma) = \int_0^\infty e^{-\sigma x} f(x) dx$$

$$\mathscr{K}(\sigma) = \int_0^\infty e^{-\sigma x} K(x) dx.$$

Applying Laplace transforms to (98) and using the convolution formula (87) leads to

$$\Phi(\sigma) - \lambda \mathscr{K}(\sigma)\Phi(\sigma) = \mathscr{F}(\sigma)$$

so that

$$\Phi(\sigma) = \frac{1}{1 - \lambda \mathscr{K}(\sigma)} \mathscr{F}(\sigma),$$

provided $1 - \lambda \mathscr{K}(\sigma) \neq 0$.

If the above has an inverse Laplace transform, one finds that

$$\phi(x) = \frac{1}{2\pi i} \int_{-i\infty+p}^{i\infty+p} e^{\sigma x} \frac{1}{1 - \lambda \mathscr{K}(\sigma)} \mathscr{F}(\sigma) \, d\sigma. \qquad (99)$$

EXAMPLE 16. Solve

$$\phi(x) - \lambda \int_0^x e^{x-y} \phi(y) \, dy = f(x). \qquad (100)$$

First we note that

$$\int_0^\infty e^{-\sigma x} e^x \, dx = \frac{1}{\sigma - 1}$$

if $\mathrm{Re}\ \sigma > 1$. Now we proceed as in Example 15 and obtain

$$\phi(x) = \frac{1}{2\pi i} \int_{-i\infty+p}^{i\infty+p} e^{\sigma x} \frac{\sigma - 1}{\sigma - (\lambda + 1)} \mathscr{F}(\sigma) \, d\sigma. \qquad (101)$$

It is possible to evaluate the above integral directly. But an easier approach is to use formula (88). We note that, if we let

$$g(x) = e^{(\lambda+1)x}$$

then

$$\mathscr{G}(\sigma) = \frac{1}{\sigma - (\lambda + 1)}$$

Now (101) can be rewritten as

$$\phi(x) = \frac{1}{2\pi i} \int_{-i\infty+p}^{i\infty+p} e^{\sigma x} \left[1 + \frac{\lambda}{\sigma - (\lambda + 1)} \right] \mathscr{F}(\sigma) \, d\sigma$$

$$= \frac{1}{2\pi i} \int_{-i\infty+p}^{i\infty+p} e^{\sigma x} \mathscr{F}(\sigma) \, d\sigma + \frac{\lambda}{2\pi i} \int_{-i\infty+p}^{i\infty+p} e^{\sigma x} \mathscr{G}(\sigma) \mathscr{F}(\sigma) \, d\sigma.$$

By means of (88), the above reduces to

$$\phi(x) = f(x) + \lambda(f * g)(x) = f(x) + \lambda \int_0^x e^{(1+\lambda)(x-y)} f(y) \, dy. \qquad (102)$$

EXAMPLE 17. We consider the Volterra equation of the first kind

$$\int_0^x \frac{\phi(y)}{(x - y)^\alpha} \, dy = f(x) \qquad (103)$$

where $0 \leq \alpha < 1$. This equation was analyzed and solved in Section 6.1 of Chapter 2. We shall now obtain a solution by the use of Laplace transform techniques.

First we see that

$$\int_0^\infty e^{-\sigma x} x^{-\alpha} \, dx = \sigma^{\alpha-1} \int_0^\infty e^{-u} u^{-\alpha} \, du = \sigma^{\alpha-1} \Gamma(1 - \alpha), \qquad (104)$$

if we use the substitution $\sigma x = u$ in the first integral. The left side of (103) is a convolution, so that by an application of the Laplace transform (103) is reduced to

$$\sigma^{\alpha-1} \Gamma(1 - \alpha) \Phi(\sigma) = \mathscr{F}(\sigma). \qquad (105)$$

Now

$$\Phi(\sigma) = \frac{\sigma^{1-\alpha}}{\Gamma(1 - \alpha)} \mathscr{F}(\sigma)$$

so that

$$\phi(x) = \frac{1}{2\pi i} \int_{-i\infty}^{i\infty} e^{\sigma x} \frac{\sigma^{1-\alpha}}{\Gamma(1 - \alpha)} \mathscr{F}(\sigma) \, d\sigma. \qquad (106)$$

As in the previous example, an indirect evaluation of (106) is more effective than a direct approach. First we observe that if

$$\int_0^\infty e^{-\sigma x} f(x) \, dx = \mathscr{F}(\sigma)$$

then

$$\int_0^\infty e^{-\sigma x} \frac{d}{dx} f(x) \, dx = e^{-\sigma x} f(x) \Big|_0^\infty + \sigma \int_0^\infty e^{-\sigma x} f(x) \, dx = \sigma \mathscr{F}(\sigma)$$

provided $f(0) = 0$. It follows that

$$\frac{1}{2\pi i} \int_{-i\infty}^{i\infty} e^{\sigma x} \sigma \mathscr{F}(\sigma) \, d\sigma = f'(x).$$

Now (106) can be rewritten as

$$\phi(x) = \frac{d}{dx} \frac{1}{\Gamma(1 - \alpha)\Gamma(\alpha)} \frac{1}{2\pi i} \int_{-i\infty}^{i\infty} e^{\sigma x} \Gamma(\alpha) \sigma^{-\alpha} \mathscr{F}(\sigma) \, d\sigma. \qquad (107)$$

In going from (103) to (105) we had effectively shown that

$$\int_0^\infty e^{-\sigma x} \int_0^x \frac{\phi(y)}{(x - y)^\alpha} \, dy = \sigma^{\alpha-1} \Gamma(1 - \alpha) \Phi(\sigma).$$

We can replace α by $1 - \alpha$ so that

$$\int_0^\infty e^{-\sigma x} \int_0^x \frac{f(y)}{(x - y)^{1-\alpha}} \, dy = \sigma^{-\alpha} \Gamma(\alpha) \mathscr{F}(\sigma)$$

and a comparison with (107) shows that

$$\phi(x) = \frac{d}{dx} \frac{1}{\Gamma(1-\alpha)\Gamma(\alpha)} \int_0^x \frac{f(y)}{(x-y)^{1-\alpha}} \, dy.$$

Finally we use the standard formula

$$\Gamma(1-\alpha)\Gamma(\alpha) = \frac{\pi}{\sin \pi\alpha}$$

to obtain

$$\phi(x) = \frac{\sin \pi\alpha}{\pi} \frac{d}{dx} \int_0^x \frac{f(y)}{(x-y)^{1-\alpha}} \, dy$$

which agrees with (41) of Chapter 2.

5. HANKEL TRANSFORMS

The theory of Fourier transforms can be extended to functions of several variables. Suppose

$$\int_{-\infty}^{\infty} \int_{-\infty}^{\infty} |f(x, y)|^2 \, dx \, dy < \infty. \tag{108}$$

We can then define

$$F(s, \sigma) = \frac{1}{2\pi} \int_{-\infty}^{\infty} \int_{-\infty}^{\infty} f(x, y) e^{i(sx+\sigma y)} \, dx \, dy, \tag{109}$$

by first taking transforms with respect to x and then with respect to y. The resultant transform is now a function of two variables s and σ. By applying the inverse transforms one finds that

$$f(x, y) = \frac{1}{2\pi} \int_{-\infty}^{\infty} \int_{-\infty}^{\infty} F(s, \sigma) e^{-i(sx+\sigma y)} \, ds \, d\sigma. \tag{110}$$

Now we shall suppose that the function $f(x, y)$ has the specialized form

$$f(x, y) = f(r) e^{in\theta}$$

where r and θ are polar coordinates. Condition (108) is now replaced by

$$\int_0^{\infty} \int_0^{\infty} |f(r)|^2 \, r \, dr \, d\theta = 2\pi \int_0^{\infty} r \, |f(r)|^2 \, dr < \infty, \tag{111}$$

and if we let $s = \rho \cos \alpha$, $\sigma = \rho \sin \alpha$, (109) and (110) reduce to

$$F(\rho, \alpha) = \frac{1}{2\pi} \int_0^{\infty} \int_0^{2\pi} f(r) e^{i[r\rho \cos (\theta-\alpha)+n\theta]} r \, dr \, d\theta \tag{112}$$

$$f(r) e^{in\theta} = \frac{1}{2\pi} \int_0^{\infty} \int_0^{2\pi} F(\rho, \alpha) e^{-ir\rho \cos (\theta-\alpha)} \rho \, d\rho \, d\alpha. \tag{113}$$

One of the standard integral representations for Bessel functions* is

$$J_n(z) = \frac{1}{2\pi} \int_0^{2\pi} e^{i[z \cos \alpha + n\alpha - n\pi/2]} \, d\alpha. \tag{114}$$

This can be used to perform one of the integrations in (112), so that

$$F(\rho, \alpha) = e^{in(\alpha + \pi/2)} \int_0^\infty rf(r)J_n(r\rho) \, dr \tag{115}$$

and similarly

$$f(r)e^{in\theta} = e^{in\theta} \int_0^\infty \rho \left(\int_0^\infty r_1 f(r_1)J_n(r_1\rho) \, dr_1 \right) J_n(r\rho) \, d\rho. \tag{116}$$

We now define

$$\mathscr{H}(f) = \int_0^\infty rf(r)J_n(r\rho) \, dr \tag{117}$$

for all functions $f(r)$ for which

$$\int_0^\infty r \, |f(r)|^2 \, dr < \infty. \tag{118}$$

From (116) it follows that

$$f(r) = \int_0^\infty \rho \mathscr{H}(f)J_n(r\rho) \, d\rho. \tag{119}$$

Equation 117 is known as the Hankel transform of f and (119) is the corresponding inverse transform. Evidently the transform is its own inverse, so that

$$\mathscr{H}^2(f) = f$$

for all f satisfying (118).

THEOREM 5. Let $L_2(r)$ denote the Hilbert space of all functions satisfying (118). The Hankel transform represents a unitary operator on that space. It is also a self-adjoint operator.

The proof of this statement is contained in the previous remarks.

EXAMPLE 16. Solve

$$\phi(x) - \lambda \int_0^\infty yJ_n(xy)\phi(y) \, dy = f(x),$$

provided that $f(x) \in L_2([0, \infty], x)$.† Writing the above in the form

$$\phi - \lambda \mathscr{H}(\phi) = f$$

---

* A detailed knowledge of Bessel functions is not necessary for an understanding of this section. In fact the symbol $J_n(z)$, can be viewed as a shorthand notation for the function on the right side of (114).

† We recall that $L_2([0, \infty], x)$ is the space of all $f(x)$ for which $\int_0^\infty x \, |f(x)|^2 \, dx < \infty$.

we see immediately that

$$\phi = \frac{f + \lambda \mathcal{H}(f)}{1 - \lambda^2} \tag{120}$$

if $\lambda^2 \neq 1$.

EXAMPLE 17.   The Hankel transforms can be used to solve differential equations in which the operator $(d/dr)r(d/dr)$ appears. Consider the equation

$$\frac{1}{r} \frac{\partial}{\partial r} r \frac{\partial u}{\partial r} = \frac{\partial u}{\partial t} \tag{121}$$

$$u(r, 0) = f(r)$$

where $\int_0^\infty r\,|f(r)|^2\,dr < \infty$. We can apply the Hankel transforms to both sides of (121). Then

$$\mathcal{H}\left(\frac{\partial u}{\partial t}\right) = \int_0^\infty r \frac{\partial u}{\partial t} J_0(\rho r)\,dr = \frac{\partial}{\partial t} \mathcal{H}(u)$$

$$\mathcal{H}\left(\frac{1}{r} \frac{\partial}{\partial r} r \frac{\partial u}{\partial r}\right) = \int_0^\infty \left(\frac{\partial}{\partial r} r \frac{\partial u}{\partial r}\right) J_0(\rho r)\,dr$$

$$= \int_0^\infty u \left(\frac{\partial}{\partial r} r \frac{\partial}{\partial r} J_0(\rho r)\right) dr = -\rho^2 \int_0^\infty r u J_0(\rho r)\,dr$$

$$= -\rho^2 \mathcal{H}(u).$$

In the above we apply two integrations by parts and recall that $J_0(\rho r)$ satisfies the differential equation.

$$J_0''(\rho r) + \frac{1}{r} J_0'(\rho r) + \rho^2 J_0(\rho r) = 0.$$

Now (121) reduces to

$$\frac{\partial}{\partial t} \mathcal{H}(u) + \rho^2 \mathcal{H}(u) = 0$$

with the initial condition

$$\mathcal{H}(u)\big|_{t=0} = \mathcal{H}(f(r)).$$

It follows that

$$\mathcal{H}(u) = e^{-\rho^2 t} \mathcal{H}(f(r))$$

so that

$$u = \mathcal{H}(e^{-\rho^2 t} \mathcal{H}(f(r))) = \int_0^\infty \rho e^{-\rho^2 t} \left(\int_0^\infty r_1 f(r_1) J_0(\rho r_1)\,dr_1\right) J_0(\rho r)\,d\rho$$

$$= \int_0^\infty r_1 f(r_1) \left(\int_0^\infty \rho e^{-\rho^2 t} J_0(\rho r_1) J_0(\rho r)\,d\rho\right) dr_1. \tag{122}$$

6. MELLIN TRANSFORMS

Another transform closely related to the other transforms treated in this chapter is the Mellin transform. In order to define it and to motivate it we shall start with the Fourier transform

$$F(f) = \frac{1}{\sqrt{2\pi}} \int_{-\infty}^{\infty} f(x)e^{isx} dx$$

$$f = \frac{1}{\sqrt{2\pi}} \int_{-\infty}^{\infty} F(f)e^{-isx} ds$$

defined on $L_2[-\infty, \infty]$. We now let $u = e^x$ so that

$$F(f) = \frac{1}{\sqrt{2\pi}} \int_{0}^{\infty} f(lnu)u^{is-1} du \tag{123}$$

$$f(lnu) = \frac{1}{\sqrt{2\pi}} \int_{-\infty}^{\infty} F(f)u^{-is} ds \tag{124}$$

provided

$$\int_{0}^{\infty} \frac{1}{u} |f(lnu)|^2 du < \infty. \tag{125}$$

We now let $s = -i\sigma$ and replace $f(lnu)$ by $\sqrt{2\pi} f(u)$ in (123), (124), and (125) and denote the right-hand side of (123) by $M(\sigma)$. Then

$$M(\sigma) = \int_{0}^{\infty} f(u)u^{\sigma-1} du \tag{126}$$

$$f(u) = \frac{1}{2\pi i} \int_{-i\infty}^{i\infty} M(\sigma)u^{-\sigma} d\sigma \tag{127}$$

provided

$$\int_{0}^{\infty} \frac{1}{u} |f(u)|^2 du < \infty. \tag{128}$$

This is very similar to what was done to define and develop the Laplace transform in (76) and (77). Equation (126) is known as the Mellin transform of f; (127) in turn is the inverse Mellin transform. By virtue of (128) the Mellin transform is defined on all functions in $L_2([0, \infty], 1/u)$ provided Re $\sigma = 0$.

In (78) and (79) it was shown that Laplace transforms can be defined for functions of exponential growth. By a similar process the definition of

Mellin transforms can be extended to a broader class of functions. Suppose that there is a real number k such that $u^k f(u) \in L_2([0, \infty], 1/u)$. Then we have

$$M(\sigma) = \int_0^\infty f(u)u^{\sigma+k-1} \, du \tag{129}$$

$$u^k f(u) = \frac{1}{2\pi i} \int_{-i\infty}^{i\infty} M(\sigma)u^{-\sigma} \, d\sigma. \tag{130}$$

We now replace $\sigma + k$ by σ in (129) and (130), which is equivalent to working with Re $\sigma = k$, instead of Re $\sigma = 0$. We then obtain the following set of formulas

$$M(\sigma) = \int_0^\infty f(u)u^{\sigma-1} \, du \tag{131}$$

$$f(u) = \frac{1}{2\pi i} \int_{-i\infty+k}^{i\infty+k} M(\sigma)u^{-\sigma} \, d\sigma \tag{132}$$

for those $f(u)$ for which

$$\int_0^\infty u^{2k-1} |f(u)|^2 \, du < \infty. \tag{133}$$

By use of Parseval's formula for the Fourier transforms, namely

$$\int_{-\infty}^\infty |F(f)|^2 \, ds = \int_{-\infty}^\infty |f(x)|^2 \, dx$$

one obtains a corresponding formula for Mellin transforms.

$$\int_0^\infty \frac{1}{u} |f(u)|^2 \, dx = \frac{1}{2\pi i} \int_{-i\infty}^{i\infty} |M(\sigma)|^2 \, d\sigma \tag{134}$$

and more generally if

$$N(\sigma) = \int_0^\infty g(u)u^{\sigma-1} \, du$$

and $g(u)$ also satisfies (128)

$$\int_0^\infty \frac{1}{u} f(u)\overline{g(u)} \, du = \frac{1}{2\pi i} \int_{-i\infty}^{i\infty} M(\sigma)\overline{N(\sigma)} \, d\sigma. \tag{135}$$

More generally if $u^k f(u) \in L_2([0, \infty], 1/u)$ and $u^k g(u)$ as well we obtain

$$\int_0^\infty u^{2k-1} f(u)\overline{g(u)} \, du = \frac{1}{2\pi i} \int_{-i\infty}^{i\infty} M(\sigma+k)\overline{N(\sigma+k)} \, d\sigma. \tag{136}$$

Evidently, many formulas that apply to Fourier transforms have their analogs for Mellin transforms. For example one can define a convolution operation by

$$(f * g)(u) = \int_0^\infty \frac{1}{v} f(v) g\left(\frac{u}{v}\right) dv. \tag{137}$$

It follows that

$$\int_0^\infty (f * g)(u) u^{\sigma-1} du = M(\sigma)N(\sigma) \tag{138}$$

in analogy to (12).

We shall now consider some applications of Mellin transforms. For example they can be used in summing certain infinite series. Let $f(u)$ and $M(\sigma)$ be a suitable function and its Mellin transform. Then

$$f(u) = \frac{1}{2\pi i} \int_{-i\infty+k}^{i\infty+k} M(\sigma) u^{-\sigma} d\sigma.$$

It follows that

$$\sum_{n=1}^\infty f(n) = \frac{1}{2\pi i} \int_{-i\infty+k}^{i\infty+k} M(\sigma) \sum_{n=1}^\infty n^{-\sigma} d\sigma = \frac{1}{2\pi i} \int_{-i\infty+k}^{i\infty+k} M(\sigma)\zeta(\sigma) d\sigma, \tag{139}$$

where the function $\zeta(\sigma)$ is defined by

$$\zeta(\sigma) = \sum_{n=1}^\infty n^{-\sigma}. \tag{140}$$

If the integration in (139) can be carried out, the corresponding sum shall have been obtained.

The zeta function $\zeta(\sigma)$ is defined for σ, such that $\mathrm{Re}\,\sigma > 1$, by the sum in (140). One can show that at $\sigma = 1$ this function has a simple pole with residue 1, so that

$$\lim_{\sigma \to 1} (\sigma - 1)\zeta(\sigma) = 1. \tag{141}$$

Another well-known fact regarding this function is its functional equation

$$\pi^{-\sigma/2}\Gamma\left(\frac{\sigma}{2}\right)\zeta(\sigma) = \pi^{-(1-\sigma)/2}\Gamma\left(\frac{1-\sigma}{2}\right)\zeta(1-\sigma). \tag{142}$$

By means of (142) one can obtain the analytic continuation of $\zeta(\sigma)$ into the entire σ plane. It also follows that $\zeta(\sigma)$ has only one singularity, namely the above-mentioned pole at $\sigma = 1$. We shall now consider an example, where the computations will not be presented in detail, since they require a more specialized knowledge of certain analytic functions.

EXAMPLE 17. Sum the series

$$S = \sum_{n=1}^\infty \frac{\cos nx}{n^2 x^2}$$

by use of (139). One can show that

$$\int_0^\infty \frac{\cos ux}{u^2 x^2} u^{\sigma-1} \, du = \frac{x^{-\sigma} 2^{\sigma-3} \sqrt{\pi} \, \Gamma((\sigma-2)/2)}{\Gamma((3-\sigma)/2)}, \qquad 2 < \mathrm{Re}\ \sigma < 3.$$

Then

$$S = \frac{1}{2\pi i} \int_{-i\infty+k}^{i\infty+k} \frac{x^{-\sigma} 2^{\sigma-3} \sqrt{\pi} \, \Gamma((\sigma-2)/2)\zeta(\sigma)}{\Gamma((3-\sigma)/2)} \, d\sigma, \qquad 2 < k < 3.$$

The above integrand has simple poles at $\sigma = 2, 1, 0$ with residues

$$\frac{\zeta(2)}{x^2}, \qquad -\frac{\pi}{2x}, \qquad \frac{1}{4}$$

respectively. By a standard result

$$\zeta(2) = \sum_{n=1}^\infty \frac{1}{n^2} = \frac{\pi^2}{6}$$

and the fact that one can close the contour on the left in the integral we obtain

$$S = \frac{\pi^2}{6x^2} - \frac{\pi}{2x} + \frac{1}{4}.$$

In dealing with the Fourier sine, Fourier cosine, and Hankel transforms we saw that they were self-reciprocal. That is, the inverse transforms coincide with the transforms. In particular the Fourier sine and cosine transforms satisfy relations like the following:

$$T(f) = \int_0^\infty f(x)K(xy) \, dx \tag{143}$$

$$f(x) = \int_0^\infty T(f)K(xy) \, dy. \tag{144}$$

More generally one can inquire, whether for a given kernel $K(xy)$ there exists an associated kernel $H(xy)$ such that if

$$T(f) = \int_0^\infty f(x)K(xy) \, dx \tag{145}$$

then

$$f(x) = \int_0^\infty T(f)H(xy) \, dy. \tag{146}$$

The Mellin transform enables one to find simple necessary conditions on two functions $K(u)$ and $H(u)$ so that (145) and (146) are satisfied.

Let

$$L(\sigma) = \int_0^\infty K(u)u^{\sigma-1}\,du$$

$$M(\sigma) = \int_0^\infty H(u)u^{\sigma-1}\,du$$

and by taking Mellin transforms of (145) and (146) find

$$\int_0^\infty T(f)y^{\sigma-1}\,dy = \int_0^\infty f(x)\left(\int_0^\infty K(xy)y^{\sigma-1}\,dy\right)dx$$

$$= \int_0^\infty f(x)x^{-\sigma}\,dx \int_0^\infty K(u)u^{\sigma-1}\,du$$

$$= L(\sigma)\int_0^\infty f(x)x^{-\sigma}\,dx$$

and

$$\int_0^\infty f(x)x^{-\sigma}\,dx = \int_0^\infty T(f)\left(\int_0^\infty H(xy)x^{-\sigma}\,dx\right)dy$$

$$= \int_0^\infty T(f)y^{\sigma-1}\,dy \int_0^\infty H(u)u^{-\sigma}\,du$$

$$= M(1-\sigma)\int_0^\infty T(f)y^{\sigma-1}\,dy.$$

In order for the last two equations to hold, it is necessary that

$$L(\sigma)M(1-\sigma) = 1. \tag{147}$$

In particular if $K(u) = H(u)$, (147) reduces to

$$L(\sigma)L(1-\sigma) = 1. \tag{148}$$

In the case of the Fourier sine transform we have

$$K(u) = \sqrt{\frac{2}{\pi}}\sin u$$

so that

$$\sqrt{}\ \int^\infty \qquad\qquad \sqrt{}\ \ )\sin\frac{\pi\sigma}{2}\ .$$

$$\text{in}\ \frac{\pi}{2}(1-\sigma)$$

$$= \frac{2}{\pi}\cdot\frac{\pi}{\sin\pi\sigma}\sin\frac{\pi}{2}\,\sigma\cos\frac{\pi\sigma}{2} = 1,$$

where we used the formula

$$\Gamma(\sigma)\Gamma(1 - \sigma) = \frac{\pi}{\sin \pi\sigma}.$$

It follows that the Fourier sine transform is self-reciprocal. Similarly for the Fourier cosine transform we have

$$L(\sigma) = \sqrt{\frac{2}{\pi}} \int_0^\infty \cos uu^{\sigma-1} \, du = \sqrt{\frac{2}{\pi}} \, \Gamma(\sigma) \cos \frac{\pi\sigma}{2}$$

and again

$$L(\sigma)L(1 - \sigma) = 1.$$

The Hankel transform is not of the form (143) or (144), but by a slight modification can be brought into this form. We merely rewrite (117) and (119) in the form

$$\sqrt{\rho} \, \mathscr{H}(f) = \int_0^\infty \sqrt{r} f(r)\sqrt{r\rho} \, J_n(r\rho) \, dr \tag{149}$$

$$\sqrt{r} f(r) = \int_0^\infty \sqrt{\rho} \, \mathscr{H}(f)\sqrt{r\rho} \, J_n(r\rho) \, d\rho \tag{150}$$

where

$$K(u) = \sqrt{u} \, J_n(u).$$

Now

$$L(\sigma) = \int_0^\infty \sqrt{u} \, J_n(u)u^{\sigma-1} \, du = \frac{2^{\sigma-\frac{1}{2}}\Gamma[(n + \sigma)/2 + \frac{1}{4}]}{\Gamma[(n - \sigma)/2 + \frac{3}{4}]}$$

so that

$$L(\sigma)L(1 - \sigma) = \frac{2^{\sigma-\frac{1}{2}}\Gamma[(n + \sigma)/2 + \frac{1}{4}]}{\Gamma[(n - \sigma)/2 + \frac{3}{4}]} \cdot \frac{2^{\frac{1}{2}-\sigma}\Gamma[(n - \sigma)/2 + \frac{3}{4}]}{\Gamma[(n + \sigma)/2 + \frac{1}{4}]} = 1$$

and again (148) is satisfied.

A discussion of the sufficiency of conditions (147) and (148) is considerably more complicated and we refer to the literature for additional information (see "Theory of Fourier Integrals" by E. C. Titchmarsh).

7. THE PROJECTION METHOD*

In Example 7 of Section 2 of this chapter the Hilbert transform was defined and its chief properties were discussed. By definition

$$H\phi = \frac{1}{\pi} \int_{-\infty}^{\infty} {}^* \frac{\phi(y)}{x - y} \, dy \tag{151}$$

* This method is due to Torsten Carleman. It was first published in 1922. See *Sur la résolution de certaines équations intégrales*, Arkiv För Matematik, Astronomi Och Fysik, Band 16, No. 26. It was reprinted in *Edition Compléte des Articles de Torsten Carleman*, Malmo 1960, pp. 141–159.

and it was shown that the operator H mapped the space $L_2[-\infty, \infty]$ into itself. Furthermore,

$$H^2\phi = -\phi \tag{152}$$

and

$$F(H\phi) = i \operatorname{sgn} s F(\phi) \tag{153}$$

from which it follows that H has an inverse, namely $-H$ and that $\|H\| = 1$.

Given any $\phi \in L_2[-\infty, \infty]$ we can define two new functions $\phi_+(x)$ and $\phi_-(x)$ as follows.

$$\phi_+(x) = \tfrac{1}{2}[\phi + iH\phi] \tag{154}$$

$$\phi_-(x) = \tfrac{1}{2}[\phi - iH\phi] \tag{155}$$

Taking Fourier transforms of the above we have

$$\begin{aligned} F(\phi_+) = \tfrac{1}{2}[F(\phi) + iF(H\phi)] &= \tfrac{1}{2}[F(\phi) - \operatorname{sgn} s F(\phi)] \\ &= 0, \quad s > 0 \\ &= F(\phi), \quad s < 0 \end{aligned} \tag{156}$$

and similarly

$$\begin{aligned} F(\phi_-) &= F(\phi), \quad s > 0 \\ &= 0, \quad s < 0. \end{aligned} \tag{157}$$

Use of the last four equations shows that

$$\phi = \phi_+ + \phi_- \tag{158}$$

$$F(\phi) = F(\phi_+) + F(\phi_-). \tag{159}$$

We see therefore that the effect of decomposing ϕ into ϕ_+ and ϕ_- in (158) when transferred to the Fourier domain is to split $F(\phi)$ into two functions, one vanishing for all $s > 0$ and the other vanishing for all $s < 0$. The behavior at $s = 0$ is unimportant since in $L_2[-\infty, \infty]$ all functions can be modified at a single point without changing their value as elements in a Hilbert space.

We can now decompose $L_2[-\infty, \infty]$ into two subspaces, namely L_2^+ and L_2^-, defined by

$$L_2^{\pm} = \{\phi \in L_2[-\infty, \infty] \mid F(\phi) = 0, s \gtrless 0\} \tag{159}$$

Evidently, for $\phi \in L_2^{\pm}$

$$H\phi = {}^-_+ i\phi. \tag{160}$$

We can view ϕ_+ as the projection of ϕ on L_2^+ and similarly ϕ_- is the projection on L_2^-. Equations 154 and 155 show that this decomposition is unique in the Hilbert space. The following relations hold

$$L_2[-\infty, \infty] = L_2^+ \cup L_2^- \tag{161}$$

$$\{0\} = L_2^+ \cap L_2^-. \tag{162}$$

(162) is a consequence of the fact that if $\phi \in L_2^+ \cap L_2^-$, $F(\phi) = 0$ for all s so that $\phi = 0$ (or a null function which is equivalent to 0 in $L_2[-\infty, \infty]$.)

EXAMPLE 18. Solve

$$\phi(x) - \frac{\lambda}{\pi} \int_{-\infty}^{\infty *} \frac{\phi(y)^{\cdot}}{x - y} \, dy = f(x) \tag{163}$$

for $f(x) \in L_2[-\infty, \infty]$. By means of (154) and (155) we can rewrite the above as

$$\phi_+ + \phi_- + i\lambda(\phi_+ - \phi_-) = f_+ + f_-$$

or equivalently as

$$(1 + i\lambda)\phi_+ - f_+ = -(1 - i\lambda)\phi_- + f_-.$$

By virtue of (162) both sides of the above must vanish so that

$$\phi_+ = \frac{f_+}{1 + i\lambda}$$

$$\phi_- = \frac{f_-}{1 - i\lambda}$$

and

$$\phi = \phi_+ + \phi_- = \frac{(1 - i\lambda)f_+ + (1 + i\lambda)f_-}{1 + \lambda^2}$$

$$\phi = \frac{f + \lambda Hf}{1 + \lambda^2} \tag{164}$$

(163) has solutions for all $\lambda \neq \pm i$.

If $\lambda = -i$ we require that $f_- = 0$ so that

$$\phi_+ = \tfrac{1}{2}f_+$$

and ϕ_- is arbitrary.

The spaces $L_2^{\pm}$ can be characterized in another way. Suppose $\phi \in L_2^+$. Then

$$F(\phi) = 0 \qquad s > 0$$

and

$$\phi = F^*F(\phi) = \frac{1}{\sqrt{2\pi}} \int_{-\infty}^{0} F(\phi)e^{-isx} \, ds. \tag{165}$$

We now consider the function

$$\phi(z) = \frac{1}{\sqrt{2\pi}} \int_{-\infty}^{0} F(\phi)e^{-isz} \, ds. \tag{166}$$

where z is a complex variable. For $y = \text{Im } z > 0$, $\phi(z)$ is an analytic function, whose boundary value on $y = 0$ is given by (165). In general it may have no analytic continuation into the lower half plane since (166) need not exist at all for $\text{Im } z < 0$.

Similarly if $\phi \in L_2^-$

$$\phi(z) = \frac{1}{\sqrt{2\pi}} \int_0^\infty F(\phi)e^{-isz}\,ds \tag{167}$$

defines a function analytic in the lower half plane.

For subsequent applications the following theorem will prove to be necessary. It shows that if a function $\phi \in L_2^+$ is multiplied by a suitable function either the product will still be in L_2^+, or by a suitable modification a function in L_2^+ will result.

THEOREM 6. Let $\phi \in L_2^+$, and let $a(z)$ where $z = x + iy$ be continuous for $y \geq 0$ and analytic for $y > 0$, except possibly for a pole of order n at $z = \zeta$. We suppose that $a(z)$ is bounded, except possibly at $z = \zeta$; that is $a(z)$ is bounded in the upper half plane outside of every neighborhood of $z = \zeta$. Then for suitable $\alpha_1, \alpha_2, \ldots, \alpha_n$

$$a(x)\phi(x) - \sum_{k=1}^n \frac{\alpha_k}{(x - \zeta)^k} \in L_2^+.$$

If $a(z)$ has no pole $a(x)\phi(x) \in L_2^+$.

PROOF. Let

$$F(\phi(x)) = \Phi(s)$$

$$F(a(x)e^{-\epsilon|x|}) = A_\epsilon(s), \qquad \epsilon > 0$$

so that by the convolution theorem*

$$F((a(x)\phi(x)e^{-\epsilon|x|}) = \frac{1}{\sqrt{2\pi}} \int_{-\infty}^\infty \Phi(\sigma)A_\epsilon(s - \sigma)\,d\sigma$$

$$= \frac{1}{\sqrt{2\pi}} \int_{-\infty}^0 \Phi(\sigma)A_\epsilon(s - \sigma)\,d\sigma \tag{168}$$

since $\Phi(s) = 0$ for $s > 0$. Eventually $\epsilon \to 0$, but for the moment it is necessary that $\epsilon > 0$ to be sure that certain integrals converge.

Now we shall examine the structure of

$$A_\epsilon(s) = \frac{1}{\sqrt{2\pi}} \int_{-\infty}^\infty a(x)e^{-\epsilon|x|+isx}\,dx.$$

We shall assume that $s > 0$, which is the case of interest and furthermore we shall assume that $\text{Re}\,\zeta > 0$. The case where $\text{Re}\,\zeta \leq 0$ can be treated with obvious modifications. By a contour integration over the second quadrant,

---

* Here we use the convolution theorem in the form $F(F^*(f)F^*(g)) = \sqrt{2\pi}\,(F_*g)$ with $= A_\epsilon(s), g = \Phi(s)$.

where $a(z)$ has no singularities we have

$$\int_{-\infty}^{0} a(x)e^{\epsilon x + isx} \, dx = -i \int_{0}^{\infty} a(iy)e^{i\epsilon y - sy} \, dy,$$

since the contribution of the integral over the infinite quarter circle vanishes. The integral over $(0, \infty)$ is evaluated by performing a contour integration over the first quadrant.

$$\int_{0}^{\infty} a(x)e^{-\epsilon x + isx} \, dx = R + i \int_{0}^{\infty} a(iy)e^{i\epsilon y - sy} \, dy.$$

R is the residue contribution from the pole at $z = \zeta$. If $a(z)$ is expressed as

$$a(z) = \sum_{k=1}^{n} \frac{a_k}{(z - \zeta)^k} + b(z)$$

where $b(z)$ is analytic at $z = \zeta$, one can easily show that

$$R = \left(\sum_{1}^{n} r_k (is - \epsilon)^{k-1} \right) e^{(is - \epsilon)\zeta},$$

for suitable constants $r_1, r_2, \ldots, r_n$. It follows that

$$\int_{-\infty}^{\infty} a(x)e^{-\epsilon|x| + isx} \, dx = R + 2 \int_{0}^{\infty} a(iy) \sin \epsilon y e^{-sy} \, dy. \qquad (169)$$

Since $|a(iy)| \leq M$ we can estimate the last integral to obtain

$$\left| \int_{0}^{\infty} a(iy) \sin \epsilon y e^{-sy} \, dy \right| \leq M\epsilon \int_{0}^{\infty} y e^{-sy} \, dy = \frac{M\epsilon}{s^2}.$$

We now return to (168) and let $\epsilon \to 0$ in (169) so that

$$F(a(x)\phi(x)) = \frac{1}{\sqrt{2\pi}} \int_{-\infty}^{0} \Phi(\sigma) \left(\sum_{1}^{n} r_k i^{k-1} (s - \sigma)^{k-1} \right) e^{i(s - \sigma)\zeta} \, d\sigma, \qquad s > 0. \quad (170)$$

Next we consider the function

$$b(z) = \sum_{k=1}^{n} \frac{\alpha_k}{(z - \zeta)^k}. \qquad (171)$$

$$F(b) = \frac{1}{\sqrt{2\pi}} \int_{-\infty}^{\infty} b(x)e^{isx} \, dx = \sqrt{2\pi} \, i \sum_{1}^{n} \frac{\alpha_k (is)^{k-1} e^{is\zeta}}{(k-1)!}, \qquad s > 0. \quad (172)$$

For $s > 0$, (170) is of the form $e^{is\zeta} p(s)$, where $p(s)$ is a polynomial of degree $n - 1$. More precisely we have

$$F(a(x)\phi(x)) = e^{is\zeta} \sum_{j=0}^{n-1} \frac{1}{j!} s^j \sum_{k=j+1}^{n} \frac{r_k i^{k-1}(k-1)!}{\sqrt{2\pi}\,(k-1-j)!} \int_{-\infty}^{0} \Phi(\sigma)(-\sigma)^{k-1-j} e^{-i\sigma\zeta} \, d\sigma.$$

A comparison of the above with (172) shows that we can choose $\alpha_1, \alpha_2, \ldots, \alpha_n$ in (171) so that

$$F(a(x)\phi(x) - b(x)) = 0, \qquad s > 0$$

and as a result

$$a(x)\phi(x) - b(x) \in L_2^+.$$

In particular if $a(z)$ has no pole in the upper half plane $a(x)\phi(x) \in L_2^+$. ∎

There is obviously a companion theorem for functions in L_2^- that are multiplied by functions having a pole in the lower half plane.

EXAMPLE 19. Solve

$$\phi(x) - \frac{\lambda}{\pi} \int_0^{\infty *} \frac{\phi(y)}{x - y} \, dy = f(x), \qquad f(x) \in L_2[0, \infty]. \tag{173}$$

We shall assume that λ^2 does not belong to the interval $(-\infty, -1]$. The reasons for this assumption will emerge in the development. We shall also assume that

$$\int_1^\infty x \, |f|^2 \, dx < \infty \qquad \text{and} \qquad \int_0^1 \frac{1}{x} |f|^2 \, dx < \infty. \tag{174}$$

These conditions are not absolutely necessary, but the most general that enable us to solve (173) for all λ, subject to the above restrictions.

In order to be able to work in $L_2[-\infty, \infty]$ and to use the projection method we shall extend $f(x)$ and $\phi(x)$ to $(-\infty, 0)$ by

$$f(x) = \phi(x) = 0, \qquad x < 0.$$

Now we can rewrite (173) as follows

$$\phi_+ + \phi_- + i\lambda(\phi_+ - \phi_-) = f, \qquad x > 0$$
$$\phi_+ + \phi_- = 0, \qquad x < 0$$

so that

$$\phi_- = -\frac{1 + \lambda i}{1 - \lambda i} \phi_+ + \frac{f}{1 - \lambda i}, \qquad x > 0$$

$$= -\phi_+, \qquad x < 0. \tag{175}$$

By defining the function

$$p(x) = \frac{1 + \lambda i}{1 - \lambda i}, \qquad x > 0$$

$$= 1, \qquad x < 0$$

(175) can be put into the more compact form

$$\phi_- = -p(x)\phi_+ + \frac{f}{1 - \lambda i}, \qquad -\infty < x < \infty. \tag{176}$$

In order to apply the projection method we should like to separate the terms in (176) into two groups, one in L_2^+, and the other in L_2^-. To accomplish this we proceed as follows. Let

$$\rho = \frac{1}{2\pi i} \log \frac{1 - \lambda i}{1 + \lambda i} \tag{177}$$

and by the restrictions imposed on λ we have

$$|\text{Re } \rho| < \tfrac{1}{2} \tag{178}$$

and

$$e^{-2\pi i\rho} = \frac{1 + \lambda i}{1 - \lambda i}.$$

We now consider the function

$$q(z) = (e^{-\pi i}z)^\rho$$

where for $z = x < 0$, $q(z) = (-x)^\rho = e^{\rho \ln(-x)}$, with $\ln (-x)$ real. Let

$$q^+(x) = \lim_{y \to +0} q(z)$$

and

$$q^-(x) = \lim_{y \to -0} q(z).$$

Then we have

$$\begin{aligned} q^+(x) &= e^{-\pi i\rho}x^\rho, && x > 0 \\ &= (-x)^\rho, && x < 0 \end{aligned} \tag{179}$$

and

$$\begin{aligned} q^-(x) &= e^{\pi i\rho}x^\rho, && x > 0 \\ &= (-x)^\rho, && x < 0 \end{aligned} \tag{180}$$

so that

$$p(x) = \frac{q^+(x)}{q^-(x)}. \tag{181}$$

We shall defer to a later section a discussion of how such functions can be found systematically. But using (181), (176) can be rewritten as

$$q^-\phi_- = -q^+\phi_+ + \frac{fq^-}{1 - \lambda i} \tag{182}$$

By virtue of (174) and (178) the term fq^- is certainly in $L_2[-\infty, \infty]$. If λ were to be kept fixed we could replace (174) by the less restrictive condition

$$\int_0^\infty x^{2\rho} |f|^2 \, dx < \infty. \tag{183}$$

Since fq^- is in $L_2[-\infty, \infty]$ it can be decomposed into two parts in L_2^+ and L_2^- respectively. Then (182) becomes

$$q^-\phi_- - \frac{(fq^-)_-}{1 - \lambda i} = -\left[q^+\phi_+ - \frac{(fq^-)_+}{1 - \lambda i}\right]. \tag{184}$$

If $q^\pm$ were both bounded we could apply theorem 6 to (184). It would follow that the left side would belong to L_2^- and the right side to L_2^+ and both would then vanish. But (179) and (180) show that $q^\pm$ become unbounded at either $x = 0$ or $x = \infty$, depending on the sign of Re ρ. We shall now assume that Re $\rho > 0$ (a similar analysis applies to Re $\rho < 0$) in which case $q^\pm$ become unbounded for large x. We now divide both sides of (184) by $x - i$, so that

$$\frac{q^-}{x - i}\phi_- - \frac{(fq^-)_-}{(x - i)(1 - \lambda i)} = -\left[\frac{q^+}{x - i}\phi_+ - \frac{(fq^-)_+}{(x - i)(1 - \lambda i)}\right]. \tag{185}$$

$\phi_- \in L_2^-$ and $q^-/(x - i)$ is bounded for all real x and analytic in the lower half plane. By theorem 6 it follows that $q^-\phi_-/(x - i) \in L_2^-$. Similarly $(fq^-)_- \in L_2^-$ and $1/(x - i)$ is bounded for real x and analytic in the lower half plane. It follows therefore that the left side of (185) is in L_2^-.

The right side of (185) is not in L_2^+, since the coefficients of ϕ_+ and $(fq^-)_+$ are bounded for real x, but have simple poles at $x = i$. By theorem 6 however, we can subtract a term of the type $\alpha/(x - i)$, for a suitable α so that the right side will be in L_α^+. Then

$$\frac{q^-}{x - i}\phi_- - \frac{(fq^-)_-}{(x - i)(1 - \lambda i)} - \frac{\alpha}{x - i}$$

$$= -\left[\frac{q^+}{x - i}\phi_+ - \frac{(fq^-)_+}{(x - i)(1 - \lambda i)}\right] - \frac{\alpha}{x - i}. \tag{186}$$

Now the right side of (186) is in L_2^+. But

$$F\left(\frac{1}{x - i}\right) = \frac{1}{\sqrt{2\pi}}\int_{-\infty}^{\infty} \frac{e^{isx}}{x - 1}dx = \sqrt{2\pi}\,ie^{-s}, \quad s > 0$$
$$= 0, \quad s < 0$$

so that the left side is in L_2^-. It follows that both sides of (186) must vanish identically and that

$$\phi_+ = \frac{(fq^-)_+}{q^+(1 - \lambda i)} - \frac{\alpha}{q^+}$$

$$\phi_- = \frac{(fq^-)_-}{q^-(1 - \lambda i)} + \frac{\alpha}{q^-}.$$

Finally we can add the above, use the explicit representation of $q^\pm$ for $x > 0$, and use (154), (155) to obtain

$$\phi = \frac{f}{1 + \lambda^2} + \frac{\lambda x^{-\rho}}{1 + \lambda^2} Hfx^\rho - 2\pi i\alpha \sin \pi\rho x^{-\rho} \tag{187}$$

We still have to determine α. To do so we consider the left side of (186). For the first term we have the obvious estimate

$$\left| \int_1^\infty \frac{q^-}{x - i} \phi_- \, dx \right| \leq K \int_1^\infty x^{\rho-1} |\phi_-| \, dx$$

$$\leq K \left\{ \int_1^\infty x^{2\rho-1} \, dx \int_1^\infty |\phi_-|^2 \, dx \right\}^{1/2} < \infty$$

and similarly

$$\left| \int_1^\infty \frac{(fq^-)_-}{x - i} \, dx \right| < \infty.$$

But since the left side of (186) vanishes identically we also require that

$$\left| \int_1^\infty \frac{\alpha}{x - i} \, dx \right| < \infty.$$

The above integral diverges for all $\alpha \neq 0$. To satisfy the above it is necessary that $\alpha = 0$. Finally our solution of (173) is given by

$$\phi(x) = \frac{f(x)}{1 + \lambda^2} + \frac{\lambda x^{-\rho}}{(1 + \lambda^2)\pi} \int_0^{\infty^*} \frac{f(y)y^\rho}{x - y} \, dy. \tag{188}$$

EXAMPLE 20. Solve

$$\frac{1}{\pi} \int_0^{\infty^*} \frac{\phi(y)}{x - y} \, dy = f(x). \tag{189}$$

The problem was already solved in Examples 8 and 11 by other methods. If we replace f by $-\lambda f$ in (173) and formally let $\lambda \to \infty$, (173) reduces to (189). We can perform the same operation in (188) and thus obtain as our solution

$$\phi(x) = \frac{1}{\pi} \int_0^{\infty^*} \sqrt{\frac{y}{\lambda}} \frac{f(y)}{x - y} \, dy. \tag{190}$$

Note that $\lim_{\lambda \to \infty} \rho = \frac{1}{2}$.

Next we shall return to the question of how a function like $p(x)$ can be represented as shown in (181). To do so we require the following preliminary theorem.

THEOREM 7. Let $\phi(\tau) \in L_2[-\infty, \infty]$ and consider the function

$$q(z) = \frac{1}{\pi i} \int_{-\infty}^{\infty} \frac{\phi(\tau)\, d\tau}{\tau - z} \tag{191}$$

where $y = \operatorname{Im} z \neq 0$. $q(z)$ is a bounded analytic function for $y > 0$ and $y < 0$. Its boundary values on $y = 0$ are given by

$$\lim_{y \to \pm 0} q(z) = q^{\pm}(x) = \pm\phi(x) + iH\phi(x). \tag{192}$$

PROOF. If $y \neq 0$

$$|q(z)| \leq \frac{1}{\pi}\left\{\int_{-\infty}^{\infty} |\phi(\tau)|^2\, d\tau \int_{-\infty}^{\infty} \frac{d\tau}{(\tau - x)^2 + y^2}\right\}^{1/2} < \infty$$

so that $q(z)$ is bounded. A similar computation shows that $q(z)$ is a continuous function of z. Now let C be a closed curve in the upper half plane. Then

$$\int_C q(z)\, dz = \frac{1}{\pi i} \int_{-\infty}^{\infty} \phi(\tau)\, d\tau \int_C \frac{dz}{\tau - z} = 0$$

since τ is outside of C. By Morera's theorem $q(z)$ must be analytic.

To determine the boundary values we shall take the Fourier transform of $q(z)$, viewing it as a function of x. Then

$$F(q(z)) = \frac{1}{\sqrt{2\pi}} \int_{-\infty}^{\infty} q(x + iy)e^{isx}\, dx = \frac{1}{\sqrt{2\pi}} \cdot \frac{1}{\pi i} \int_{-\infty}^{\infty} \phi(\tau)\, d\tau \int_{-\infty}^{\infty} \frac{e^{isx}}{\tau - z}\, dx.$$

If $s > 0$ we can close the contour over the upper half plane, and for $s < 0$ over the lower half plane. For $y > 0$ we find

$$F(q(z)) = 0, \qquad s > 0$$

$$= \frac{2}{\sqrt{2\pi}} e^{sy} \int_{-\infty}^{\infty} \phi(\tau)e^{is\tau}\, d\tau, \qquad s < 0.$$

Letting $y \to 0_+$ we finally have

$$F(q^+(x)) = 0, \qquad s > 0$$
$$= 2F(\phi), \qquad s < 0.$$

Similarly

$$F(q^-(x)) = -2F(\phi), \qquad s > 0$$
$$= 0, \qquad s < 0,$$

so that

$$F(q^+ + q^-) = -2 \operatorname{sgn} s F(\phi)$$
$$F(q^+ - q^-) = 2F(\phi).$$

Recalling that
$$F(H\phi) = i \operatorname{sgn} s F(\phi)$$
we finally have
$$q^+ + q^- = 2iH\phi \qquad (193)$$
$$q^+ - q^- = \cdot 2\phi \qquad (194)$$

and these are equivalent to (192). ■

With this theorem it is possible to present a method for finding the decomposition of (181).

THEOREM 8. Let $p(x)$ be such that

$$\lim_{|x| \to \infty} p(x) = 1$$

and define $\log p(x)$ such that

$$\lim_{x \to \infty} \log p(x) = 0.$$

Suppose furthermore that*

$$\lim_{x \to -\infty} \log p(x) = 0$$

and that $\log p(x) \in L_2[-\infty, \infty]$. Then there exists a bounded function $q(z)$ analytic for $\operatorname{Im} z \neq 0$ such that

$$p(x) = \frac{q^+(x)}{q^-(x)} \qquad (195)$$

where

$$q^\pm(x) = \lim_{y \to \pm 0} q(z).$$

PROOF.   We apply theorem 7 with $\phi(x)$ replaced by $\frac{1}{2} \log p(x)$. Then we can define q^+ and q^- by

$$\log p = \log q^+ - \log q^-$$
$$iH \log p = \log q^+ + \log q^-$$

and

$$\log q(z) = \frac{1}{2\pi i} \int_{-\infty}^{\infty} \frac{\log p(\tau)}{\tau - z} \, d\tau. \qquad (196)$$

By the assumptions on $\log p(x)$, (196) defines a suitable function $\log q(z)$ with boundary values

$$q^\pm = \exp \tfrac{1}{2}\{iH \log p \pm \log p\}$$

---

* Since $p(x) \to 1$ as $|x| \to \infty$ we can certainly select that branch of the logarithm function for which $\lim_{x \to \infty} \log p(x) = 0$. Once, however, this branch is selected, the best we can say about $\lim_{x \to -\infty} \log p(x)$ is that it must be $2n\pi i$ for some integer n. As a hypothesis in this theorem we require that $n = 0$.

and it follows that

$$\frac{q^+}{q^-} = p(x). \quad \blacksquare$$

EXAMPLE 20. Define $p(x)$ by

$$p(x) = a, \qquad |x| < 1$$
$$= 1, \qquad |x| > 1$$

where a is not on the negative real axis in the complex plane; that is, $a \notin (-\infty, 0]$. Then $p(x)$ satisfies the conditions of theorem 8. Now

$$\log q(z) = \frac{1}{2\pi i} \int_{-\infty}^{\infty} \frac{\log p(\tau)}{\tau - z} d\tau = \frac{\log a}{2\pi i} \int_{-1}^{1} \frac{d\tau}{\tau - z} = \frac{\log a}{2\pi i} \log \left(\frac{z-1}{z+1}\right)$$

and for $|x| > 1$ we find

$$q^\pm = \left(\frac{x-1}{x+1}\right)^{\log a / 2\pi i}.$$

For $|x| < 1$ we obtain

$$q^\pm = \left(\frac{1-x}{1+x}\right)^{\log a / 2\pi i} a^{\pm 1/2}. \tag{197}$$

EXAMPLE 21. Solve

$$\phi(x) - \frac{\lambda}{\pi} \int_{-1}^{1*} \frac{\phi(y)}{x - y} dy = f(x), \qquad f(x) \in L_2[-1, 1] \tag{198}$$

where $\lambda^2 \notin (-\infty, -1]$, and

$$\int_{-1}^{1} \frac{|f(x)|^2}{1 - x^2} dx < \infty.$$

(198) is posed as a problem on $L_2[-1, 1]$, but we shall extend it to the space $L_2[-\infty, \infty]$, by defining

$$\phi(x) = f(x) = 0, \qquad |x| > 1.$$

Now, as in Example 19, we rewrite (198) as

$$\phi_+ + \phi_- + \lambda i(\phi_+ - \phi_-) = f, \qquad |x| < 1.$$
$$\phi_+ + \phi_- = 0, \qquad |x| > 1.$$

These can be combined into a single equation

$$\phi_- = -p(x)\phi_+ + \frac{f}{1 - \lambda i} \tag{199}$$

where

$$p(x) = \frac{1 + \lambda i}{1 - \lambda i}, \qquad |x| < 1$$

$$= 1, \qquad |x| > 1.$$

We can now apply theorem 8 and the results of Example 21, to write $p(x)$ as

$$p(x) = \frac{q^+}{q^-}$$

$$q^\pm = \left(\frac{x-1}{x+1}\right)^\rho, \qquad |x| > 1 \qquad (200)$$

$$= \left(\frac{1-x}{1+x}\right)^\rho e^{\pm i\pi\rho}, \qquad |x| < 1$$

and where

$$\rho = \frac{1}{2\pi i} \log \frac{1 + \lambda i}{1 - \lambda i}.$$

By the hypotheses on λ, $|\mathrm{Re}\,\rho| < \frac{1}{2}$. (199) can now be rewritten as follows

$$q^-\phi_- - \frac{(fq^-)_-}{1 - \lambda i} = -\left[q^+\phi_+ - \frac{(fq^-)_+}{1 - \lambda i}\right]. \qquad (201)$$

We shall now assume that $\mathrm{Re}\,\rho \geq 0$. The case $\mathrm{Re}\,\rho < 0$ can be treated in a similar manner. $q^\pm$ become unbounded at $x = -1$, but in order to apply theorem 6 we require boundedness. This can be accomplished by multiplying (201) by $(x + 1)/(x - i)$ for example. Equation (201) is then replaced by

$$\frac{x + 1}{x - i} q^-\phi_- - \frac{[(x + 1)/(x - i)](fq^-)_-}{1 - \lambda i}$$

$$= -\left[\frac{x + 1}{x - i} q^+\phi_+ - \frac{[(x + 1)/(x - i)](fq^-)_+}{1 - \lambda i}\right]. \qquad (202)$$

Theorem 6 now tells us that the left side is in L_2^-, but the right side is not in L_2^+, because of the pole at $x = i$. We can, however, subtract a term $\alpha/(x - i)$ from both sides of (202) such that the right side is in L_2^+. Since the term $\alpha/(x - i)$ is in L_2^- the left side will still be in L_2^-.

$$\frac{x + 1}{x - i} q^-\phi_- - \frac{[(x + 1)/(x - i)](fq^-)_-}{1 - \lambda i} - \frac{\alpha}{x - i}$$

$$= -\left[\frac{x + 1}{x - i} q^+\phi_+ - \frac{[(x + 1)/(x - i)](fq^-)_+}{1 - \lambda i} + \frac{\alpha}{x - i}\right]. \qquad (203)$$

$L_2^+ \cap L_2^- = \{0\}$ so that both sides of (203) vanish, enabling us to find $\phi^{\pm}$ in terms of α.

$$q^-\phi_- = \frac{(fq^-)_-}{1 - \lambda i} + \frac{\alpha}{x + 1}.$$

$$q^+\phi_+ = \frac{(fq^-)_+}{1 - \lambda i} - \frac{\alpha}{x + 1}$$

One can easily verify that

$$\int_{-1}^{1} \left| q^-\phi_- - \frac{(fq^-)_-}{1 - \lambda i} \right| dx < \infty.$$

The first of the above equations therefore implies that

$$\int_{-1}^{1} \left| \frac{\alpha}{x + 1} \right| dx < \infty$$

so that $\alpha = 0$. Now we finally can solve for ϕ.

$$\phi = \frac{f}{1 + \lambda^2} + \frac{2\lambda}{\pi(1 + \lambda^2)} \left(\frac{1 + x}{1 - x}\right)^\rho \int_{-1}^{1^*} \frac{f(y)}{x - y} \left(\frac{1 - y}{1 + y}\right)^\rho dy. \qquad (204)$$

EXAMPLE 22. Solve

$$\frac{1}{\pi} \int_{-1}^{1^*} \frac{\phi(y)}{x - y} dy = f(x), \qquad f(x) \in L_2[-1, 1] \qquad (205)$$

under the same restrictions on λ and $f(x)$ as in Example 21. This problem was treated in Example 13 by another method. If $f(x)$ is replaced by $-\lambda f(x)$ in (198) and $\lambda \to \infty$, (205) is obtained. We perform the same limiting process in (204). Note that $\lim_{\lambda \to \infty} \rho = \frac{1}{2}$. Then

$$\phi = -\frac{1}{\pi} \sqrt{\frac{1 + x}{1 - x}} \int_{-1}^{1^*} \frac{f(y)}{x - y} \sqrt{\frac{1 - y}{1 + y}} dy. \qquad (206)$$

8. THE WIENER–HOPF TECHNIQUE I*

In a study of the distribution of temperature in a stellar atmosphere in a radioactive equilibrium the following integral equation arises.

$$\phi(x) - \frac{1}{2} \int_0^\infty \left\{ \int_{|x-y|}^\infty \frac{e^{-t}}{t} dt \right\} \phi(y) \, dy = 0 \qquad (207)$$

* The origin of this work may be found in N. Wiener and E. Hopf, *Über eine Klasse singularer Integralgleichungen*, Sitzungsberichte der Preussischen Akademie, Mathematisch-Physikalische Klasse, 1931, p. 696.

The above is a special case of the more general equation

$$\phi(x) - \int_0^\infty K(x - y)\phi(y)\, dy = f(x), \qquad f(x) \in L_2[0, \infty] \qquad (208)$$

Were the integration extended over the interval $(-\infty, \infty)$, (208) could be solved by use of Fourier transforms. Such cases were treated in Section 2 of this chapter. In order to treat Eq. 207, and more generally (208), N. Wiener and E. Hopf discovered a method to which this section will be devoted. The projection method, discussed in the previous section, will be used to solve equations of the type (208), that have become known as Wiener–Hopf equations.

Equation 208 can be extended to $(-\infty, \infty)$ by letting

$$\phi(x) = f(x) = 0, \qquad x < 0. \qquad (209)$$

Now we define a new function $g(x)$ by

$$\begin{aligned} g(x) &= -\int_0^\infty K(x - y)\phi(y)\, dy, \qquad &x < 0 \\ &= 0, \qquad &x > 0 \end{aligned} \qquad (210)$$

so that (208) can be rewritten in the form

$$\phi(x) - \int_{-\infty}^\infty K(x - y)\phi(y)\, dy = f(x) + g(x). \qquad (211)$$

For $x > 0$ (211) reduces to (208) and for $x < 0$, it reduces to an identity by use of (210). $g(x)$ is of course not known, since it depends on $\phi(x)$, the solution of (208). However, we can formally apply Fourier transforms to (211). Recognizing the integral in (211) as a convolution, we obtain

$$F(\phi) - \sqrt{2\pi}\, F(K)F(\phi) = F(f) + F(g). \qquad (212)$$

In order to derive (212) we assume that all functions that we are dealing with do belong to $L_2[-\infty, \infty]$.

Rather than work with (212), it will be more convenient to work with the inverse of the Fourier transform. By applying F^* to (211) we obtain

$$F^*(\phi) - \sqrt{2\pi}\, F^*(K)F^*(\phi) = F^*(f) + F^*(g).$$

or equivalently

$$p(s)F^*(\phi) - F^*(f) = F^*(g) \qquad (213)$$

where

$$p(s) = 1 - \sqrt{2\pi}\, F^*(K). \qquad (214)$$

To solve (213) we shall use the projection method of the previous section.

In other words we shall decompose every term in (213) into two parts, one in $L_2{}^+$ and one in $L_2{}^-$. Now

$$F(F^*(\phi)) = \phi = 0 \qquad \text{for} \quad x < 0$$

by (209), so that $F^*(\phi) \in L_2{}^-$. Similarly $F^*(f) \in L_2{}^-$ and $F^*(g) \in L_2{}^+$ so that (213) can be rewritten as

$$p(s)F_-^*(\phi) - F_-^*(f) = F_+^*(g).$$

Our next step requires that we write $p(s)$ in the form

$$p(s) = \frac{q^-(s)}{q^+(s)} \tag{215}$$

by applying theorem 8.* In order to be able to apply that theorem $p(s)$ must satisfy the requirements,

$$\lim_{|s| \to \infty} p(s) = 1 \tag{216}$$

and

$$\log p(s) \in L_2[-\infty, \infty]. \tag{217}$$

The above conditions impose additional constraints on the kernel $K(x)$ in (208). Equation (216) will certainly be satisfied by

$$\lim_{|s| \to \infty} F^*(K) = 0 \tag{218}$$

In general, if $K(x) \in L_2[-\infty, \infty]$, (218) will not hold. Equation (218) therefore adds a constraint to the general condition that $K(x) \in L_2[-\infty, \infty]$. One can give sufficient conditions on $K(x)$ in order for (218) to hold. One such condition is the following

$$\int_{-\infty}^{\infty} |K(x)| \, dx < \infty. \tag{219}$$

If (219) holds, according to the Riemann–Lebesgue theorem,† (218) will also hold.

If (218) holds we can define $\log p(s)$ such that

$$\lim_{s \to \infty} \log p(s) = 0.$$

* In theorem 8 we showed how to represent $p(s)$ in the form $[q^+(s)/q^-(s)]$. Equation (215) can be found by applying theorem 8 to $1/p(s)$ rather than to $p(s)$.
† For a full statement and proof of this theorem see, for example, R. R. Goldberg, *Fourier Transforms*, Cambridge University Press, London, 1961, p. 7.

But the above need not hold as $s \rightarrow -\infty$. It could happen that

$$\lim_{s \to -\infty} \log p(s) = 2n\pi i \tag{220}$$

for some integer n. For the moment we shall assume that $n = 0$ in (220). Now we see that for $|s|$ sufficiently large

$$|F^*(K)| < \epsilon$$

by (218). Then

$$|\log p(s)| = |\log (1 - \sqrt{2\pi}\, F^*(K))| \leq \frac{\sqrt{2\pi}\, |F^*(K)|}{1 - \sqrt{2\pi}\, |F^*(K)|} \leq \frac{\sqrt{2\pi}\, |(F^*(K)|}{1 - \epsilon}$$

and since $F^*(K) \in L_2[-\infty, \infty]$ we can conclude that $\log p(s) \in L_2[-\infty, \infty]$. The case where $n \neq 0$ in (220) will be treated in a later section.

Theorem 8 enables us to write

$$p(s) = \frac{q^-(s)}{q^+(s)}$$

where

$$q^-(s) = \sqrt{p(s)} \exp\left\{-\frac{i}{2} H \log p(s)\right\} \tag{221}$$

$$q^+(s) = \frac{1}{\sqrt{p(s)}} \exp\left\{-\frac{i}{2} H \log p(s)\right\}. \tag{222}$$

Finally (214) can be written as

$$q^- F^*_-(\phi) - q^+ F^*_-(f) = q^+ F^*_+(g) \tag{223}$$

where q^- and q^+ are boundary values of functions analytic and bounded in the lower and upper half planes respectively. By theorem 6, $q^- F^*_-(\phi) \in L_2^-$ and $q^+ F^*_+(g) \in L_2^+$. The remaining function $q^+ F^*_-(f)$ can be written as

$$q^+ F^*_-(f) = (q^+ F^*_-(f))_+ + (q^+ F^*_-(f))_-$$

where by (154) and (155)

$$(q^+ F^*_-(f))_\pm = \tfrac{1}{2}[q^+ F^*_-(f) \pm iH(q^+ F^*_-(f))].$$

Equation 223 now becomes

$$q^- F^*_-(\phi) - (q^+ F^*_-(f))_- = (q^+ F^*_-(f))_+ + q^+ F^*_+(g). \tag{224}$$

In (224) the left side is in L_2^- and the right side in L_2^+, and since $L_2^+ \cap L_2^- = \{0\}$, both vanish. Then

$$F^*_-(\phi) = \frac{(q^+ F^*_-(f))_-}{q^-} \tag{225}$$

But

$$F_-^*(\phi) = F^*(\phi), \qquad F_-^*(f) = F^*(f)$$

so that the solution of (208) is given by

$$\phi = F\left(\frac{(q^+F^*(f))_-}{q^-}\right). \tag{226}$$

It is clear from this construction that the solution found in (226) is unique. For $f(x) = 0$ (225) shows that only $\phi(x) = 0$ can satisfy the original equation (208).

EXAMPLE 23. Consider the function

$$p(x) = \frac{x^2 + a^2}{x^2 + 1}, \qquad \text{Re } a > 0 \tag{227}$$

and find q^- and q^+ such that

$$p(x) = \frac{q^-(x)}{q^+(x)}.$$

In a simple example of this type the above factorization can often be produced by inspection. We shall, however, carry out the general procedure described by Eqs. (221) and (222). Then $p(x)$ clearly satisfies conditions (216) and (217) so that the general procedure applies. The condition Re $a > 0$ is needed precisely to satisfy these conditions. Equation (217) holds in view of the fact that $p(x) = 1 + O(1/x^2)$ for large $|x|$, and also both the numerator and the denominator of $p(x)$ have each one zero in the upper half plane and one zero in the lower half plane.

In order to calculate $H \log p$ we shall first calculate $F(\log p)$, then use the fact that

$$F(H \log p) = i \, \text{sgn} \, s F(\log p)$$

and finally apply F^* to the above. Now

$$F(\log p) = \frac{1}{\sqrt{2\pi}} \int_{-\infty}^{\infty} \log \frac{x^2 + a^2}{x^2 + 1} e^{isx} \, dx.$$

First we shall differentiate the above with respect to the parameter a and evaluate the resultant integral by a residue integration. We find

$$\frac{\partial}{\partial a} F(\log p) = \frac{1}{\sqrt{2\pi}} \int_{-\infty}^{\infty} \frac{2a}{x^2 + a^2} e^{isx} \, dx$$

$$= \begin{array}{ll} \sqrt{2\pi} \, e^{-as}, & s > 0 \\ \sqrt{2\pi} \, e^{as}, & s < 0 \end{array}$$

$$= \sqrt{2\pi} \, e^{-a|s|}.$$

To find $F(\log p)$ we integrate the above and evaluate the constant of integration so that

$$F(\log p)|_{a=1} = 0$$

since in that case $p(x) = 1$. It follows that

$$F(\log p) = \sqrt{2\pi} \, \frac{e^{-|s|} - e^{-a|s|}}{|s|}$$

$$= \sqrt{2\pi} \, \frac{e^{-|s|} - e^{-a|s|}}{s} \, \text{sgn } s$$

and

$$F(H \log p) = \sqrt{2\pi} \, i \, \frac{e^{-|s|} - e^{-a|s|}}{s} \, .$$

Now

$$H \log p = i \int_{-\infty}^{\infty} \frac{e^{-|s|} - e^{-a|s|}}{s} \, e^{-isx} \, ds$$

and we shall differentiate with respect to a. Then

$$\frac{\partial}{\partial a} H \log p = i \int_{-\infty}^{\infty} e^{-a|s|} \, \frac{|s|}{s} \, e^{-isx} \, ds = 2 \int_{0}^{\infty} e^{-as} \sin sx \, ds = \frac{2x}{x^2 + a^2} \, .$$

By an integration one can find that

$$H \log p = 2 \left[\tan^{-1} \frac{a}{x} - \tan^{-1} \frac{1}{x} \right]$$

Finally

$$q^- = \sqrt{p(x)} \exp \left\{ -\frac{i}{2} H \log p \right\} = \sqrt{p(x)} \, \frac{x - ia}{\sqrt{x^2 + a^2}} \, \frac{x + i}{\sqrt{x^2 + 1}} = \frac{x - ia}{x - i}$$

(228)

and

$$q^+ = \frac{1}{\sqrt{p(x)}} \exp \left\{ -\frac{i}{2} H \log p \right\} = \frac{x + i}{x + ia} \, .$$

(229)

Evidently these satisfy (227) and are also bounded and analytic in their respective half planes.

EXAMPLE 24. Solve

$$\phi(x) - \lambda \int_{0}^{\infty} e^{-|x-y|} \phi(y) \, dy = f(x), \qquad f(x) \in L_2[0, \infty],$$

(230)

on $L_2[0, \infty]$. Following the general procedure we extend $\phi(x)$ and $f(x)$ so that

$$\phi(x) = f(x) = 0, \qquad x < 0$$

and define $g(x)$ by

$$g(x) = -\lambda \int_0^\infty e^{-|x-y|}\,\phi(y)\,dy = -\lambda e^x \int_0^\infty e^{-y}\phi(y)\,dy, \qquad x < 0$$

$$= 0, \qquad\qquad x > 0.$$

Equation (230) now becomes

$$\phi(x) - \lambda \int_{-\infty}^\infty e^{-|x-y|}\phi(y)\,dy = f(x) + g(x) \tag{231}$$

and we can apply F^* to obtain (as in (213))

$$\frac{s^2 + 1 - 2\lambda}{s^2 + 1}\, F^*_-(\phi) - F^*_-(f) = F^*_+(g). \tag{232}$$

The function $p(s)$ in (232) is given by

$$p(s) = \frac{s^2 + a^2}{s^2 + 1}, \qquad a^2 = 1 - 2\lambda. \tag{233}$$

In order to be able to produce the required factorization we need to have $\mathrm{Re}\, a > 0$ or equivalently $\lambda \notin [\tfrac12, \infty)$. By Example 23 we have

$$p(s) = \frac{q^-}{q^+}$$

where

$$q^- = \frac{s - ia}{s - i}, \qquad q^+ = \frac{s + ia}{s + i}.$$

Equation (232) can now be rewritten as

$$\frac{s - ia}{s - i}\, F^*_-(\phi) - \frac{s + i}{s + ia}\, F^*_-(f) - \frac{\alpha}{s + ia} = \frac{s + i}{s + ia}\, F^*_+(g) - \frac{\alpha}{s + ia}.$$

By theorem 6 the first term on the left is in L_2^- and by a suitable choice of α the remaining two terms are also in L_2^-. The right side is in L_2^+ and it follows that both sides vanish. We can now solve for $F^*_-(\phi)$ and finally for ϕ to obtain

$$\phi = F\!\left\{ F^*(f) + \frac{2\lambda}{s^2 + 1 - 2\lambda}\, F^*(f) + \frac{s - i}{s^2 + 1 - 2\lambda}\, \alpha \right\}.$$

Now

$$F(F^*(f)) = f$$

and since

$$F^*\!\left(\frac{\sqrt{2\pi}}{2a}\, e^{-a|x|} \right) = \frac{1}{s^2 + a^2}$$

we have by the convolution theorem

$$F\left(\frac{2\lambda}{s^2 + 1 - 2\lambda} F^*(f)\right) = \frac{\lambda}{a} \int_{-\infty}^{\infty} e^{-a|x-y|} f(y)\, dy$$

$$= \frac{\lambda}{a} \int_0^{\infty} e^{-a|x-y|} f(y)\, dy$$

since $f(x) = 0$ for $x < 0$. Finally, by a residue integration we find

$$F\left(\frac{s-i}{s^2 + a^2}\right) = \frac{\sqrt{2\pi} i}{2a} (a \operatorname{sgn} x - 1) e^{-a|x|}.$$

Then

$$\phi(x) = f(x) + \frac{\lambda}{a} \int_0^{\infty} e^{-a|x-y|} f(y)\, dy + \beta(a \operatorname{sgn} x - 1) e^{-a|x|},$$

where β is a multiple of α. To determine β we use the condition that

$$\phi(x) = 0, \qquad x < 0$$

and then

$$\beta = \frac{\lambda}{a(a+1)} \int_0^{\infty} e^{-ay} f(y)\, dy$$

and

$$\phi(x) = f(x) + \frac{\lambda}{\sqrt{1-2\lambda}} \int_0^{\infty} e^{-\sqrt{1-2\lambda}|x-y|} f(y)\, dy$$

$$+ \frac{\lambda(\sqrt{1-2\lambda}-1)}{1-2\lambda+\sqrt{1-2\lambda}} e^{-\sqrt{1-2\lambda}\, x} \int_0^{\infty} e^{-\sqrt{1-2\lambda}\, y} f(y)\, dy, \qquad x > 0.$$

$$(234)$$

There are physical applications where one is interested in finding solutions of (208) that grow exponentially. To study such cases we can replace $\phi(x)$ by $e^{\alpha x}\psi(x)$ and $f(x)$ by $e^{\alpha x}k(x)$. Then (208) becomes

$$\psi(x) - \int_0^{\infty} K(x-y) e^{-\alpha(x-y)} \psi(y)\, dy = k(x). \qquad (235)$$

If the kernel $K(x-y) e^{-\alpha(x-y)}$ and the function $k(x)$ have all the required properties, (235) can be solved as before and its solution $\psi(x) \in L_2[0, \infty]$. The function $\phi(x) = e^{\alpha x}\psi(x)$ will then satisfy (208), but be of exponential growth.

EXAMPLE 25. Solve

$$\phi(x) - \lambda \int_0^{\infty} e^{-|x-y|} \phi(y)\, dy = f(x),$$

such that $\phi(x)$ is of exponential growth. Equation (235) for this case becomes

$$\psi(x) - \lambda \int_0^\infty e^{-|x-y|-\alpha(x-y)}\psi(y)\,dy = k(x)$$

and we proceed exactly as in Example 24. Of course α has to be restricted so that $0 \leq \alpha < 1$. Let

$$\psi(x) = k(x) = 0, \qquad x < 0$$

$$g(x) = -\lambda \int_0^\infty e^{-|x-y|-\alpha(x-y)}\psi(y)\,dy, \qquad x < 0$$

$$= 0, \qquad\qquad\qquad\qquad\qquad x > 0.$$

Then

$$\psi(x) - \lambda \int_{-\infty}^\infty e^{-|x-y|-\alpha(x-y)}\psi(y)\,dy = k(x) + g(x)$$

and applying F^* we have

$$\frac{(s-i\alpha)^2 + 1 - 2\lambda}{(s-i\alpha)^2 + 1} F_-^*(\psi) - F_-^*(k) = F_+^*(g). \tag{236}$$

For $\alpha = 0$ (236) reduces to (232). Here

$$p(s) = \frac{(s-i\alpha)^2 + 1 - 2\lambda}{(s-i\alpha)^2 + 1} \tag{237}$$

and for $0 \leq \alpha < 1$ the denominator has one zero each in the upper and lower half plane. In order to satisfy the condition that

$$\log p(s) \in L_2[-\infty, \infty]$$

the zeros of the numerator of (237) must also be so distributed that one is in the upper and one in the lower half plane. This requirement imposes certain constraints on λ. The case where both zeros of the numerator lie in the same half plane will be considered in the next section. Now

$$(s-i\alpha)^2 + 1 - 2\lambda = [s - i(\alpha + a)][s - i(\alpha - a)]$$

where $a^2 = 1 - 2\lambda$ and we suppose that

$$\text{Im } i(\alpha + a) > 0$$

$$\text{Im } i(\alpha - a) < 0.$$

It is easy to see that

$$p(s) = \frac{q^-(s)}{q^+(s)}$$

$$q^-(s) = \frac{s - i(\alpha + a)}{s - i(\alpha + 1)}, \qquad q^+(s) = \frac{s - i(\alpha - a)}{s - i(\alpha - 1)}.$$

Equation (236) can now be rewritten as

$$\frac{s - i(\alpha + a)}{s - i(\alpha + 1)} F_-^*(\psi) - \frac{s - i(\alpha - 1)}{s - i(\alpha - a)} F_-^*(k) - \frac{\beta}{s - i(\alpha - a)}$$

$$= \frac{s - i(\alpha - 1)}{s - i(\alpha - a)} F_+^*(g) - \frac{\beta}{s - i(\alpha - a)}.$$

By theorem 6 the first term on the left is in L_2^-, and by a suitable choice of β the remaining two terms are also in L_2^-. The right side is in L_2^+ and hence both sides vanish identically. Now

$$\psi = F\left\{F^*(k) + \frac{2\lambda}{(s - i\alpha)^2 + a^2} F^*(k) + \frac{\beta[s - i(\alpha + 1)]}{(s - i\alpha)^2 + a^2}\right\}$$

and by operations similar to those carried out in Example 24

$$\psi(x) = k(x) + \frac{\lambda}{a} \int_0^\infty e^{-a|x-y| - \alpha(x-y)} k(y)\, dy$$

$$+ \frac{\lambda(a - 1)}{a(a + 1)} e^{-(a+\alpha)x} \int_0^\infty e^{-(a-\alpha)y} k(y)\, dy$$

Finally, we let $\psi(x) = e^{-\alpha x}\phi(x)$, $k(x) = e^{-\alpha x}f(x)$ and obtain

$$\phi(x) = f(x) + \frac{\lambda}{\sqrt{1 - 2\lambda}} \int_0^\infty e^{-\sqrt{1-2\lambda}|x-y|} f(y)\, dy$$

$$+ \frac{\lambda(\sqrt{1 - 2\lambda} - 1)}{\sqrt{1 - 2\lambda}(\sqrt{1 - 2\lambda} + 1)} e^{-\sqrt{1-2\lambda}x} \int_0^\infty e^{-\sqrt{1-2\lambda}y} f(y)\, dy, \quad (238)$$

Formally (234) and (238) are identical. The difference, however, is in the class of functions to which $\phi(x)$ and $f(x)$ belong. Equation (234) holds for the space $L_2[0, \infty]$. In (238) $f(x)$ and $\phi(x)$ may be such that $e^{-\alpha x} \cdot f(x)$ and $e^{-\alpha x}\phi(x)$ belong to $L_2[0, \infty]$ provided that $\mathrm{Re}\,\sqrt{1 - 2\lambda} - \alpha > 0$.

9. THE WIENER–HOPF TECHNIQUE II

In the previous section we treated the following integral equation

$$\phi(x) - \int_0^\infty K(x - y)\phi(y)\, dy = f(x) \qquad (239)$$

subject to certain constraints on $K(x)$. We required that $K(x) \in L_2[-\infty, \infty]$

and also that

$$\lim_{|s| \to \infty} F^*(K) = 0. \tag{240}$$

Under these conditions we can define $\log p(s)$, where

$$p(s) = 1 - F^*(K)$$

such that

$$\lim_{s \to \infty} \log p(s) = 0.$$

From (240) it follows that

$$\lim_{s \to \infty} \log |p(s)| = 0 \tag{241}$$

but because of the change in the argument of $p(s)$ as we traverse the s axis from $+\infty$ to $-\infty$ it may happen that

$$\lim_{s \to -\infty} \log p(s) = 2n\pi i \tag{242}$$

where n is an integer. To see this we write

$$p(s) = |p(s)| \, e^{i\theta(s)}.$$

We have selected that branch of the logarithm for which

$$\lim_{s \to \infty} \theta(s) = 0$$

and we know that

$$\lim_{|s| \to \infty} p(s) = 1.$$

From the above we see that

$$\lim_{s \to -\infty} p(s) = \lim_{s \to -\infty} |p(s)| \, e^{i\theta(s)} = \lim_{s \to -\infty} e^{i\theta(s)} = 1$$

and necessarily

$$\lim_{s \to -\infty} \theta(s) = 2n\pi.$$

This establishes (242), where n is known as the index of $I - K(x)$.

In the previous section we assumed that $n = 0$. Under this condition $\log p(s) \in L_2[-\infty, \infty]$ and by means of theorem 8 we were able to represent $p(s)$ in the form $q^-(s)/q^+(s)$. $q^{\pm}(s)$ represent the boundary values on the real axis of two functions $q^{\pm}(z)$ that are analytic in the upper and lower half planes respectively. If $n \neq 0$ such a decomposition of $p(s)$ does not exist. This section will be devoted to a discussion of the Wiener–Hopf equation (239) when the index $n \neq 0$.

When $n = 0$ it was shown that (239) had unique solutions in $L_2[0, \infty]$ for $f(x) \in L_2[0, \infty]$. Questions of existence and uniqueness are much more delicate when $n \neq 0$ and for this reason we shall first treat the homogeneous case and come back to inhomogeneous equations later.

(a) The Homogeneous Equation, $n > 0$

In steps (208)–(214) the following equation was developed

$$p(s)F_-^*(\phi) - F_-^*(f) = F_+^*(g)$$

and for $f = 0$ it reduces to

$$p(s)F_-^*(\phi) = F_+^*(g). \tag{243}$$

We assume that index $I - K(x) = n > 0$; that is, if

$$\lim_{s \to \infty} \log p(s) = 0$$

it follows that

$$\lim_{s \to -\infty} \log p(s) = 2n\pi i.$$

We shall show that for this case (243) has a unique solution, namely $\phi(x) = 0$.

Consider the function

$$\tau(s) = \frac{s - i}{s + i} \tag{244}$$

for which

$$\log \tau(s) = -2i \tan^{-1} 1/s.$$

To define the above uniquely we specify that

$$\lim_{s \to \infty} \tan^{-1} 1/s = 0$$

so that

$$\lim_{s \to -\infty} -2 \tan^{-1} 1/s = -2\pi$$

For the function $\log \tau^n(s)$ we have accordingly

$$\lim_{s \to \infty} \log \tau^n(s) = 0$$

$$\lim_{s \to -\infty} \log \tau^n(s) = -2n\pi i.$$

It follows therefore that

$$\log \tau^n(s)p(s) \in L_2[-\infty, \infty]. \tag{245}$$

We can now solve (243) by multiplying (243) by $\tau^n(s)$.

$$\tau^n(s)p(s)F_-^*(\phi) = \tau^n(s)F_+^*(g) \tag{246}$$

where $\tau^n(s)p(s)$ satisfies the conditions of theorem 8 and can be decomposed into $q^-(s)/q^+(s)$. By multiplying (246) by $q^+(s)$ we obtain

$$q^-(s)F_-^*(\phi) = \tau^n(s)q^+(s)F_+^*(g). \tag{247}$$

The left side of the above is in L_2^- and the right side in L_2^+, since $\tau(s)$ has no poles in the upper half plane. Now both sides vanish and we have

$$F_-^*(\phi) = 0$$

so that $\phi(x) = 0$.

We have succeeded in showing that (243) has a unique solution. From this result we can conclude that the inhomogeneous equation (239) has at most one solution. For if there were two we would have (using symbolic notation)

$$\phi_1 - K\phi_1 = f$$
$$\phi_2 - K\phi_2 = f$$

so that

$$(\phi_1 - \phi_2) - K(\phi_1 - \phi_2) = 0.$$

But the above is (243) and we just saw that (243) has precisely one solution, namely $\phi(x) = 0$. It follows, therefore, that (239) has at most one solution for $n > 0$.

Whether a solution of (239) will exist for all $f(x)$ or not will be discussed in a later section. It will be shown that (239) is not solvable for arbitrary $f(x) \in L_2[0, \infty]$. Necessary and sufficient conditions on $f(x)$ for a solution to exist will be derived.

(b) The Homogeneous Equation, $n < 0$

We can follow the same steps as in Section (a) up to (247). But since $n < 0$, $\tau^n(s)$ will have poles in the upper half plane. The left side is still in L_2^-, but the right side is not in L_2^+. We can, by theorem 6, subtract a term from both sides such that the right side will be in L_2^+.

$$q^-(s)F_-^*(\phi) - \sum_{k=1}^{|n|} \frac{\alpha_k}{(s-i)^k} = \tau^n(s)q^+(s)F_+^*(g) - \sum_{k=1}^{|n|} \frac{\alpha_k}{(s-i)^k}$$

For a suitable choice of $\alpha_1, \alpha_2, \ldots, \alpha_{|n|}$ the right side is now in L_2^+. It is easy to verify that the left side is in L_2^- so that both sides vanish. Now

$$F_-^*(\phi) = \frac{1}{q^-(s)} \sum_{k=1}^{|n|} \frac{\alpha_k}{(s-i)^k}$$

$$\phi(x) = F\left[\frac{1}{q^-(s)} \sum_{k=1}^{|n|} \frac{\alpha_k}{(s-i)^k}\right] \tag{248}$$

Similarly

$$g = F\left[\frac{1}{\tau^n(s)q^+(s)} \sum_{k=1}^{|n|} \frac{\alpha_k}{(s-i)^k}\right] \tag{249}$$

and it is easy to see that $F^*(\phi) \in L_2^-$, $F^*(g) \in L_2^+$ and that these solutions satisfy (243).

Evidently for $n < 0$, (243) has nonunique solutions and the number of linearly independent ones are $|n|$ in number. In fact, for every choice of $\alpha_1, \alpha_2, \ldots, \alpha_{|n|}$ (248) and (249) do satisfy the homogeneous Wiener–Hopf equation.

The function $\tau^n(s)$ was needed in order for the product $\tau^n(s)p(s)$ to be such that $\log \tau^n(s)p(s) \in L_2[-\infty, \infty]$. But there is an infinite choice of such functions. For example $((s - ia)/(s + ia))^n$, for Re $a > 0$, has the same property. One might well ask whether the solutions (248) and (249) might not somehow depend on such a choice. Suppose $\rho(s)$ is a second such function so that in place of (246) we have

$$\rho(s)p(s)F_-^*(\phi) = \rho(s)F_+^*(g).$$

Now let $\sigma(s) = \rho(s)/\tau^n(s)$ and replace $\rho(s)$ by $\tau^n(s)\sigma(s)$ in the above equation. Evidently both sides have a common factor of $\sigma(s)$ and we are led back to (246).

EXAMPLE 26. Solve

$$\phi(x) - \lambda \int_0^\infty e^{-|x-y|}\phi(y) \, dy = 0 \tag{250}$$

for $\phi(x)$ such that $e^{-\alpha x}\phi(x) \in L_2[0, \infty]$. The corresponding inhomogeneous case was discussed in Example 25. Letting $\phi(x) = e^{\alpha x}\psi(x)$ and proceeding as before we are led to the equation

$$\frac{(s - i\alpha)^2 + 1 - 2\lambda}{(s - i\alpha)^2 + 1} F_-^*(\psi) = F_+^*(g) \tag{251}$$

as in (236). We shall suppose that $0 \leq \alpha < 1$ and that Re $a = $ Re $\sqrt{1 - 2\lambda} > 0$. Now the denominator of

$$p(s) = \frac{(s - i\alpha)^2 + a^2}{(s - i\alpha)^2 + 1} = \frac{[s - i(\alpha + a)][s - i(\alpha - a)]}{[s - i(1 + \alpha)][s + i(1 - \alpha)]}$$

has one zero in the upper half plane and one in the lower half plane. If, furthermore,

$$\text{Re } \alpha + a > 0, \qquad \text{Re } \alpha - a < 0$$

the numerator has the same property and the index $n = 0$. In this case (250) has the unique solution $\phi(x) = 0$.

Suppose however that $|a|$ is so small that Re $(\alpha \pm a) > 0$. In this case the numerator of $p(s)$ has two zeros in the upper half plane and the denominator has one zero each in the lower and upper half planes. The index is essentially a topological property of $p(s)$ and does not depend on the actual location of the zeros and poles of $p(s)$. Furthermore, the index is a continuous function of the zeros and poles of $p(s)$, but it is also an integer. As a consequence, the index is a fixed constant that remains invariant under continuous changes of the zeros and poles of $p(s)$, as long as they do not cross the real axis. There, $\log p(s)$ will have discontinuities.

Now if $p(s)$ has two zeros and one pole in the upper half plane and one pole in the lower half plane, its index will coincide with the index of $[(s - i)^2/(s - i)(s + i)] = (s - i)/(s + i)$. The latter, as was seen in (244), has index -1. We now multiply (251) by $[s + i(1 - \alpha)]/[s - i(\alpha - a)]$ (which is equivalent to multiplying by $(s + i)/(s - i)$ so far as the index is concerned) and obtain

$$\frac{s - i(\alpha + a)}{s - i(1 + \alpha)} F_-^*(\psi) - \frac{\beta}{s - i(\alpha - a)}$$
$$= \frac{s + i(1 - \alpha)}{s - i(\alpha - a)} F_+^*(g) - \frac{\beta}{s - i(\alpha - a)},$$

after subtracting $\beta/[s - i(\alpha - a)]$ from both sides. The left side is now in L_2^- and the right in L_2^+. It follows that

$$\psi(x) = \beta F\left[\frac{s - i(1 + \alpha)}{(s - i(\alpha + a))(s - i(\alpha - a))}\right]$$
$$= \frac{\beta\sqrt{2\pi} i}{2a}[(a - 1)e^{-(\alpha+a)x} + (a + 1)e^{-(\alpha-a)x}]$$

where β is arbitrary. Finally

$$\phi(x) = \beta'[(\sqrt{1 - 2\lambda} - 1)e^{-\sqrt{1-2\lambda}x} + (\sqrt{1 - 2\lambda} + 1)e^{\sqrt{1-2\lambda}x}] \quad (252)$$

where β' is arbitrary. In this case we have a one dimensional solution space of the homogeneous equation (250).

(c) The Inhomogeneous Equation, $n < 0$

We now reconsider the equation

$$\phi(x) - \int_0^\infty K(x - y)\phi(y) \, dy = f(x) \quad (253)$$

and suppose that the index of $I - K = n < 0$. After the standard manipulations, we have

$$p(s)F_-^*(\phi) - F_-^*(f) = F_+^*(g) \tag{254}$$

and we multiply the above by $\tau^n(s)$, where $\tau(s) = (s - i)/(s + i)$. Next we can, by theorem 7, write $\tau^n(s)p(s)$ in the form $q^-(s)/q^+(s)$ so that the above equation can be rewritten in the form

$$q^-(s)F_-^*(\phi) - \tau^n(s)q^+(s)F_-^*(f) = \tau^n(s)q^+(s)F_+^*(g).$$

The second term on the left can be written as a sum of two terms, one in L_2^- and one in L_2^+ [see (154) and (155)]. The first term on the left is in L_2^-. $\tau^n(s)$ has a pole of order $|n|$ at $s = i$ (n is negative) so that the right side is not in L_2^+. But we can subtract a suitable term to bring it into L_2^+. We then obtain

$$q^-(s)F_-^*(\phi) - (\tau^n(s)q^+(s)F_-^*(f))_- - \sum_{k=1}^{|n|} \frac{\alpha_k}{(s-i)^k}$$

$$= \tau^n(s)q^+(s)F_+^*(g) + (\tau^n(s)q^+(s)F_+^*(f))_+ - \sum_{k=1}^{|n|} \frac{\alpha_k}{(s-i)^k}. \tag{255}$$

An inspection of the above shows that the left side is in L_2^- and the right side in L_2^+. Solving for ϕ and g we obtain

$$\phi = F\left\{\frac{1}{q^-(s)}\left[(\tau^n(s)q^+(s)F_-^*(f))_- + \sum_{k=1}^{|n|} \frac{\alpha_k}{(s-i)^k}\right]\right\} \tag{256}$$

$$g = F\left\{\frac{1}{\tau^n(s)q^+(s)}\left[-(\tau^n(s)q^+(s)F_-^*(f))_+ + \sum_{k=1}^{|n|} \frac{\alpha_k}{(s-i)^k}\right]\right\}. \tag{257}$$

An inspection of (256) and (257) shows that regardless how $\alpha_1, \alpha_2, \ldots, \alpha_{|n|}$ are selected $F^*(\phi)$ and $F^*(g)$ are in L_2^- and L_2^+ respectively and satisfy (254). It follows that for $n < 0$, the Wiener–Hopf equation (253) has solutions for all $f(x) \in L_2[0, \infty]$. However these solutions are not unique. The difference of two solutions must lie in the $|n|$ dimensional solution space of the corresponding homogeneous equation.

EXAMPLE 27. Solve the equation

$$\phi(x) - \lambda \int_0^\infty e^{-|x-y|}\phi(y)\,dy = f(x) \tag{257}$$

for $f(x)$ and $\phi(x)$ of exponential growth. We proceed as in Example 25 to Eq. (236).

$$\frac{(s - i\alpha)^2 + a^2}{(s - i\alpha)^2 + 1} F_-^*(\psi) - F_-^*(k) = F_+^*(g)$$

where $a^2 = 1 - 2\lambda$ and $\operatorname{Re} a > 0$. In Example 26 it was shown that if $0 \leq \alpha < 1$ and $\operatorname{Re}(\alpha \pm a) > 0$, the index of $I - K(x)$, $n = -1$. To derive the formula equivalent to (255) we first multiply by $[s - i(1 - \alpha)]/[s - i(\alpha - a)]$ and subtract $\beta/[s - i(\alpha - a)]$. Then

$$\frac{s - i(\alpha + a)}{s - i(1 + \alpha)} F_-^*(\psi) - \frac{s + i(1 - \alpha)}{s - i(\alpha - a)} F_-^*(k) - \frac{\beta}{s - i(\alpha - a)}$$

$$= \frac{s + i(1 - \alpha)}{s - i(\alpha - a)} F_+^* g - \frac{\beta}{s - i(\alpha - a)}$$

and solving for $\psi(x)$ we find

$$\psi(x) = F\left\{ F^*(k) + \frac{2\lambda}{(s - i\alpha)^2 + a^2} F^*(k) + \beta \frac{s - i(1 + \alpha)}{(s - i\alpha)^2 + a^2} \right\}$$

$$= k(x) - \frac{2\lambda}{a} \int_0^x e^{-\alpha(x-y)} \sinh a(x - y) k(y)\, dy$$

$$+ \beta'[(a - 1)e^{-(\alpha+a)x} + (a + 1)e^{-(\alpha-a)x}] \qquad \text{for} \quad x > 0.$$

If we now let $\psi(x) = e^{-\alpha x}\phi(x)$, $k(x) = e^{-\alpha x} f(x)$ we finally have

$$\phi(x) = f(x) - \frac{2\lambda}{\sqrt{1 - 2\lambda}} \int_0^x \sinh \sqrt{1 - 2\lambda}(x - y) f(y)\, dy$$

$$+ \beta'[(\sqrt{1 - 2\lambda} - 1)e^{-\sqrt{1-2\lambda}\,x} + (\sqrt{1 - 2\lambda} + 1)e^{\sqrt{1-2\lambda}\,x}], \qquad x > 0$$

where β' is arbitrary. For $x < 0$, $\phi(x) = 0$.

(d) The Inhomogeneous Equation, $n > 0$

In Lemma 2 of Chapter 1 it was shown that

$$N(L) = R(L^*)^\perp. \tag{258}$$

Here, L is a bounded operator on a Hilbert space and $N(L)$ is its null space. L^* is the operator adjoint to L, and $R(L^*)^\perp$ is the orthogonal complement of the range of L^*. In Chapter 1 we worked in the finite dimensional spaces, but the proof for bounded operators on general Hilbert spaces is essentially the same.

In the Wiener–Hopf equation the integral operator

$$K\phi = \int_0^\infty K(x - y)\phi(y)\, dy \tag{259}$$

occurs. We shall show that it is a bounded operator of $L_2[0, \infty]$. As before we view it as an operator on the Hilbert space $L_2[-\infty, \infty]$, by extending

$\phi(x)$ to $(-\infty, 0]$ by setting $\phi(x) = 0$ for $x < 0$. Then by taking Fourier transforms of (259) we have

$$F(K\phi) = \sqrt{2\pi}\, F(K)F(\phi)$$

from which, by the Cauchy–Schwarz inequality, we obtain

$$\|F(K\phi)\| \le \sqrt{2\pi}\, \|F(K)\|\, \|F(\phi)\|.$$

In Section 1 of this chapter it was shown that F is a unitary operator, so that

$$\|K\phi\| \le \sqrt{2\pi}\, \|\phi\| \left(\int_{-\infty}^{\infty} |K(x)|^2\, dx\right)^{1/2} \tag{260}$$

The operator K defined by (259) is therefore bounded.

It had been shown earlier that for a bounded integral operator

$$K\phi = \int_a^b K(x, y)\phi(y)\, dy$$

the adjoint operator is given by

$$K^*\phi = \int_a^b \overline{K(y, x)}\phi(y)\, dy.$$

For the operator defined by (259) we find that

$$K^*\phi = \int_0^\infty \overline{K(y - x)}\phi(y)\, dy. \tag{261}$$

As a consequence of the preceding remarks we can now draw the following conclusion. A necessary and sufficient condition for the Wiener–Hopf equation

$$\phi(x) - \int_0^\infty K(x - y)\phi(y)\, dy = f(x) \tag{262}$$

to have a solution is that

$$f(x) \in R(I - K) = N(I - K^*)^\perp.$$

This is equivalent to saying that in order for (262) to have a solution, $f(x)$ must be orthogonal to all functions $\psi(x) \in L_2[0, \infty]$ satisfying

$$\psi(x) - \int_0^\infty \overline{K(y - x)}\psi(y)\, dy = 0. \tag{263}$$

As has been shown in previous sections (262) can be reduced to

$$p(s)F_-^*(\phi) - F_-^*(f) = F_+^*(g) \tag{264}$$

where

$$p(s) = 1 - \sqrt{2\pi}\, F^*(K(x)). \tag{265}$$

By a similar process (263) can be reduced to an equation of the form

$$[1 - \sqrt{2\pi}\, F^*\overline{(K(-x))}]F_-^*(\psi) = F_+^*(h) \tag{266}$$

where

$$h(x) = -\int_{-\infty}^{0} \overline{K(y-x)}\psi(y)\,dy, \qquad x < 0$$

$$= 0, \qquad\qquad\qquad\qquad x > 0.$$

A simple calculation shows that

$$F^*\overline{(K(-x))} = \frac{1}{\sqrt{2\pi}}\int_{-\infty}^{\infty} \overline{K(-x)}e^{-isx}\,dx = \overline{F^*(K(x))}$$

so that (266) can be rewritten as

$$\overline{p(s)}F_-^*(\psi) = F_+^*(h), \tag{267}$$

where $\overline{p(s)}$ is the conjugate of the same function $p(s)$ that appears in (254). The index of $I - K(x)$ is related to the change of the argument of $p(s)$ as s decreases from $+\infty$ to $-\infty$. If that change is $2n\pi i$, the index of $I - K(x)$ is n. Similarly, the argument of $\overline{p(s)}$, as s decreases from $+\infty$ to $-\infty$, will increase by $-2n\pi i$ so that the index of $I - \overline{K(-x)}$ is $-n$. For $n > 0$ the index associated with (267) is negative and, by the results of Section (b), (267) will have n linearly independent solutions. Now (264) will have solutions only for those $f(x)$ that are orthogonal to the n independent solutions of (267).

If the index of (264) were negative, the index of (267) would be positive and (267) would have only the null solution by Section (a). In this case, (264) would have solutions for all $f(x) \in L_2[0, \infty]$, which reconfirms the results of Section (c). This argument only assures the existence of solutions, but not their uniqueness.

To solve (264) we multiply by $\tau^n(s)$, where $\tau(s) = (s - i)/(s + i)$ to obtain

$$\tau^n(s)p(s)F_-^*(\phi) - \tau^n(s)F_-^*(f) = \tau^n(s)F_+^*(g).$$

If $\tau^n(s)p(s) = q^-(s)/q^+(s)$ the above reduces to

$$q^-(s)F_-^*(\phi) - (\tau^n(s)q^+(s)F_-^*(f))_- = (\tau^n(s)q^+(s)F_-^*(f))_+ + \tau^n(s)q^+(s)F_+^*(g).$$

Evidently both sides must vanish so that

$$(F_-^*\phi) = \frac{[\tau^n(s)q^+(s)F_-^*(f)]_-}{q^-(s)}$$

$$F_+^*(g) = \frac{-[\tau^n(s)q^+(s)F_-^*(f)]_+}{\tau^n(s)q^+(s)}$$

It is clear that $F_-^*(\phi)$ as constructed above is in L_2^-. But $F_+^*(g)$ may not be in L_2^+, because the denominator contains the term $(s - i)^n$, so that there is a pole in the upper half plane. A solution will exist only for those $f(x)$ orthogonal to all solutions of (263). In that case

$$\phi = F\left\{\frac{1}{q^-(s)} [\tau^n(s)q^+(s)F^*(f)]_-\right\} \tag{268}$$

$$g = F\left\{\frac{-1}{\tau^n(s)q^+(s)} [\tau^n(s)q^+(s)F^*(f)]_+\right\}. \tag{269}$$

It was pointed out earlier that the above solution, when it exists, is unique.

EXAMPLE 28. Solve the following equation and obtain conditions on $f(x)$ for a solution to exist.

$$\phi(x) - \tfrac{1}{2}\int_0^\infty e^{-|x-y|+\frac{1}{2}(x-y)}\phi(y)\,dy = f(x), \qquad f(x) \in L_2[0, \infty]. \tag{270}$$

The above corresponds to the problem treated in Example 25 with $\lambda = \tfrac{1}{2}$. There we sought solutions of exponential growth. By taking $\alpha = -\tfrac{1}{2}$ (exponential decay) in that example, we arrive at (270). The standard manipulations lead to

$$\frac{[s + (i/2)]^2}{[s - (i/2)][s + (3i/2)]} F_-^*(\phi) - F_-^*(f) = F_+^*(g) \tag{271}$$

The index associated with the above problem is $n = 1$. In this case we find

$$\tau(s) = \frac{s - (i/2)}{s + (i/2)}, \qquad q^-(s) = 1, \qquad q^+(s) = \frac{s + (3i/2)}{s + (i/2)}$$

so that we obtain

$$F_-^*(\phi) - \frac{[s + (3i/2)][s - (i/2)]}{[s + (i/2)]^2} F_-^*(f) = \frac{[s + (3i/2)][s - (i/2)]}{[s + (i/2)]^2} F_+^*(g)$$

or equivalently

$$F_-^*(\phi) - F_-^*(f) - \frac{1}{[s + (i/2)]^2} F_-^*(f) = \frac{[s + (3i/2)][s - (i/2)]}{[s + (i/2)]^2} F_+^*(g). \tag{272}$$

The first two terms on the left are in L_2^-, and the term on the right is in L_2^+. The third term on the left can be decomposed into a term in L_2^- and L_2^+ as follows. A standard computation shows that

$$F\left(\frac{1}{\sqrt{2\pi}[s + (i/2)]^2}\right) = \frac{1}{2\pi}\int_{-\infty}^\infty \frac{e^{isx}}{[s + (i/2)]^2}\,ds = 0, \qquad x > 0$$

$$= xe^{x/2}, \qquad x < 0.$$

We can now define the function

$$k(x) = xe^{x/2}, \qquad x < 0$$
$$= 0, \qquad x > 0$$

so that

$$F^*\left[\int_{-\infty}^{\infty} k(x - y)f(y)\,dy\right] = \frac{1}{[s + (i/2)]^2}\,F_-^*(f).$$

Now let

$$h_1(x) = \int_x^{\infty} (x - y)e^{(x-y)/2}f(y)\,dy, \qquad x > 0$$

$$= 0, \qquad x < 0$$

$$h_2(x) = 0, \qquad x > 0 \qquad\qquad (273)$$

$$= -\int_0^{\infty} (x - y)e^{(x-y)/2}f(y)\,dy, \qquad x < 0 \qquad (\text{N.B. } f(x) = 0,\, x < 0)$$

so that

$$h_1(x) - h_2(x) = \int_{-\infty}^{\infty} k(x - y)f(y)\,dy.$$

Clearly

$$F^*(h_1) = F_-^*(h_1) \in L_2^-$$

$$F^*(h_2) = F_+^*(h_2) \in L_2^+$$

and (272) can now be rewritten as

$$F_-^*(\phi) - F_-^*(f) - F_-^*(h_1) = -F_+^*(h_2) + \frac{[s - (3i/2)][s - (i/2)]}{[s + (i/2)]^2}\,F_+^*(g).$$

From the above we finally have

$$\phi(x) = f(x) + \int_x^{\infty} (x - y)e^{(x-y)/2}f(y)\,dy, \qquad x > 0 \qquad (274)$$

$$g(x) = F\left[\frac{[s + (i/2)]^2}{[s + (3i/2)][s - (i/2)]}\,F_+^*(h_2)\right]. \qquad (275)$$

We are not, as yet, assured that $F^*(g) \in L_2^+$. In fact a computation shows that

$$F^*(h_2) = \frac{-1}{\sqrt{2\pi}[s + (i/2)]^2}\left\{\int_0^{\infty} e^{-(y/2)}f(y)\,dy - i\left(s + \frac{i}{2}\right)\int_0^{\infty} ye^{-(y/2)}f(y)\,dy\right\}$$

and

$$g = -\frac{1}{\sqrt{2\pi}} F\left[\frac{1}{[s + (3i/2)][s - (i/2)]}\right.$$

$$\left. \times \left\{\int_0^\infty e^{-(y/2)}f(y)\,dy - i\left(s + \frac{i}{2}\right)\int_0^\infty ye^{-(y/2)}f(y)\,dy\right\}\right]. \quad (276)$$

As a consequence of the pole at $s = i/2$ in (276) $F^*(g)$ may not be in L_2^+. But if $f(x)$ satisfies the condition

$$\int_0^\infty (1 + y)e^{-(y/2)}f(y)\,dy = 0 \quad (277)$$

(276) reduces to

$$g = \frac{i}{\sqrt{2\pi}} F\left[\frac{1}{s + (3i/2)}\int_0^\infty e^{-(y/2)}f(y)\,dy\right] \quad (278)$$

and now $F^*(g) \in L_2^+$.

We now consider the adjoint homogeneous equation associated with (270), namely

$$\psi(x) - \tfrac{1}{2}\int_0^\infty e^{-|x-y|-\frac{1}{2}(x-y)}\psi(y)\,dy = 0. \quad (279)$$

By the Wiener–Hopf technique the above is reduced to, as in (267),

$$\frac{[s - (i/2)]^2}{[s + (i/2)][s - (3i/2)]} F_-^*(\psi) = F_+^*(h).$$

Its index $n = -1$, and we obtain

$$\frac{s - (i/2)}{s - (3i/2)} F_-^*(\psi) - \frac{\alpha}{s - (i/2)} = \frac{s + (i/2)}{s - (i/2)} F_+^*(h) - \frac{\alpha}{s - (i/2)}.$$

The left side is in L_2^- and the right side will be in L_2^+ for a suitable choice of α. Then

$$F_-^*(\psi) = \frac{\alpha[s - (3i/2)]}{[s - (i/2)]^2}$$

and

$$\psi(x) = \alpha'(1 + x)e^{-x/2}, \quad x > 0$$
$$= 0, \quad\quad\quad\quad\quad x < 0. \quad (280)$$

Finally, as has been noted before, (280) satisfies (279) for arbitrary α'.

Condition (277) was necessary to ensure that $F^*(g) \in L_2^+$. This condition is nothing but a statement of the fact that $f(x)$ is orthogonal to all solutions of (279). If $f(x)$ does not meet this condition $F^*(g) \notin L_2^+$ and accordingly (270) will have no solution.

10. WIENER–HOPF EQUATIONS OF THE FIRST KIND

In the preceding two sections, Wiener–Hopf equations of the second kind were reduced to equations of the type [see (213)].

$$p(s)F^*_-(\phi) - F^*_-(f) = F^*_+(g).$$

Our success in solving the above depended on our ability to represent $p(s)$ as a ratio $q^-(s)/q^+(s)$, as is done in (215). That such a factorization exists is a consequence of theorem 8. In order to apply this theorem we require that $\lim_{|s|\to\infty} p(s) = 1$. In some ways a similar technique can be applied to equations of the first kind as well. However, the coefficients $p(s)$ will not in general be such that theorem 8 can be applied. As a consequence $p(s)$ cannot be factored by a systematic procedure and the investigator's ingenuity must be invoked. Unfortunately, many important physical problems lead to equations of the first kind. For a thorough discussion of many of these problems the reader is referred to the book by B. Noble, listed in the bibliography.

EXAMPLE 29. Solve

$$\int_0^\infty e^{-|x-y|}h(y)\,dy = x^n e^{-x} \tag{281}$$

on $L_2[0, \infty]$. We now extend this problem to $L_2[-\infty, \infty]$, by letting

$$
\begin{aligned}
h(x) &= 0, & x &< 0 \\
f(x) &= x^n e^{-x}, & x &> 0 \\
&= 0, & x &< 0 \\
g(x) &= 0, & x &> 0 \\
&= \int_0^\infty e^{-|x-y|}h(y)\,dy = e^x \int_0^\infty e^{-y}h(y)\,dy, & x &< 0
\end{aligned}
$$

so that (281) can be rewritten as

$$\int_{-\infty}^\infty e^{-|x-y|}h(y)\,dy = f(x) + g(x). \tag{282}$$

To the above we can now apply the inverse Fourier transform F^*. To do so we note that

$$F^*(e^{-|x|}) = \frac{1}{\sqrt{2\pi}}\int_{-\infty}^\infty e^{-|x|-isx}\,dx = \frac{\sqrt{2/\pi}}{1+s^2} = \frac{\sqrt{2/\pi}}{(s+i)(s-i)}$$

and

$$\frac{1}{\sqrt{2\pi}}\int_{-\infty}^\infty f(x)e^{-isx}\,dx = \frac{1}{\sqrt{2\pi}}\int_0^\infty x^n e^{-(1+is)x}\,dx = \frac{(1/\sqrt{2\pi})(-i)^{n+1}n!}{(s-i)^{n+1}}.$$

After an application of the operation F^* (282) can now be rewritten as

$$\frac{\sqrt{2/\pi}}{(s+i)(s-i)} F_-^*(h) - \frac{(1/\sqrt{2\pi})(-i)^{n+1}n!}{(s-i)^{n+1}} = F_+^*(g). \tag{283}$$

The second term on the left in (283) is in L_2^- and the right side is in L_2^+. The first term on the left has a pole at $s = -i$ and therefore is not in L_2^-. That pole can be removed by multiplying (283) by $(s+i)/(s-i)$ so that

$$\frac{\sqrt{2/\pi}}{(s-i)^2} F_-^*(h) - \frac{(1/\sqrt{2\pi})(-i)^{n+1}n!(s+i)}{(s-i)^{n+2}} = \frac{s+i}{s-i} F_+^*(g).$$

Now the entire left side is in L_2^-, but the right side is not in L_2^+ because of the pole at $s = i$. By theorem 6 one can subtract a term of the form $(\alpha/s - i)$ from both sides so that the right side will be in L_2^+. Such a term is in L_2^- as was seen in the previous examples. After this step we have

$$\frac{\sqrt{2/\pi}}{(s-i)^2} F_-^*(h) - \frac{(1/\sqrt{2\pi})(-i)^{n+1}n!(s+i)}{(s-i)^{n+2}} - \frac{\alpha}{s-i}$$

$$= \frac{s+i}{s-i} F_+^*(g) - \frac{\alpha}{s-i}.$$

Since $L_2^+ \cap L_2^- = \{0\}$ we can conclude that

$$F_-^*(h) = \frac{(-i)^{n+1}n!(s+i)}{2(s-i)^n} + \sqrt{\frac{\pi}{2}}\alpha(s-i) \tag{284}$$

$$F_+^*(g) = \frac{\alpha}{s+i} \tag{285}$$

In order for the right side of (284) to be in L_2^- it is necessary that $n \geq 2$ and $\alpha = 0$; otherwise it is not in $L_2[-\infty, \infty]$. For $n = 1$ the right side of (284) will not be in L_2^- and (281) will have no solution. We see that

$$F_-^*(h) = \frac{(-i)^{n+1}n!(s+i)}{2(s-i)^n}, \qquad F_+^*(g) = 0, \qquad n \geq 2$$

satisfy (283) and thus lead to a solution of (281). Finally

$$h(x) = F(F_-^*(h)) = \frac{(-i)^{n+1}n!}{2\sqrt{2\pi}} \int_{-\infty}^{\infty} \frac{s+i}{(s-i)^n} e^{isx} ds$$

$$= 0, \qquad x < 0$$

$$= \sqrt{\frac{\pi}{2}} [2nx^{n-1} - n(n-1)x^{n-2}]e^{-x}, \qquad n \geq 2 \tag{286}$$

11. DUAL INTEGRAL EQUATIONS

The projection method can be used to solve so-called dual integral equations. We shall devote this section to an investigation of the following equations.

$$\int_0^\infty f(u)J_\nu(\rho u)\,du = g(\rho), \qquad 0 < \rho < 1, \quad \nu > -\tfrac{1}{2}, \qquad (287)$$

$$\int_0^\infty u^\alpha f(u)J_\mu(\rho u)\,du = 0, \qquad 1 < \rho < \infty, \,\mu > -\tfrac{1}{2}. \qquad (288)$$

If (287) were to hold for all $0 < \rho < \infty$ it could be solved directly by means of Hankel transforms, and the same is true for (288). It is precisely because each of these equations holds for a different ρ interval that these equations constitute a system of dual equations.

In order to bring (287) and (288) into forms to which Fourier transforms can be applied we shall make the following changes of variable

$$u = e^{-y}, \qquad \rho = e^x \qquad (289)$$

and then multiply (287) by $e^{(1-k)x}$ and (288) by $e^{\alpha x}$; the choice of k will be taken up shortly. We then obtain

$$\int_{-\infty}^\infty e^{-ky}f(e^{-y})J_\nu(e^{x-y})e^{(1-k)(x-y)}\,dy = e^{(1-k)x}g(e^x), \qquad -\infty < x < 0 \quad (290)$$

$$\int_{-\infty}^\infty e^{\alpha(x-y)}e^{-ky}f(e^{-y})J_\mu(e^{x-y})e^{(1-k)(x-y)}\,dy = 0, \qquad x > 0. \qquad (291)$$

Equations (290) and (291) are now in the form of convolutions; however, the right sides are defined only for $x < 0$ and $x > 0$ respectively. k will now be selected so that

$$\int_{-\infty}^\infty |J_\nu(e^x)e^{(1-k)x}|^2\,dx < \infty, \qquad \int_{-\infty}^\infty |e^{\alpha x}J_\mu(e^x)e^{(1-k)x}|^2\,dx < \infty. \quad (292)$$

To analyze the integrals in (292) certain facts about Bessel functions are required.* The standard series for Bessel functions is given by

$$J_\nu(x) = \left(\frac{x}{2}\right)^\nu \sum_{n=0}^\infty \frac{(ix/2)^{2n}}{n!\,\Gamma(n+\nu+1)} \qquad (293)$$

From (293) it follows that for small x, $J_\nu(x)$ behaves like x^ν. For large x we

---

* In this section a considerable knowledge about Bessel functions and gamma functions will be required. Pertinent information about these may be found, for example, in H. Hochstadt, *The Functions of Mathematical Physics*, John Wiley and Sons, New York, 1971.

have the following asymptotic estimate

$$J_\nu(x) = \sqrt{\frac{2}{\pi x}} \cos \left(x - \frac{\nu \pi}{2} - \frac{\pi}{4} \right) + 0 \left(\frac{1}{x} \right).$$

We now return to the first integral in (292) and let $u = e^x$, so that

$$\int_{-\infty}^{\infty} |J_\nu(e^x) e^{(1-k)x}|^2 \, dx = \int_0^{\infty} |J_\nu(u)|^2 \, u^{1-2k} \, du$$

For small u the integrand behaves like $u^{2\nu+1-2k}$. In order for the integral to converge near $u = 0$ we require that $k < 1 + \nu$. For large u the integrand behaves like u^{-2k} and for convergence we require that $2k > 1$. A similar analysis applies to the second integral. Finally k will be so selected that

$$\nu + 1 > k > \tfrac{1}{2}, \qquad \alpha + \mu + 1 > k > \alpha + \tfrac{1}{2}, \tag{294}$$

provided, of course, that ν, μ, and α are such that we can select a k satisfying (294). Equation (292) will be satisfied if k is selected as above. Of course, such a choice of k can be made only if $\nu > -\tfrac{1}{2}$, $\mu > -\tfrac{1}{2}$, and the intervals in (294) overlap.

Now we can define the following functions:

$$
\begin{aligned}
h_+(x) &= e^{(1-k)x} g(e^x), & x &< 0 \\
&= 0, & x &> 0 \qquad (295)\\
h_-(x) &= 0, & x &< 0
\end{aligned}
$$

$$= \int_{-\infty}^{\infty} e^{-ky} f(e^{-y}) J_\nu(e^{x-y}) e^{(1-k)(x-y)} \, dy, \qquad x > 0 \tag{296}$$

$$m_+(x) = \int_{-\infty}^{\infty} e^{-\alpha y} e^{-ky} f(e^{-y}) J_\mu(e^{x-y}) e^{(1-k)(x-y)} \, dy, \qquad x < 0,$$

$$= 0, \qquad x > 0. \tag{297}$$

By the use of these functions we rewrite (290) and (291) as

$$\int_{-\infty}^{\infty} w(y) J_\nu(e^{x-y}) e^{(1-k)(x-y)} \, dy = h_+(x) + h_-(x), \qquad -\infty < x < \infty \tag{298}$$

$$\int_{-\infty}^{\infty} e^{\alpha(x-y)} w(y) J_\mu(e^{x-y}) e^{(1-k)(x-y)} \, dy = m_+(x), \qquad -\infty < x < \infty \tag{299}$$

where

$$w(y) = e^{-ky} f(e^{-y}) \tag{300}$$

In an appendix to this section we shall show that

$$F^*(J_\nu(e^x)e^{\beta x}) = \frac{1}{\sqrt{2\pi}} \int_{-\infty}^{\infty} J_\nu(e^x)e^{\beta x}e^{-isx} \, dx$$

$$= \frac{2^{\beta-is-1}\Gamma[(\nu + \beta - is)/2]}{\sqrt{2\pi}\ \Gamma[(\nu - \beta + is + 2)/2]}, \tag{301}$$

for $J_\nu(e^x)e^{\beta x} \in L_2[-\infty, \infty]$.

If we apply the operation F^* to (298) and (299) we obtain

$$\frac{F^*(w)2^{-is-k}\Gamma[(\nu + 1 - k - is)/2]}{\Gamma[(\nu + 1 - k + is)/2]} = F^*(h_+) + F^*(h_-), \tag{302}$$

$$\frac{F^*(w)2^{-is+\alpha-k}\Gamma[(\mu + \alpha + 1 - k - is)/2]}{\Gamma[(\mu - \alpha + 1 + k + is)/2]} = F^*(m_+) \tag{303}$$

In order to perform these operations we must make the assumption on the function $g(\rho)$ in (287), that $h_+(x) \in L_2[-\infty, \infty]$. Note that $F^*(h_+)$ and $F^*(m_+)$ are in L_2^+ and $F^*(h_-)$ is in L_2^-, by virtue of their definitions in (295), (296), and (297). To solve (302) and (303) for $F^*(w)$ we first eliminate $F^*(w)$ between these two equations to obtain

$$\frac{F^*(m_+)2^{-\alpha}\Gamma[(\nu + 1 - k - is)/2]\Gamma[(\mu - \alpha + 1 + k + is)/2]}{\Gamma[(\nu + 1 + k + is)/2]\Gamma[(\mu + \alpha + 1 - k - is)/2]}$$
$$= F^*(h_+) + F^*(h_-). \tag{304}$$

In order to solve (304) we shall try to regroup the terms in such a way that the left side will be in L_2^+ and the right side in L_2^-. In order to complete this program we require certain facts about gamma functions.

$$\Gamma(z) = \int_0^\infty e^{-t}t^{z-1} \, dt = \int_0^1 e^{-t}t^{z-1} \, dt + \int_1^\infty e^{-t}t^{z-1} \, dt, \qquad \operatorname{Re} z > 0.$$

The second integral above represents an entire function of z. The first integral can be evaluated as follows

$$\int_0^1 e^{-t}t^{z-1} \, dt = \sum_{n=0}^{\infty} \frac{(-1)^n}{n!} \int_0^1 t^{n+t-1} \, dt = \sum_{n=0}^{\infty} \frac{(-1)^n}{n!\,(n+z)}$$

so that

$$\Gamma(z) = \sum_{n=0}^{\infty} \frac{(-1)^n}{n!\,(n+z)} + \int_1^\infty e^{-t}y^{z-1} \, dt. \tag{305}$$

From (305) we observe that $\Gamma(z)$ is a meromorphic function with simple poles at $z = 0, -1, -2, -3, \ldots$. The residue at $z = -n$ is given by

$(-1)^n/n!$. There is a standard asymptotic formula for $\Gamma(z)$ for large $|z|$ known as Stirling's formula. It is given by

$$\ln \Gamma(z) = (z - \tfrac{1}{2}) \ln z - z + \tfrac{1}{2} \ln 2\pi + 0\left(\frac{1}{z}\right). \qquad (306)$$

From the above one can show that for large $|z|$ we have approximately

$$\frac{\Gamma(a + z)}{\Gamma(b + z)} \approx z^{a-b}. \qquad (307)$$

Equation (304) can now be rewritten as

$$\frac{F^*(m_+)2^{-\alpha}\Gamma[(\nu + 1 - k - is)/2]}{\Gamma[(\mu + \alpha + 1 - k - is)/2]} = \frac{\Gamma[(\nu + 1 + k + is)/2]}{\Gamma[(\mu - \alpha + 1 + k + is)/2]} F^*(h_+)$$

$$+ \frac{\Gamma[(\nu + 1 + k + is)/2]}{\Gamma[(\mu - \alpha + 1 + k + is)/2]} F^*(h_-). \qquad (308)$$

The coefficient of $F^*(m_+)$ has simple poles at $s = -i(2n + \nu + 1 - k)$ where $n = 0, 1, 2, \ldots$ by the use of (305). Inspection of (294) shows that all these poles are in the lower half plane. This same term has zeros at $s = -i(2n + \mu + \alpha + 1 - k)$ where $n = 0, 1, 2, \ldots$ and all these are in the lower half plane also. We shall now make the additional assumptions that

$$\nu - \mu \leq -|\alpha| \qquad (309)$$

Now by means of (307) the coefficient of $F^*(m_+)$ satisfies

$$\left|\frac{\Gamma[(\nu + 1 - k - is)/2]}{\Gamma[(\mu + \alpha + 1 - k - is)/2]}\right| \leq K |s|^{(\nu-\mu-\alpha)/2}$$

for large $|s|$, for a suitable constant K. It follows that if ν, μ, α satisfy (309) the coefficient of $F^*(m_+)$ is bounded for real s and is analytic in the upper half plane. Therefore the left side of (308) is in L_2^+.

We now inspect the second term on the right of (308). It has poles at $s = i(2n + \nu + 1 + k)$, $n = 0, 1, 2, \ldots$, and all these are in the upper half plane. We also have

$$\left|\frac{\Gamma[(\nu + 1 + k + is)/2]}{\Gamma[(\mu - \alpha + 1 + k + is)/2]}\right| \leq K |s|^{(\nu-\mu+\alpha)/2}$$

and again this is bounded for real s, by use of (309). Since F_-^* (h) is in L_2^- and its coefficient is bounded for real s and analytic in the lower half plane the second term on the right of (308) is in L_2^-. The first term on the right can be decomposed into two terms, one in L_2^- and the other in L_2^+. Then

$$\frac{F^*(m_+)2^{-\alpha}\Gamma[(\nu + 1 - k - is)/2]}{\Gamma[(\mu + \alpha + 1 - k - is)/2]} - \left[\frac{\Gamma[(\nu + 1 + k + is)/2]}{\Gamma[(\mu - \alpha + 1 + k + is)/2]} F^*(h_+)\right]_+$$

$$= \frac{\Gamma[(\nu + 1 + k + is)/2]}{\Gamma[(\mu - \alpha + 1 + k + is)/2]} F^*(h_-)$$

$$+ \left[\frac{\Gamma[(\nu + 1 + k + is)/2]}{\Gamma[(\mu - \alpha + 1 + k + is)/2]} F^*(h_+)\right]_-. \quad (310)$$

Now the left side is in L_2^+ and the right in L_2^-. Accordingly, both vanish. We then find that

$$F^*(m_+) = \frac{2^{\alpha}\Gamma[(\mu + \alpha + 1 - k - is)/2]}{\Gamma[(\nu + 1 - k - is)/2]}$$

$$\times \left[\frac{\Gamma[(\nu + 1 + k + is)/2]}{\Gamma[(\mu - \alpha + 1 + k + is)/2]} F^*(h_+)\right]_+. \quad (311)$$

Finally to find $F^*(w)$ we return to (303), and use (311). Then

$$F^*(w) = \frac{2^{k+is}\Gamma[(\mu - \alpha + 1 + k + is)/2]}{\Gamma[(\nu + 1 - k - is)/2]}$$

$$\times \left[\frac{\Gamma[(\nu + 1 + k + is)/2]}{\Gamma[(\mu - \alpha + 1 + k + is)/2]} F^*(h_+)\right]_+. \quad (312)$$

(312) was derived under the following assumptions.

$$\int_{-\infty}^{0} |e^{(1-k)x}g(e^x)|^2\, dx < \infty$$

$$\nu - \mu \le -|\alpha| \quad (313)$$

$$\nu + 1 > k > \tfrac{1}{2}, \qquad \alpha + \mu + 1 > k > \alpha + \tfrac{1}{2}.$$

Equation (312) can be brought into a more usable form from which an explicit solution for $f(x)$ in terms of $g(\rho)$ can be found. First we compute $F^*(h_+)$.

$$F^*(h_+) = \frac{1}{\sqrt{2\pi}} \int_{-\infty}^{0} e^{(1-k)t}g(e^t)e^{-ist}\, dt$$

and letting $e^t = V$ we obtain

$$F^*(h_+) = \frac{1}{\sqrt{2\pi}} \int_{0}^{1} V^{-is-k}g(V)\, dV.$$

Now we see that

$$
\left[\frac{\Gamma[(\nu + 1 + k + is)/2]}{\Gamma[(\mu - \alpha + 1 + k + is)/2]} F^*(h_+) \right]_+
$$

$$
= \frac{1}{\sqrt{2\pi}} \int_0^1 V^{-k} g(V) \left[\frac{\Gamma[(\nu + 1 + k + is)/2]}{\Gamma[(\mu - \alpha + 1 + k + is)/2]} V^{-is} \right]_+ dV \quad (314)
$$

and we shall calculate

$$
\left[\frac{\Gamma[(\nu + 1 + k + is)/2]}{\Gamma[(\mu - \alpha + 1 + k + is)/2]} V^{-is} \right]_+ \equiv [\phi]_+. \quad (315)
$$

To accomplish this we recall that for $\phi(x) \in L_2[-\infty, \infty]$ we have

$$
F(\phi_+) = F(\phi), \quad s < 0
$$
$$
= 0, \quad s > 0
$$

It follows that

$$
\int_{-\infty}^{\infty} \left[\frac{\Gamma[(\nu + 1 + k + i\xi)/2]}{\Gamma[(\mu - \alpha + 1 + k + i\xi)/2]} V^{-i\xi} \right]_+ e^{is\xi} d\xi = 0, \quad s > 0
$$

$$
= \int_{-\infty}^{\infty} \frac{\Gamma[(\nu + 1 + k + i\xi)/2]}{\Gamma[(\mu - \alpha + 1 + k + i\xi)/2]} V^{-i\xi} e^{is\xi} d\xi, \quad s < 0
$$

To evaluate the above integral we write it in the form

$$
\int_{-\infty}^{\infty} \frac{\Gamma[(\nu + 1 + k + i\xi)/2]}{\Gamma[(\mu - \alpha + 1 + k + i\xi)/2]} e^{i\xi(s - \ln V)} d\xi. \quad (316)
$$

For $s < \ln V$, $e^{i\xi(s - \ln V)}$ decreases exponentially in the lower half of the ξ plane. The integrand is analytic in the lower half plane and by closing the contour about the lower half plane we see that the integral vanishes for $s < \ln V$.

For $\ln V < s < 0$ we can evaluate (316) by closing the contour over the upper half plane and summing over the residues due to the poles of the gamma function in the numerator. Then

$$
\frac{1}{\sqrt{2\pi}} \int_{-\infty}^{\infty} \frac{\Gamma[(\nu + 1 + k + i\xi)/2]}{\Gamma[(\mu - \alpha + 1 + k + i\xi)/2]} e^{i\xi(s - \ln V)} d\xi
$$

$$
= \frac{4\pi}{\sqrt{2\pi}} \sum_{n=0}^{\infty} \frac{(-1)^n (e^{-2s} V^2)^{n + [(\nu + 1 + k)/2]}}{n! \, \Gamma[(\mu - \alpha - \nu)/2 - n]}
$$

$$
= \frac{2\sqrt{2\pi} (e^{-s} V)^{\nu + 1 + k}}{\Gamma[(\mu - \alpha - \nu)/2]} (1 - e^{-2s} V^2)^{(\mu - \alpha - \nu - 2)/2}, \quad \ln V < s < 0 \quad (317)
$$

where the last step is accomplished by means of the binomial theorem. Finally, using

$$[\phi]_+ = \frac{1}{\sqrt{2\pi}} \int_{-\infty}^{0} F(\phi)e^{-is\xi}\, ds$$

we obtain

$$\left[\frac{\Gamma[(\nu + 1 + k + i\xi)/2]}{\Gamma[(\mu - \alpha + 1 + k + i\xi)/2]} V^{-i\xi} \right]_+$$

$$= \frac{1}{\sqrt{2\pi}} \int_{\ln V}^{0} \frac{2\sqrt{2\pi}(e^{-s}V)^{\nu+1+k}}{\Gamma[(\mu - \alpha - \nu)/2]} (1 - e^{-2s}V^2)^{(\mu-\alpha-\nu-2)/2} e^{-is\xi}\, ds$$

The above can now be inserted in (314). After that one can interchange the order of integration, let $V = e^s \rho$ and finally let $u = e^s$. The final result after all these manipulations becomes

$$\left[\frac{\Gamma[(\nu + 1 + k + is)/2]}{\Gamma[(\mu - \alpha + 1 + k + is)/2]} F^*(h_+) \right]_+$$

$$= \frac{\sqrt{2/\pi}}{\Gamma[(\mu - \alpha - \nu)/2]} \int_0^1 u^{-is-k}\, du \int_0^1 \rho^{\nu+1} g(V\rho)(1 - \rho^2)^{(\mu-\nu-\alpha-2)/2}\, d\rho. \quad (318)$$

We now insert (318) in (312). To compute $w(x)$ we have to evaluate

$$F\left[\frac{2^{k+is}\Gamma[(\mu - \alpha + 1 + k + is)/2]u^{-is-k}}{\Gamma[(\nu + 1 - k - is)/2]} \right]$$

In the appendix to this section we shall show that the above reduces to

$$2\sqrt{2\pi}\, e^{-kx} \left(\frac{ue^{-x}}{2} \right)^{(\mu-\alpha-\nu+2)/2} J_{(\nu+\mu-\alpha)/2}(ue^{-x}). \quad (319)$$

The above can now be used in (312) and

$$w(x) = \frac{2^{(2+\nu+\alpha-\mu)/2} e^{-x[k+(\mu-\alpha-\nu+2)/2]}}{\Gamma[(\mu - \alpha - \nu)/2]} \int_0^1 V^{(\mu-\alpha-\nu+2)/2} J_{(\nu+\mu-\alpha)/2}(Ve^{-x})\, dV$$

$$\times \int_0^1 \rho^{1+\nu} g(V\rho)(1 - \rho^2)^{(\mu-\nu-\alpha-2)/2}\, d\rho. \quad (320)$$

Finally, we can use (300) to obtain

$$f(x) = \frac{2^{(2+\nu+\alpha-\mu)/2} x^{(\mu-\alpha-\nu+2)/2}}{\Gamma[(\mu - \alpha - \nu)/2]} \int_0^1 V^{(\mu-\alpha-\nu+2)/2} J_{(\nu+\mu-\alpha)/2}(Vx)\, dV$$

$$\times \int_0^1 \rho^{1+\nu} g(V\rho)(1 - \rho^2)^{(\mu-\nu-\alpha-2)/2}\, d\rho. \quad (321)$$

Equation (321) represents the solution of (287) and (288) subject to the conditions listed in (313).

Equations (287) and (288) may still have solutions under other conditions on μ, ν, and α than those listed in (313). In order for a solution to exist, however, more stringent conditions may have to be imposed on $g(\rho)$. For example, we shall consider the case when $\mu = \nu = 0$ and $\alpha = 1$. Then we consider the equations

$$\int_0^\infty f(u)J_0(\rho u)\,du = g(\rho), \qquad 0 < \rho < 1, \tag{322}$$

$$\int_0^\infty uf(u)J_0(\rho u)\,du = 0, \qquad 1 < \rho < \infty, \tag{323}$$

(298) and (299) now become

$$\int_{-\infty}^\infty w(y)J_0(e^{x-y})e^{(1-k)(x-y)}\,dy = h_+(x) + h_-(x), \qquad -\infty < x < \infty \tag{324}$$

$$\int_{-\infty}^\infty e^{(x-y)}w(y)J_0(e^{x-y})e^{(1-k)(x-y)}\,dy = m_+(x), \qquad -\infty < x < \infty \tag{325}$$

$$w(y) = e^{-ky}f(e^{-y}) \tag{326}$$

where $h_+(x)$, $h_-(x)$, and $m_+(x)$ are as defined in (295), (296), and (297). Inspection of (294) shows that we can apply the F^* operator to (324), but not to (325) since k cannot be selected so that both inequalities in (294) are satisfied. We therefore rewrite (325) as

$$\int_{-\infty}^\infty e^{-y}w(y)J_0(e^{x-y})e^{(1-k)(x-y)}\,dy = e^{-x}m_+(x), \qquad -\infty < x < \infty \tag{327}$$

Applying F^* to the above and using (301) we find

$$\frac{1}{\sqrt{2\pi}} \int_{-\infty}^\infty w(y)e^{-iy(s-i)}\,dy\, \frac{2^{-k-is}\Gamma[(1-k-is)/2]}{\Gamma[(1+k+is)/2]}$$

$$= \frac{1}{\sqrt{2\pi}} \int_{-\infty}^0 m_+(x)e^{-ix(s-i)}\,dx \tag{328}$$

We have to assume that both $e^{-y}w(y)$ and $e^{-x}m_+(x)$ are in $L_2[-\infty, \infty]$. It will be convenient to use the following notation in this case.

$$F_s^*(w) = \frac{1}{\sqrt{2\pi}} \int_{-\infty}^\infty w(y)e^{-isy}\,dy$$

to bring out the s dependence. In this notation (328) can be rewritten as

$$F^*_{s-i}(w) \frac{2^{-k-is}\Gamma[(1 - k - is)/2]}{\Gamma[(1 + k + is)/2]} = F^*_{s-i}(m_+). \tag{329}$$

We now apply F^* to (324) to obtain

$$F^*_s(w) \frac{2^{-k-is}\Gamma[(1 - k - is)/2]}{\Gamma[(1 + k + is)/2]} = F^*(h_+) + F^*(h_-). \tag{330}$$

It is desirable to have a second equation for $F^*_s(w)$, rather than (329). This can be accomplished by replacing s by $s + i$ in (329). Then

$$F^*_s(w)2^{1-k-is} \frac{\Gamma[(2 - k - is)/2]}{\Gamma[(k + is)/2]} = F^*_s(m_+). \tag{331}$$

Now we can eliminate $F^*(w)$ from (330) and (331) to obtain

$$F^*_s(m_+) \frac{\Gamma[(1 - k - is)/2]\Gamma[(k + is)/2]}{2\Gamma[(1 + k + is)/2]\Gamma[(2 - k - is)/2]} = F^*(h_+) + F^*(h_-). \tag{332}$$

Equation (332) is identical to (304) for the case $\mu = \nu = 0$, $\alpha = 1$. However, (304) was derived under different hypotheses. In analogy to (308) we now write

$$\frac{F^*_s(m_+)\Gamma[(1 - k - is)/2]}{2\Gamma[(2 - k - is)/2]}$$

$$= \frac{\Gamma[(1 + k + is)/2]}{\Gamma[(k + is)/2]} F^*(h_+) + \frac{\Gamma[(1 + k + is)/2]}{\Gamma[(k + is)/2]} F^*(h_-). \tag{333}$$

Our next task is to regroup the terms in (333) so that one side is in L_2^+ and the other in L_2^-. The coefficient of $F^*_s(m_+)$ has poles in the lower half plane and is analytic in the upper half plane. According to (307) we also have

$$\left| \frac{\Gamma[(1 - k - is)/2]}{\Gamma[(2 - k - is)/2]} \right| \leq K |s|^{-\frac{1}{2}}$$

so that the coefficient of $F^*_s(m_+)$ is bounded for real s. It follows that the left side of (333) is in L_2^+. The coefficient $F^*(h_-)$ is analytic in the lower half plane and satisfies the inequality

$$\left| \frac{\Gamma[(1 + k + is)/2]}{\Gamma[(k + is)/2]} \right| \leq K |s|^{\frac{1}{2}}$$

on the axis. The second term on the right will therefore be in L_2^- only if

$F^*(h_-)$ vanishes sufficiently fast for large $|s|$ so that

$$\frac{\Gamma[(1 + k + is)/2]}{\Gamma[(k + is)/2]} F^*(h_-) \in L_2[-\infty, \infty].$$

The function $F^*(h_-)$ is unknown, and whether it satisfies these conditions will depend on the inhomogeneous term $F^*(h_+)$, which of course depends on the inhomogeneous term $g(\rho)$ in (322).

If we now assume that $F^*(h_+)$ is sufficiently small for large s so that the first term on the right of (333) is in $L_2[-\infty, \infty]$ we obtain

$$F_s^*(m_+) = \frac{2\Gamma[(2 - k - is)/2]}{\Gamma[(1 - k - is)/2]} \left[\frac{\Gamma[(1 + k + is)/2]}{\Gamma[(k + is)/2]} F^*(h_+) \right]_+. \tag{334}$$

The above agrees with (311), but as seen requires additional hypotheses on $F^*(h_+)$. Finally using (331) we find

$$F_s^*(w) = \frac{2^{k+is}\Gamma[(k + is)/2]}{\Gamma[(1 - k - is)/2]} \left[\frac{\Gamma[(1 + k + is)/2]}{\Gamma[(k + is)/2]} F^*(h_+) \right]_+. \tag{335}$$

To make (335) more tractable we shall assume that $g(e^x)$ has the special form

$$g(e^x) = -2\int_x^0 e^{-(1-k)u} p(e^u)\, du, \qquad x < 0,$$

$$= 0, \qquad\qquad\qquad x > 0. \tag{336}$$

The advantages of this choice will become evident in the succeeding steps. Then

$$h_+(x) = e^{(1-k)x} g(e^x) = -2\int_x^0 e^{(1-k)(x-u)} p(e^u)\, du, \qquad x < 0$$

$$= 0, \qquad\qquad\qquad\qquad x > 0, \tag{337}$$

and

$$F^*(h_+) = \frac{2}{-1 + k + is} F^*(p). \tag{338}$$

We now use one of the standard identities for the gamma function, namely

$$\Gamma(1 + z) = z\Gamma(z)$$

to obtain

$$\frac{\Gamma[(1 + k + is)/2]}{\Gamma[(k + is)/2]} F^*(h_+) = \frac{\Gamma[(1 + k + is)/2]2/(-1 + k + is)}{\Gamma[(k + is)/2]} F^*(p)$$

$$= \frac{\Gamma[(-1 + k + is)/2]}{\Gamma[(k + is)/2]} F^*(p). \tag{339}$$

One can now proceed exactly as in the derivation of (318) to show that

$$\left[\frac{\Gamma[(-1 + k + is)/2]}{\Gamma[(k + is)/2]} F^*(p)\right]_+$$

$$= \frac{\sqrt{2}}{\pi} \int_0^1 u^{-is-k}\, du \int_0^u p(V)V^{k-2}\left[\left(1 - \frac{V^2}{u^2}\right)^{-\frac{1}{2}} - 1\right] dV. \quad (340)$$

Finally one can determine $w(x)$ to show that

$$w(x) = \frac{2}{\pi} \int_0^1 (ue^{-x})^k u^{-1} \cos{(ue^{-x})}\, du \int_0^1 p(\rho u)\rho^{k-2}[(1 - \rho^2)^{\frac{1}{2}} - 1]\, d\rho \quad (341)$$

and

$$f(x) = \frac{2}{\pi} \int_0^1 u^{k-1} \cos{(ux)}\, du \int_0^1 p(\rho u)\rho^{k-2}[(1 - \rho^2)^{-\frac{1}{2}} - 1]\, d\rho. \quad (342)$$

Inspection of (336) shows that

$$g'(x) = 2x^{k-2}p(x) \quad (343)$$

so that (342) can be written as

$$f(x) = \frac{1}{\pi} \int_0^1 u \cos{(ux)}\, du \int_0^1 g'(\rho u)[(1 - \rho^2)^{-\frac{1}{2}} - 1]\, d\rho. \quad (344)$$

The above solution is valid provided $g(x)$ and $p(x)$ are related via (343) and $p(e^u) \in L_2[-\infty, 0]$ as a function of u.

EXAMPLE 30. In numerous physical problems, for example in elasticity theory and potential theory, one encounters the following dual integral equations

$$\int_0^\infty f(u)J_0(\rho u)\, du = 1, \qquad 0 < \rho < 1, \quad (345)$$

$$\int_0^\infty uf(u)J_0(\rho u)\, du = 0, \qquad 1 < \rho < \infty. \quad (346)$$

Formally these look like (322), (323) with $g(\rho) = 1$. For this case (see (295))

$$h_+(x) = e^{(1-k)x}, \qquad x < 0$$
$$= 0, \qquad x > 0$$

so that

$$F^*(h_+) = \frac{1}{\sqrt{2\pi}} \int_{-\infty}^0 e^{(1-k)x - isx}\, dx = \frac{1}{\sqrt{2\pi}(1 - k - is)}.$$

Equation (335) now takes the form

$$F^*(w) = \frac{2^{k+is}\Gamma[(k + is)/2]}{\Gamma[(1 - k - is)/2]}\left[\frac{\Gamma[(1 + k + is)/2]}{\sqrt{2\pi}\,\Gamma[(k + is)/2](1 - k - is)}\right]_+. \quad (347)$$

The function in the bracket above behaves for large $|s|$ like $|s|^{-\frac{1}{2}}$. It is therefore not in $L_2[-\infty, \infty]$ and the preceding theory cannot be applied.

In order to produce a solution, which it turns out has physical significance, we can treat the above as a limiting case. We consider

$$\left[\frac{\Gamma[(1 + k + is)/2]}{\sqrt{2\pi}\,\Gamma[(k + is)/2](1 - k - is)(1 - i\epsilon s)}\right]_+$$

to which we can apply the preceding theory and then let $\epsilon \to 0$. A calculation similar to those performed earlier shows that the above reduces to

$$\frac{1}{\sqrt{2}\,\pi(1 - k - is)(1 - \epsilon + \epsilon k)}$$

$$-\frac{\Gamma\{[1 + k + (1/\epsilon)]/2\}\epsilon}{\sqrt{2\pi}\,\Gamma\{[k + (1/\epsilon)]/2\}[1 - k - (1/\epsilon)](1 - \epsilon + \epsilon k)}$$

and as $\epsilon \to 0$ the above reduces to

$$\frac{1}{\sqrt{2}\,\pi(1 - k - is)}$$

so that

$$F^*(w) = \frac{2^{k+is}\Gamma[(k + is)/2]}{\sqrt{2}\,\pi\Gamma[(1 - k - is)/2](1 - k - is)}.$$

Next we can perform a standard residue integration to determine $w(x)$ and finally $f(x)$. The resultant infinite series can be summed by inspection and we find

$$f(x) = \frac{2}{\pi}\frac{\sin x}{x}. \tag{348}$$

APPENDIX

In (301) and (319) we used integrations equivalent to

$$F^*(J_\nu(\rho e^x)e^{\beta x}) = \frac{2^{\beta-is-1}\rho^{is-\beta}\Gamma[(\nu + \beta - is)/2]}{\sqrt{2\pi}\,\Gamma[(\nu - \beta + is + 2)/2]}. \tag{349}$$

To prove (349) we shall compute

$$F\left(\frac{2^{\beta-is-1}\rho^{is-\beta}\Gamma[(\nu + \beta - is)/2]}{\sqrt{2\pi}\,\Gamma[(\nu - \beta + is + 2)/2]}\right)$$

$$= \frac{1}{2\pi}\int_{-\infty}^{\infty}\frac{2^{\beta-is-1}\rho^{is-\beta}\Gamma[(\nu + \beta - is)/2]}{\Gamma[(\nu - \beta + is + 2)/2]}e^{isx}\,ds.$$

The above integrand has poles in the lower half plane at $s = -i(2n + \nu + \beta)$,

$n = 0, 1, 2, \ldots$. We can close the contour over the lower half plane and a residue integration yields

$$e^{\beta x} \left(\frac{e^x \rho}{2} \right)^\nu \sum_{n=0}^{\infty} \frac{(-1)^n (e^x \rho/2)^{2n}}{n! \, \Gamma(n + 1 + \nu)} .$$

A comparison with (293) shows that the above is in fact $e^{\beta x} J_\nu(\rho e^x)$ thereby proving (349).

EXERCISES

1. Show that the Fourier transform of an integrable function is integrable.
2. Carry out the details of the computations summarized in (19), (20), and (21).
3. Calculate

$$F(e^{-|x|}) = \frac{1}{\sqrt{2\pi}} \int_{-\infty}^{\infty} e^{-|x|+isx} \, dx.$$

4. In Section 2, Example 2,

$$u = \int_{-\infty}^{\infty} \frac{e^{-y^2/4t}}{2\sqrt{\pi t}} g(x - y) \, dy.$$

Show that u, u_x, and u_{xx} are in $L_2[-\infty, \infty]$.

5. In Section 2, it was shown that $\|K\| \leq 2$, where

$$K\phi = \int_{-\infty}^{\infty} e^{-|x-y|} \phi(y) \, dy.$$

Show that $\|K\| = 2$. (Hint: Either find a $\phi(x) \in L_2[-\infty, \infty]$ such that $\|K\phi\| = 2 \|\phi\|$, or find a sequence $\{\phi_n\}$ such that $\|K\phi_n\|/\|\phi_n\| \to 2$.)

6. Calculate

$$F\left(\frac{1}{\sqrt{2t}} e^{-x^2/t} \right) = \frac{1}{\sqrt{2\pi}} \int_{-\infty}^{\infty} \frac{1}{\sqrt{2t}} e^{-x^2/t+isx} \, dx.$$

7. Analyze in detail

$$\phi(x) - \frac{\lambda}{\pi} \int_0^{\infty} \frac{\phi(y)}{x + y} \, dy = f(x).$$

from Example 6 in Section 2 to obtain

$$\phi(e^{2\xi}) e^\xi = \frac{1}{\sqrt{\pi}} \int_{-\infty}^{\infty} \frac{e^{ix\xi} \cosh{(\pi/2)s} F(g)}{\cosh{(\pi/2)s} - \lambda} \, ds$$

8. In Example 8, Section 2, show that

$$\left\| \frac{1}{\pi} \int_0^\infty \frac{\phi(y)}{x - y} \, dy \right\| \leq \|\phi\|.$$

9. Let

$$K^*\phi = - \frac{1}{\pi} \int_0^{\pi*} \frac{\sin x}{\cos x - \cos y} \, \phi(y) \, dy.$$

Show that

$$K^* \cos nx = \sin nx, \quad n = 0, 1, 2, \ldots$$

10. In Example 12, Section 2 show that

$$\left\| \frac{1}{\pi} \int_0^{\infty*} \left(\frac{1}{x + y} - \frac{1}{x - y} \right) \phi(y) \, dy \right\| \leq \|\phi\|.$$

11. Let $k(x)$ be such that $\int_{-\infty}^\infty |k(x)| \, dx < \infty$. Define the operator $Kf = \int_{-\infty}^\infty k(x - y) f(y) \, dy$ in $L_2[-\infty, \infty]$. Show that K is bounded.

12. Solve the heat flow equation

$$\frac{\partial u}{\partial t} = \frac{\partial^2 u}{\partial x^2}$$

subject to the auxiliary conditions

$$\frac{\partial u}{\partial x}(0, t) = \frac{\partial u}{\partial x}(L, t) = 0$$

$$u(x, 0) = f(x)$$

13. Solve

$$\frac{\partial^2 u}{\partial t^2} = \frac{\partial^2 u}{\partial x^2}$$

$$u(x, 0) = f(x), \qquad \frac{\partial u}{\partial t}(x, 0) = 0$$

for $-\infty < x < \infty$, $t \geq 0$, $f(x) \in L_2[-\infty, \infty]$ using Fourier transforms with respect to x.

14. Show that

$$\int_0^\infty \cos u u^{\sigma-1} \, du = \Gamma(\sigma) \cos \frac{\pi\sigma}{2}$$

$$\int_0^\infty \sin u u^{\sigma-1} \, du = \Gamma(\sigma) \sin \frac{\pi\sigma}{2}.$$

15. Prove in detail theorem 5 in Section 5.

16. Use Laplace transforms with respect to t to solve

$$\frac{\partial^2 u}{\partial t^2} = \frac{\partial^2 u}{\partial x^2}, \qquad u(x, 0) = f(x), \qquad \frac{\partial u}{\partial t}(x, 0) = 0$$

$$u(0, t) = u(L, t) = 0.$$

17. Solve

$$\int_0^x \frac{\phi(y)}{(x - y)^\alpha} \, dy = f(x), \qquad 0 \le \alpha < 1$$

$$f(x) \in L_2[0, \infty]$$

18. Solve $(1/\pi) \int_0^{*\infty} [\phi(y)/(x - y)] \, dy = f(x)$ on $L_2[0, \infty]$. (Hint: Treat it as a limiting case of

$$\phi(x) - \frac{\lambda}{\pi} \int_0^{*\infty} \frac{\phi(y)}{x - y} \, dy = -\lambda f(x) \quad \text{as} \quad \lambda \to \infty).$$

19. Let $p(x) = (x^2 + a^2)(x^2 + b^2)/(x^2 + 1)^2$ where a and b are real. Decompose $p(x)$ in the form

$$p(x) = \frac{q^+(x)}{q^-(x)} \quad \text{in accordance with theorem 8.}$$

20. Derive Eq. (266) from (263) in detail.
21. Show that $F^*(\overline{K(-x)}) = \overline{F^*(K(x))}$.

6

THE FREDHOLM
THEORY

SUMMARY

This chapter is devoted to the classical Fredholm theory. For self-adjoint compact operators, little new information is obtained. But for non-self-adjoint operators, one obtains strong results regarding the existence of eigenvalues. The techniques employed rely heavily on function theoretic techniques. Finally, integral operators with positive kernels (not to be confused with positive operators) are discussed and it is shown that they must have at least one positive eigenvalue.

1. INTRODUCTION

In Section 3 of Chapter 1 a Fredholm equation of the second kind

$$\phi(x) - \lambda \int_0^1 K(x, y)\phi(y)\,dy = f(x) \tag{1}$$

was studied via a finite difference approximation. In order for such an approximation to make sense we shall assume in this chapter that $K(x, y)$ and $f(x)$ are continuous functions. Then (1) can be reduced to the algebraic system

$$\phi\left(\frac{j}{n}\right) - \lambda \sum_{i=1}^{n} \frac{1}{n} K\left(\frac{j}{n}, \frac{i}{n}\right)\phi\left(\frac{i}{n}\right) = f\left(\frac{j}{n}\right), \qquad j = 1, 2, \ldots, n. \tag{2}$$

234

Let

$$D_n(\lambda) = \begin{vmatrix} 1 - \dfrac{\lambda}{n} K\left(\dfrac{1}{n},\dfrac{1}{n}\right) & -\dfrac{\lambda}{n} K\left(\dfrac{1}{n},\dfrac{2}{n}\right) & \cdots & -\dfrac{\lambda}{n} K\left(\dfrac{1}{n},1\right) \\[2mm] -\dfrac{\lambda}{n} K\left(\dfrac{2}{n},\dfrac{1}{n}\right) & 1 - \dfrac{\lambda}{n} K\left(\dfrac{2}{n},\dfrac{2}{n}\right) & \cdots & -\dfrac{\lambda}{n} K\left(\dfrac{2}{n},1\right) \\[2mm] \cdot & \cdot & \cdot \\ \cdot & \cdot & \cdot \\ \cdot & \cdot & \cdot \\[2mm] -\dfrac{\lambda}{n} K\left(1,\dfrac{1}{n}\right) & -\dfrac{\lambda}{n} K\left(1,\dfrac{2}{n}\right) & \cdots & 1 - \dfrac{\lambda}{n} K(1,1) \end{vmatrix} \tag{3}$$

and

$$D_n(i, j; \lambda)$$

$$= \begin{vmatrix} 1 - \dfrac{\lambda}{n} K\left(\dfrac{1}{n},\dfrac{1}{n}\right) & \cdots & -\dfrac{\lambda}{n} K\left(\dfrac{1}{n},\dfrac{i-1}{n}\right) & 0 & \cdots & -\dfrac{\lambda}{n} K\left(\dfrac{1}{n},1\right) \\[2mm] -\dfrac{\lambda}{n} K\left(\dfrac{2}{n},\dfrac{1}{n}\right) & \cdots & -\dfrac{\lambda}{n} K\left(\dfrac{2}{n},\dfrac{i-1}{n}\right) & 0 & \cdots & -\dfrac{\lambda}{n} K\left(\dfrac{2}{n},1\right) \\[2mm] \cdot & & \cdot & & & \cdot \\ \cdot & & \cdot & & & \cdot \\ \cdot & & \cdot & & & \cdot \\[2mm] -\dfrac{\lambda}{n} K\left(\dfrac{j}{n},\dfrac{1}{n}\right) & \cdots & -\dfrac{\lambda}{n} K\left(\dfrac{j}{n},\dfrac{i-1}{n}\right) & 1 & \cdots & -\dfrac{\lambda}{n} K\left(\dfrac{j}{n},1\right) \\[2mm] \cdot & & \cdot & \cdot & & \cdot \\ \cdot & & \cdot & \cdot & & \cdot \\ \cdot & & \cdot & \cdot & & \cdot \\[2mm] -\dfrac{\lambda}{n} K\left(1,\dfrac{1}{n}\right) & \cdots & -\dfrac{\lambda}{n} K\left(1,\dfrac{i-1}{n}\right) & 0 & \cdots & 1 - \dfrac{\lambda}{n} K(1,1) \end{vmatrix}$$

$$\tag{4}$$

where $D_n(\lambda)$ is the determinant of the system (2) and $D_n(i, j; \lambda)$ agrees with $D_n(\lambda)$ except for the ith column where all but the jth term is 0 and the jth term is 1. Then

$$\phi\left(\frac{i}{n}\right) = \frac{1}{D_n(\lambda)} \sum_{j=1}^{n} D_n(i, j; \lambda) f\left(\frac{j}{n}\right) \tag{5}$$

represents the solution of (2), providing that $D_n(\lambda) \neq 0$.

It will be shown that in the limit as $n \to \infty$ $D_n(\lambda)$ and $D_n(i, j; \lambda)$ tend to limits $D(\lambda)$ and $D(x, y; \lambda)$ respectively. Then (5) will reduce to a solution of (1). For self-adjoint operators this method provides relatively little new information. The operator is clearly compact and if also self-adjoint the methods of Chapter 3 suffice. Of course, the present approach has an elegance worth noting. It has, however, one overwhelming advantage. It works even when the operator is not self-adjoint. In the latter eventuality the methods of Chapter 3 break down.

2. THE FREDHOLM THEORY

We shall now study the structure of $D_n(\lambda)$, defined in (3). Clearly, $D_n(\lambda)$ is a polynomial of degree n in λ. Then

$$D_n(\lambda) = \sum_{k=0}^{n} \frac{a_k(n)}{k!} \lambda^k$$

and

$$a_k(n) = \left(\frac{d}{d\lambda}\right)^k D_n(\lambda)\big|_{\lambda=0}.$$

To evaluate the above, we note that

$$\frac{d}{d\lambda} D_n(\lambda) = (\partial_1 + \partial_2 + \cdots + \partial_n) D_n(\lambda)$$

where ∂_i is a differentiation operator with respect to λ, acting only on the terms in the ith column. This is nothing but the differentiation formula for a product. Now

$$(\partial_1 + \partial_2 + \cdots + \partial_n)^k = \sum_{\alpha_1+\alpha_2+\cdots+\alpha_k=k} \frac{k!}{\alpha_1! \, \alpha_2! \cdots \alpha_n!} \partial_1^{\alpha_1} \partial_2^{\alpha_2} \ldots \partial_n^{\alpha_n}$$

by the multinomial theorem, so that

$$a_k(n) = \sum_{\alpha_1+\alpha_2+\cdots+\alpha_n=k} \frac{k!}{\alpha_1! \, \alpha_2! \cdots \alpha_n!} \partial_1^{\alpha_1} \partial_2^{\alpha_2} \cdots \partial_n^{\alpha_n} D_n(\lambda)\big|_{\lambda=0}$$

However, each column of $D_n(\lambda)$ is linear in λ so that every term where some $\alpha_i \geq 2$ vanishes. Now the sum representing $a_k(n)$ can be simplified to

$$a_k(n) = \sum_{\substack{\Sigma\alpha_1=k \\ \alpha_i=0 \text{ or } 1}} k! \, \partial_1^{\alpha_1} \partial_2^{\alpha_2} \cdots \partial_n^{\alpha_n} D_n(\lambda)\big|_{\lambda=0}$$

$$= \sum_{p} \partial_{i_1} \partial_{i_2} \cdots \partial_{i_k} \partial_{i_{k+1}}^0 \partial_{i_{k+2}}^0 \cdots \partial_{i_n}^0 D_n(\lambda)\big|_{\lambda=0} \tag{6}$$

where the above sum runs over all permutations of the subscripts $1, 2, \ldots, n$.

Finally we observe that

$$\partial_1 \partial_2 \cdots \partial_k D_n(\lambda)\big|_{\lambda=0}$$

$$= \begin{vmatrix}
-\dfrac{1}{n} K\!\left(\dfrac{1}{n},\dfrac{1}{n}\right) & -\dfrac{1}{n} K\!\left(\dfrac{1}{n},\dfrac{2}{n}\right) & \cdots & -\dfrac{1}{n} K\!\left(\dfrac{1}{n},\dfrac{k}{n}\right) & 0 & \cdots & 0 \\[2mm]
-\dfrac{1}{n} K\!\left(\dfrac{2}{n},\dfrac{1}{n}\right) & -\dfrac{1}{n} K\!\left(\dfrac{2}{n},\dfrac{2}{n}\right) & \cdots & -\dfrac{1}{n} K\!\left(\dfrac{2}{n},\dfrac{k}{n}\right) & 0 & \cdots & 0 \\[2mm]
\vdots & \vdots & & \vdots & \vdots & & \vdots \\[2mm]
-\dfrac{1}{n} K\!\left(\dfrac{k}{n},\dfrac{1}{n}\right) & -\dfrac{1}{n} K\!\left(\dfrac{k}{n},\dfrac{2}{n}\right) & \cdots & -\dfrac{1}{n} K\!\left(\dfrac{k}{n},\dfrac{k}{n}\right) & 0 & \cdots & 0 \\[2mm]
-\dfrac{1}{n} K\!\left(\dfrac{k+1}{n},\dfrac{1}{n}\right) & -\dfrac{1}{n} K\!\left(\dfrac{k+1}{n},\dfrac{2}{n}\right) & \cdots & -\dfrac{1}{n} K\!\left(\dfrac{k+1}{n},\dfrac{k}{n}\right) & 1 & \cdots & 0 \\[2mm]
\vdots & \vdots & & \vdots & \vdots & & \vdots \\[2mm]
-\dfrac{1}{n} K\!\left(\dfrac{n}{n},\dfrac{1}{n}\right) & -\dfrac{1}{n} K\!\left(\dfrac{n}{n},\dfrac{2}{n}\right) & \cdots & -\dfrac{1}{n} K\!\left(\dfrac{n}{n},\dfrac{k}{n}\right) & 0 & \cdots & 1
\end{vmatrix}$$

and similarly for the other terms in (6). It follows that

$$a_k(n) = \frac{(-1)^k}{n^k} \sum_{i_1,i_2,\ldots,i_k}
\begin{vmatrix}
K\!\left(\dfrac{i_1}{n},\dfrac{i_1}{n}\right) & \cdots & K\!\left(\dfrac{i_1}{n},\dfrac{i_k}{n}\right) \\[2mm]
\vdots & & \vdots \\[2mm]
K\!\left(\dfrac{i_k}{n},\dfrac{i_1}{n}\right) & \cdots & K\!\left(\dfrac{i_k}{n},\dfrac{i_k}{n}\right)
\end{vmatrix}. \tag{7}$$

We now wish to investigate what happens in the limit as $n \to \infty$. Clearly

$$a_0(n) = 1$$

and

$$a_1(n) = \frac{-1}{n} \sum_{i_1=1}^{n} K\!\left(\frac{i_1}{n},\frac{i_1}{n}\right). \tag{8}$$

It follows that

$$\lim_{n \to \infty} a_1(n) = -\int_0^1 K(x,x)\,dx \tag{9}$$

since (8) is a Riemann sum that approximates (9) and $K(x, x)$ is a continuous function. Now

$$a_2(n) = \frac{1}{n^2} \sum_{i_1, i_2} \begin{vmatrix} K\left(\frac{i_1}{n}, \frac{i_1}{n}\right) & K\left(\frac{i_1}{n}, \frac{i_2}{n}\right) \\ K\left(\frac{i_2}{n}, \frac{i_1}{n}\right) & K\left(\frac{i_2}{n}, \frac{i_2}{n}\right) \end{vmatrix}$$

and as before

$$\lim_{n \to \infty} a_2(n) = \int_0^1 \int_0^1 \begin{vmatrix} K(x_1, x_1) & K(x_1, x_2) \\ K(x_2, x_1) & K(x_2, x_2) \end{vmatrix} dx_1 \, dx_2.$$

To investigate the general case it will prove convenient to introduce the following notation.

$$K\begin{pmatrix} x_1 & x_2 & \cdots & x_n \\ y_1 & y_2 & \cdots & y_n \end{pmatrix} = \begin{vmatrix} K(x_1, y_1) & K(x_1, y_2) & \cdots & K(x_1, y_n) \\ K(x_2, y_1) & K(x_2, y_2) & \cdots & K(x_2, y_n) \\ \cdot & \cdot & & \cdot \\ \cdot & \cdot & & \cdot \\ \cdot & \cdot & & \cdot \\ K(x_n, y_1) & K(x_n, y_2) & \cdots & K(x_n, y_n) \end{vmatrix} \quad (10)$$

Using this notation we see that

$$\lim_{n \to \infty} a_1(n) = -\int_0^1 K\begin{pmatrix} x \\ x \end{pmatrix} dx$$

$$\lim_{n \to \infty} a_2(n) = \int_0^1 \int_0^1 K\begin{pmatrix} x_1 & x_2 \\ x_1 & x_2 \end{pmatrix} dx_1 \, dx_2$$

and more generally

$$\lim_{n \to \infty} a_k(n) = (-1)^k \int_0^1 \int_0^1 \cdots \int_0^1 K\begin{pmatrix} x_1 & x_2 & \cdots & x_k \\ x_1 & x_2 & \cdots & x_k \end{pmatrix} dx_1 \, dx_2 \ldots dx_k \quad (11)$$

so that finally

$$D(\lambda) = \lim_{n \to \infty} D_n(\lambda)$$

$$= \sum_{k=0}^{\infty} \frac{(-1)^k \lambda^k}{k!} \int_0^1 \int_0^1 \cdots \int_0^1 K\begin{pmatrix} x_1 & x_2 & \cdots & x_k \\ x_1 & x_2 & \cdots & x_k \end{pmatrix} dx_1 \, dx_2 \ldots dx_k. \quad (12)$$

To investigate the convergence of (12) we use the Hadamard inequality, derived in Chapter 1. Then, if

$$|K(x, y)| \le M$$

we have

$$\left| K \begin{pmatrix} x_1 & x_2 & \cdots & x_k \\ y_1 & y_2 & \cdots & y_k \end{pmatrix} \right| \leq M^k k^{k/2} \tag{13}$$

and

$$|D(\lambda)| \leq \sum_{k=0}^{\infty} \frac{(|\lambda| \, M k^{1/2})^k}{k!} \tag{14}$$

The series in (14) converges for all λ since, by the ratio test,

$$\lim_{k \to \infty} \frac{(|\lambda| \, M(k+1)^{1/2})^{k+1}/(k+1)!}{(|\lambda| \, M k^{1/2})^k/k!} = \lim_{k \to \infty} \frac{|\lambda| \, M[1 + (1/k)]^{k/2}}{(k+1)^{1/2}} = 0.$$

It follows that $D(\lambda)$ as a function of λ converges for all λ and therefore is an entire function of λ.

Then $D(\lambda)$ is known as the Fredholm determinant. It will be seen that its zeros are the eigenvalues of the integral operator in (1). Since $D(\lambda)$ is an entire function it can have only a finite number of zeros in a bounded region of the λ plane. Accordingly, if $D(\lambda)$ has an infinite number of zeros they must accumulate at infinity.

Next, we examine the expression $D_n(i, j; \lambda)$ defined in (4). For $i = j$ it is of the same type as $D_{n-1}(\lambda)$ in (3) so that

$$\lim_{n \to \infty} D_n(i, i; \lambda) = \lim_{n \to \infty} D_n(\lambda) = D(\lambda). \tag{15}$$

It is of course possible to obtain the limit of $D_n(i, j; \lambda)$ as was done in the case of $D_n(\lambda)$. An alternate procedure, however, is the following. Equation (5) can be rewritten as

$$\phi\left(\frac{i}{n}\right) = \frac{D_n(i, i; \lambda)}{D_n(\lambda)} f\left(\frac{i}{n}\right) + \frac{1}{D_n(\lambda)} \sum_{\substack{j=1 \\ j \neq i}}^{n} D_n(i, j; \lambda) f\left(\frac{j}{n}\right).$$

Using (15) one can take a formal limit of the above as $n \to \infty$, to obtain

$$\phi(x) = f(x) + \int_0^1 \frac{D(x, y; \lambda)}{D(\lambda)} f(y) \, dy. \tag{16}$$

Equation (16) shall be verified independently of the above limiting procedure, and $D(x, y; \lambda)$ shall be calculated directly.

We now insert (16) in (1) to obtain

$$\int_0^1 \left[D(x, y; \lambda) - \lambda D(\lambda) K(x, y) - \lambda \int_0^1 K(x, z) D(z, y; \lambda) \, dz \right] f(y) \, dy = 0. \tag{17}$$

In order for (17) to hold for all continuous $f(y)$ the term in the bracket must vanish, since we are dealing with continuous functions. Then

$$D(x, y; \lambda) = \lambda D(\lambda)K(x, y) + \lambda \int_0^1 K(x, z)D(z, y; \lambda)\, dz. \qquad (18)$$

We now expand

$$D(\lambda) = \sum_{n=0}^{\infty} c_n \frac{(-\lambda)^n}{n!} \qquad (19)$$

$$D(x, y; \lambda) = \sum_{n=0}^{\infty} C_n(x, y) \frac{(-\lambda)^n}{n!} \qquad (20)$$

and insert these in (18). It will be shown subsequently that the summation in (20) is uniformly convergent so that the interchange of summation and integration is permissible. A comparison of coefficients shows that

$$C_0(x, y) = 0$$
$$C_{n+1}(x, y) = -(n + 1)\left[c_n K(x, y) + \int_0^1 K(x, z)C_n(z, y)\, dz \right]. \qquad (21)$$

It follows that

$$C_1(x, y) = -K(x, y)c_0 = -K(x, y)$$

and

$$C_2(x, y) = -2\left[K(x, y) \int_0^1 K(x_1, x_1)\, dx_1 - \int_0^1 K(x, x_1)K(x_1, y)\, dx_1 \right]$$

$$= -2 \int_0^1 K\begin{pmatrix} x & x_1 \\ y & y_1 \end{pmatrix} dx_1.$$

More generally we will show that

$$C_{n+1}(x, y) = -(n + 1) \int_0^1 \int_0^1 \cdots \int_0^1 K\begin{pmatrix} x & x_1 & \cdots & x_n \\ y & x_1 & \cdots & x_n \end{pmatrix} dx_1\, dx_2 \cdots dx_n \qquad (22)$$

using the notation defined in (10). Expanding the determinant in (22) by minors with respect to the first row we obtain

$$K\begin{pmatrix} x & x_1 & \cdots & x_n \\ y & x_1 & \cdots & x_n \end{pmatrix}$$

$$= K(x, y)K\begin{pmatrix} x_1 & \cdots & x_n \\ x_1 & \cdots & x_n \end{pmatrix} - K(x, x_1)\begin{pmatrix} x_1 & x_2 & \cdots & x_n \\ y & x_2 & \cdots & x_n \end{pmatrix}$$

$$- K(x, x_2)K\begin{pmatrix} x_1 & x_2 & \cdots & x_n \\ x_1 & y & \cdots & x_n \end{pmatrix} - \cdots - K(x, x_n)K\begin{pmatrix} x_1 & x_2 & \cdots & x_n \\ x_1 & x_2 & \cdots & y \end{pmatrix}.$$

In alternate terms certain columns were interchanged to produce a sequence of minus signs. We now integrate both sides and note that the last n terms on the right all produce the same integral. Then by use of (22) for the case n we find

$$\int_0^1 \int_0^1 \cdots \int_0^1 K \begin{pmatrix} x & x_1 & \cdots & x_n \\ y & x_1 & \cdots & x_n \end{pmatrix} dx_1\, dx_2 \cdots dx_n$$

$$= K(x, y) \int_0^1 \int_0^1 \cdots \int_0^1 K \begin{pmatrix} x_1 & \cdots & x_n \\ x_1 & \cdots & x_n \end{pmatrix} dx_1\, dx_2 \cdots dx_n$$

$$+ \int_0^1 K(x, x_1) \left[-n \int_0^1 \int_0^1 \cdots \int_0^1 K \begin{pmatrix} x_1 & x_2 & \cdots & x_n \\ y & x_2 & \cdots & x_n \end{pmatrix} dx_2 \cdots dx_n \right] dx_1$$

$$= c_n K(x, y) + \int_0^1 K(x, x_1) C_n(x_1, y)\, dx_1.$$

It follows that (22) satisfies (21). As in the case of $D(\lambda)$ it follows that $D(x, y; \lambda)$ is an entire function in λ for fixed values of x and y.

We can summarize the preceding discussion in the following theorem.

THEOREM 1. Let $K(x, y)$ be a continuous kernel for all $0 \leq x, y \leq 1$. The integral operator

$$(I - \lambda K)\phi = \phi(x) - \lambda \int_0^1 K(x, y)\phi(y)\, dy$$

defined on $L_2[0, 1]$ has an inverse given by

$$f(x) + \int_0^1 \frac{D(x, y; \lambda)}{D(\lambda)} f(y)\, dy$$

where $D(x, y; \lambda)$ and $D(\lambda)$ are given in (20) and (12), provided $D(\lambda) \neq 0$. Both of these functions are entire functions of λ and the zeros of $D(\lambda)$, if infinite in number, accumulate at infinity.

Originally the operator K was studied when it acts on continuous functions $f(x)$, in order to use finite difference approximations. It can however, be applied to functions in $L_2[0, 1]$. Similarly, the operator defined in (16) was shown to be inverse to $I - \lambda K$ when acting on continuous functions. But (16) too can be extended to functions in $L_2[0, 1]$ so that $I - \lambda K$ and $(I - \lambda K)^{-1}$ are defined on $L_2[0, 1]$, providing that $D(\lambda) \neq 0$, of course.

In Chapter 2 it was shown that for small λ

$$(I - \lambda K)^{-1} f = \sum_0^\infty \lambda^n \int_0^1 K_n(x, y) f(y)\, dy \tag{23}$$

where $K_n(x, y)$ is defined recursively by

$$K_{n+1}(x, y) = \int_0^1 K(x, z) K_n(z, y) \, dz.$$

These facts enable us to prove a theorem regarding the structure of $D(\lambda)$.

THEOREM 2.

$$-\lambda D'(\lambda) = \int_0^1 D(x, x; \lambda) \, dx \tag{25}$$

$$\frac{D'(\lambda)}{D(\lambda)} = -\sum_{n=1}^\infty \lambda^{n-1} \int_0^1 K_n(x, x) \, dx. \tag{26}$$

Equation (26) holds for all $|\lambda|$ sufficiently small; that is for all $|\lambda| < |\lambda_0|$ where λ_0 is a zero of $D(\lambda)$ closest to the origin.

PROOF. Equation (25) is a consequence of a direct integration of (20). Using (22) we have

$$\int_0^1 C_n(x, x) \, dx = -nc_n$$

so that

$$\int_0^1 D(x, x; \lambda) \, d\lambda = -\sum_{n=1}^\infty \frac{(-\lambda)^n}{(n-1)!} c_n$$

and also

$$D'(\lambda) = \sum_{n=1}^\infty \frac{(-1)^n \lambda^{n-1}}{(n-1)!} c_n.$$

A direct comparison of the last two series yields (25).

Use of (23) and the inverse operator given in theorem 1 shows that

$$f(x) + \int_0^1 \frac{D(x, y; \lambda)}{D(\lambda)} f(y) \, dy = f(x) + \sum_{n=1}^\infty \lambda^n \int_0^1 K_n(x, y) f(y) \, dy$$

provided $|\lambda|$ is sufficiently small. It follows that

$$\frac{D(x, y; \lambda)}{D(\lambda)} = \sum_{n=1}^\infty \lambda^n K_n(x, y)$$

By letting $x = y$ in the above and integrating one obtains, by use of (25)

$$\frac{D'(\lambda)}{D(\lambda)} = -\sum_{n=1}^\infty \lambda^{n-1} \int_0^1 K_n(x, x) \, dx$$

which is of course (26). ■

A consequence of this theorem is the following. Suppose K is a Volterra operator so that

$$K(x, y) = 0, \qquad y > x.$$

Then

$$K_2(x, y) = \int_0^1 K(x, z)K(z, y)\, dz$$

$$= \int_y^x K(x, z)K(z, y)\, dz, \qquad x > y$$

$$= 0, \qquad\qquad\qquad\qquad \text{otherwise}$$

so that

$$K_2(x, x) = 0$$

and more generally

$$K_n(x, x) = 0, \qquad n \geq 2.$$

Then, using (26)

$$\frac{D'(\lambda)}{D(\lambda)} = -\int_0^1 K(x, x)\, dx$$

it follows that

$$D(\lambda) = \exp\left(-\lambda \int_0^1 K(x, x)\, dx\right).$$

Hence $D(\lambda)$ can never vanish and a Volterra operator has no eigenvalues.

We will now show that the zeros of $D(\lambda)$ are indeed the eigenvalues of the integral operator.

THEOREM 3. Every zero of $D(\lambda)$ is an eigenvalue of the integral operator K, and in turn every eigenvalue of K is a zero of $D(\lambda)$.

PROOF. Suppose λ_0 is an eigenvalue and we expand $D(\lambda)$ and $D(x, y; \lambda)$ about that point. Then

$$D(\lambda) = \sum_{n=r}^{\infty} d_n(\lambda - \lambda_0)^n, \qquad d_r \neq 0$$

and

$$D(x, y; \lambda) = \sum_{n=s}^{\infty} d_n(x, y)(\lambda - \lambda_0)^n, \qquad d_s(x, y) \not\equiv 0.$$

Inserting these in (25) we have

$$-\lambda \sum_{n=r}^{\infty} n d_n(\lambda - \lambda_0)^{n-1} = \sum_{n=s}^{\infty} (\lambda - \lambda_0)^n \int_0^1 d_n(x, x)\, dx$$

so that

$$s \leq r - 1.$$

It follows that $D(x, y; \lambda)$ cannot vanish identically. Now the above expansion can be inserted in (18) to obtain

$$\sum_{n=s}^{\infty} d_n(x, y)(\lambda - \lambda_0)^n = \lambda \sum_{n=r}^{\infty} d_n(\lambda - \lambda_0)^n K(x, y)$$

$$+ \lambda \sum_{n=s}^{\infty} (\lambda - \lambda_0)^n \int_0^1 K(x, z) \, d_n(z, y) \, dz$$

One can now divide the above by $(\lambda - \lambda_0)^s$ and then let $\lambda = \lambda_0$, resulting in

$$d_s(x, y) = \lambda_0 \int_0^1 K(x, z) \, d_s(z, y) \, dz. \tag{27}$$

Since $d_s(x, y) \not\equiv 0$ we can select a value of y, say y_0, so that

$$d_s(x, y_0) = g(x) \not\equiv 0$$

and $g(x)$ is an eigenfunction of the integral operator.

In turn, if λ_0 is an eigenvalue of K, $I - \lambda_0 K$ will not have an inverse. The inverse operator defined in (16) can fail to exist only if $D(\lambda)$ vanishes. Hence, if $I - \lambda_0 K$ has no inverse, $D(\lambda_0) = 0$ so that all eigenvalues are zeros of $D(\lambda)$. ■

One can develop the Fredholm method considerably further and use it to prove the Fredholm alternative. The latter was already proved in theorem 22 in Chapter 3. Rather than do this it will prove to be more valuable to investigate results that are specific to the Fredholm theory instead of results that are more conveniently derived by general operator theoretical techniques. The next sections will be devoted to an investigation of the analytic properties of the Fredholm determinant.

3. ENTIRE FUNCTIONS

We shall devote this section to a review of the pertinent facts about entire functions. No proofs will be provided. For proofs and additional details we refer to any treatise on analytic functions, in particular, to Chapter 14 of Analytic Function Theory, Vol. 2, by E. Hille.

Let $f(z)$ denote an entire function; that is an analytic function

$$f(z) = \sum_{n=0}^{\infty} a_n z^n$$

where the series converges for all finite values of z. By $M(r)$ we denote its maximum modulus, namely,

$$M(r) = \max_{0 \le \theta \le 2\pi} |f(re^{i\theta})|.$$

The order ρ of $f(z)$ is now defined to be

$$\rho = \limsup_{r \to \infty} \frac{\log \log M(r)}{\log r} . \tag{28}$$

It is intuitively clear that the order ρ is related to the asymptotic properties of $f(z)$ via

$$|f(z)| \leq a e^{b|z|^\rho}$$

for suitable positive constants a and b.

One can also deduce the order of an entire function $f(z)$ from the structure of its Taylor coefficients

$$\rho = \limsup_{n \to \infty} \frac{n \log n}{-\log |a_n|} \tag{29}$$

Suppose $\{r_n\}$ is an increasing sequence of positive numbers such that

$$\sum_{n=1}^{\infty} \frac{1}{r_n^\alpha} < \infty \qquad \text{for all} \quad \alpha > \sigma$$

and that σ is the smallest number for which the above statement is true. Then σ is known as the exponent of convergence of the above series.

Next we introduce the set of functions $E(z, p)$. These are sometimes known as the Weierstrass prime factors.

$$E(z, 0) = 1 - z$$
$$E(z, p) = (1 - z) \exp \left\{ z + \tfrac{1}{2} z + \cdots + \frac{1}{p} z^p \right\}, \qquad p \geq 1. \tag{30}$$

Note that

$$\log E(z, p) = \log (1 - z) + z + \tfrac{1}{2} z^2 + \cdots + \frac{1}{p} z^p = -\sum_{k=p+1}^{\infty} \frac{1}{k} z^k$$

so that

$$E(z, p) = \exp \left\{ -\sum_{k=p+1}^{\infty} \frac{1}{k} z^k \right\} = 1 - \frac{z^{p+1}}{p + 1} + \cdots . \tag{31}$$

Now let $\{z_n\}$ be a set in the complex plane that has no finite point of accumulation and let σ be the exponent of convergence of

$$\sum_{n=1}^{\infty} \frac{1}{|z_n|^\alpha} \tag{32}$$

where σ is supposed to be finite by hypothesis. We shall now define an integer p as follows.

a. if σ is not an integer $p = [\sigma]$ where $[\sigma]$ denotes the greatest integer less than or equal to σ.

b. if σ is an integer and (32) converges for $\alpha = \sigma$, $p = \sigma - 1$.

c. if σ is an integer and (32) does not converge for $\alpha = \sigma$, $p = \sigma$.

It follows that the infinite product

$$P(z) = \prod_{n=1}^{\infty} E\left(\frac{z}{z_n}, p\right) \tag{33}$$

converges for all finite z and therefore represents an entire function. To show that it converges we use (31) and see that (33) converges like

$$\prod_{n=1}^{\infty} \left(1 - \frac{1}{p+1}\left|\frac{z}{z_n}\right|^{p+1}\right).$$

But the above product converges if and only if

$$\sum_{n=1}^{\infty} \frac{1}{p+1}\left|\frac{z}{z_n}\right|^{p+1}$$

converges. The exponent of convergence of the above series is σ and by the definition of p the series must converge. One can show that the entire function $P(z)$ is of order σ.

We are now in a position to state Hadamard's factorization theorem.

THEOREM 4. Let $f(z)$ be an entire function of finite order ρ. Then $f(z)$ can be represented in the form

$$f(z) = e^{g(z)} z^k \prod_{n=1}^{\infty} E\left(\frac{z}{z_n}, p\right). \tag{34}$$

In (34) k is a non-negative integer and $g(z)$ is a polynomial of degree $q \le \rho$. The set $\{z_n\}$ represents the zeros of $f(z)$, each z_n appearing as often in (34) as is warranted by its multiplicity. The product in (34) is an entire function of order σ where necessarily $\sigma \le \rho$. If ρ is not an integer $\sigma = \rho$ and $q \le [\rho] \le \rho$. If ρ is an integer at least one of σ and q is equal to ρ, therefore p is related to σ as in (33).

A useful corollary to the above theorem is the following.

COROLLARY. Let $f(z)$ be an entire function of finite order $\rho < 1$. Then $f(z)$ is either a polynomial, or else it must have an infinite number of zeros and

$$f(z) = z^k \prod_{n=1}^{\infty} \left(1 - \frac{z}{z_n}\right).$$

These facts about entire functions will enable us to obtain a great deal of information regarding the structure of the Fredholm determinant.

4. THE ANALYTIC STRUCTURE OF $D(\lambda)$

The Fredholm determinant $D(\lambda)$, as given in (12), is an entire function of λ. This follows from the estimate (14). If

$$D(\lambda) = \sum_{n=0}^{\infty} a_n \lambda^n$$

then

$$|a_n| \leq \frac{(Mn^{1/2})^n}{n!}$$

By the use of the Stirling asymptotic formula

$$n! = \sqrt{2\pi n} \left(\frac{n}{e}\right)^n \left[1 - \frac{1}{12n} + \cdots\right]$$

and (29), we see that $\rho \leq 2$. If $\{\lambda_n\}$ is the set of eigenvalues of the integral operator K, and therefore also the set of zeros of $D(\lambda)$ we have by (34)

$$D(\lambda) = e^{a\lambda + b\lambda^2} \prod_{1}^{\infty} E\left(\frac{\lambda}{\lambda_n}, p\right). \tag{35}$$

$\lambda = 0$ is certainly not an eigenvalue so that the term λ^k does not appear and we use the fact that $D(0) = 1$. In accordance with theorem 4 we have $\sigma \leq 2$. If $\sigma = 2$, $p = 2$, or $p = 3$ depending on whether $\sum (1/|\lambda_n|^2)$ does or does not converge. We shall show that this sum does converge so that $p = 2$.

In order to show that $p = 2$ we require a lemma about matrices.

LEMMA. Let $A = (a_{ij})$ be an $n \times n$ matrix and $\{\mu_i\}$ the set of its eigenvalues. Then

$$\sum_{i=1}^{n} |\mu_i|^2 \leq \sum_{i,j=1}^{n} |a_{ij}|^2.$$

Equality in the above is attained if and only if A is a normal matrix.

The above result is due to I. Schur. (For a proof of it we refer to *Introduction to Linear Algebra* by M. Marcus and H. Minc, The Macmillan Co., New York, 1965, p. 169.)

THEOREM 5. Let $Kf = \int_0^1 K(x, y) f(y) \, dy$, where $K(x, y)$ is continuous. The eigenvalues of this operator satisfy

$$\sum_{i=1}^{\infty} \frac{1}{|\lambda_i|^2} \leq \int_0^1 \int_0^1 |K(x, y)|^2 \, dx \, dy.$$

PROOF. For a self-adjoint operator the above result was proved in theorem 14 of Chapter 3. In fact for that case equality holds in the above inequality.

The expression for $D_n(\lambda)$, defined by (3), can be written as

$$D_n(\lambda) = |I - \lambda K_n|$$

where K_n is the $n \times n$ matrix $K_n = [(1/n)K(i/n, j/n)]$. Denote now the zeros of $D_n(\lambda)$ by $\{\lambda_i^{(n)}\}$. Use of the lemma now shows that

$$\sum_{i=1}^{n} \frac{1}{|\lambda_i^{(n)}|^2} \leq \sum_{i,j=1}^{n} \frac{1}{n^2} \left| K\left(\frac{i}{n}, \frac{j}{n}\right) \right|^2.$$

In the limit as $n \to \infty$ the above reduces to the desired result. ■
Now, in accordance with theorem 4, $p = 2$ in (35) so that

$$D(\lambda) = e^{a\lambda + b\lambda^2} \prod_{1}^{\infty} \left(1 - \frac{\lambda}{\lambda_n}\right) e^{\lambda/\lambda_n} \tag{36}$$

The conclusions of theorem 5 can be refined considerably, as the following corollary shows.

COROLLARY.   Let $Kf = \int_0^1 K(x, y)f(y)\, dy$, where $K(x, y)$ is continuous for $0 \leq x, y \leq 1$. The eigenvalues of this operator satisfy

$$\sum_{i=1}^{\infty} \frac{1}{\lambda_i^2} = \int_0^1 \int_0^1 K(x, y)K(y, x)\, dy\, dx = \int_0^1 K_2(x, x)\, dx. \tag{37}$$

PROOF.   As in the proof of theorem 5 we have the matrices

$$K_n = \left(\frac{1}{n} K\left(\frac{i}{n}, \frac{j}{n}\right)\right)$$

and

$$K_n^2 = \left(\sum_{k=1}^{n} \frac{1}{n^2} K\left(\frac{i}{n}, \frac{k}{n}\right) K\left(\frac{k}{n}, \frac{j}{n}\right)\right).$$

Now, by a standard theorem for matrices, (that the trace of a matrix is equal to the sum of the eigenvalues)

$$\text{trace } K_n^2 = \sum_{i=1}^{n} \sum_{k=1}^{n} \frac{1}{n^2} K\left(\frac{i}{n}, \frac{k}{n}\right) K\left(\frac{k}{n}, \frac{i}{n}\right) = \sum_{i=1}^{n} \frac{1}{(\lambda_i^{(n)})^2}.$$

In the limit as $n \to \infty$ the above becomes

$$\sum_{i=1}^{\infty} \frac{1}{\lambda_i^2} = \int_0^1 \int_0^1 K(x, y)K(y, x)\, dy\, dx = \int_0^1 K_2(x, x)\, dx. \quad ■$$

In fact the above proof can be adapted to show that

$$\sum_{i=1}^{\infty} \frac{1}{\lambda_i^n} = \int_0^1 K_n(x, x)\, dx, \qquad n \geq 2. \tag{38}$$

This result was derived earlier in theorem 14 of Chapter 3, but there the operator K was taken to be self-adjoint. In (38) we did not require self-adjointness, but did require that $K(x, y)$ be continuous.

THEOREM 6. Let $Kf = \int_0^1 K(x, y)f(y)\, dy$, where K is self-adjoint and $K(x, y)$ is continuous. Then the operator K has at least one nonzero eigenvalue.

PROOF. Suppose that $D(\lambda)$ has no zeros, or equivalently K has no nonzero eigenvalues. Then (35) reduces to

$$D(\lambda) = e^{a\lambda + b\lambda^2}$$

so that

$$\frac{D'(\lambda)}{D(\lambda)} = a + 2b\lambda.$$

Combining this result with (26) we have

$$-\sum_{n=1}^{\infty} \lambda^{n-1} \int_0^1 K_n(x, x)\, dx = a + 2b\lambda. \tag{39}$$

If K is self-adjoint then $K(x, y) = \overline{K(y, x)}$ and it follows that the functions $K_n(x, y)$, the iterated kernels, have also the symmetry property $K_n(x, y) = \overline{K_n(y, x)}$. It follows that

$$K_4(x, x) = \int_0^1 K_2(x, z)K_2(z, x)\, dz = \int_0^1 |K_2(x, z)|^2\, dz.$$

Using (39) we have

$$\int_0^1 K_4(x, x)\, dx = \int_0^1 \int_0^1 |K_2(x, z)|^2\, dz\, dx = 0$$

so that $K_2(x, z)$ is a null function. But a continuous null function necessarily vanishes so that

$$K_2(x, z) \equiv 0.$$

Similarly

$$K_2(x, x) = \int_0^1 |K(x, z)|^2\, dz = 0$$

so that $K(x, z) \equiv 0$. But the latter is clearly false so that $D(\lambda)$ has to vanish at least once. ∎

We shall now derive some results that enable us to refine the structure of $D(\lambda)$ even further.

THEOREM 7. Let $D(\lambda)$ be the Fredholm determinant of an integral operator K with a continuous kernel $K(x, y)$ defined on $L_2[0, 1]$. Then

$$D(\lambda) = e^{a\lambda} \prod_{1}^{\infty} \left(1 - \frac{\lambda}{\lambda_i}\right) e^{\lambda/\lambda_i} \tag{40}$$

PROOF.   Let

$$F(\lambda) = \prod_1^\infty \left(1 - \frac{\lambda}{\lambda_i}\right) e^{\lambda/\lambda_i}$$

so that

$$\frac{F'(\lambda)}{F(\lambda)} = -\sum_{i=1}^\infty \sum_{n=1}^\infty \frac{\lambda^n}{\lambda_i^{n+1}} = -\sum_{n=2}^\infty \lambda^{n-1} \int_0^1 K_n(x, x)\, dx$$

by means of (38). If we compare $F'(\lambda)/F(\lambda)$ to $D'(\lambda)/D(\lambda)$ we have

$$\frac{F'(\lambda)}{F(\lambda)} - \frac{D'(\lambda)}{D(\lambda)} = \int_0^1 K(x, x)\, dx$$

so that

$$D(\lambda) = \exp\left(-\lambda \int_0^1 K(x, x)\, dx\right) F(\lambda) \qquad (41)$$

which is of course (40) with $a = -\int_0^1 K(x, x)\, dx$. ∎

In many problems of mathematical physics it is important to know whether an integral operator does or does not have eigenvalues. The following theorem can be of great value in reaching such conclusions.

THEOREM 8.   Let $D(\lambda)$ be the Fredholm determinant of an integral operator K, with a continuous kernel. A necessary and sufficient condition for K having at least one nonzero eigenvalue is that for at least one $n \geq 2$, $\int_0^1 K_n(x, x)\, dx \neq 0$.

PROOF.   If K has no eigenvalues then

$$\frac{D'(\lambda)}{D(\lambda)} = -\sum_{n=1}^\infty \lambda^{n-1} \int_0^1 K_n(x, x)\, dx = -\int_0^1 K(x, x)\, dx$$

by the use of (41). In that case

$$\int_0^1 K_n(x, x)\, dx = 0, \qquad n \geq 2. \qquad (42)$$

It follows that if at least one of these integrals does not vanish $D(\lambda)$ must have a zero. Conversely if (42) holds $D(\lambda)$ has no zeros. ∎

EXAMPLE 1.   In laser theory the following integral equation arises

$$\phi(x) - \lambda \int_{-1}^1 e^{iH(x-y)^2} \phi(y)\, dy = 0. \qquad (43)$$

Here, H is a physical parameter, that is real in applications. The eigenvalues and eigenfunctions of (43) correspond to the resonant modes of the system. If (43) had no eigenvalues one of two possibilities would arise. Either the laser cannot work or the mathematical model (43) is an incorrect model. Now we consider the expression

$$\int_{-1}^{1} K_2(x, x)\, dx = \int_{-1}^{1} \int_{-1}^{1} e^{2iH(x-y)^2}\, dx\, dy \tag{44}$$

For $H = 0$ this integral does not vanish. For H in general, the integral is an entire function of H and hence has at most a countable number of H values for which it can vanish. We conclude that except for at most a countable number of H values, at least one eigenvalue exists.

We shall turn our attention once more to self-adjoint operators and in particular to positive operators. These were discussed at great length in Section 5 of Chapter 3. There it was shown in Corollary 2 to theorem 17 that

$$\int_{0}^{1} K(x, x)\, dx = \sum_{i=1}^{\infty} \mu_i = \sum_{i=1}^{\infty} \frac{1}{\lambda_i}. \tag{45}$$

With this result we can refine theorem 7 considerably.

THEOREM 9. Let K be a positive integral operator on $L_2[0, 1]$ with a continuous kernel. Then

$$D(\lambda) = \prod_{n=1}^{\infty} \left(1 - \frac{\lambda}{\lambda_n}\right). \tag{46}$$

PROOF. We have by theorem 7

$$D(\lambda) = \exp\left(-\lambda \int_{0}^{1} K(x, x)\, dx\right) \prod_{n=1}^{\infty} \left(1 - \frac{\lambda}{\lambda_n}\right) e^{\lambda/\lambda_n}.$$

In view of (45) $\prod_{n=1}^{\infty} [1 - (\lambda/\lambda_n)]$ converges so that

$$D(\lambda) = \exp -\left\{\lambda\left[\int_{0}^{1} K(x, x)\, dx - \sum_{n=1}^{\infty} \frac{1}{\lambda_n}\right]\right\} \prod_{n=1}^{\infty} \left(1 - \frac{\lambda}{\lambda_n}\right).$$

Combining the above with (45) leads immediately to the result

$$D(\lambda) = \prod_{n=1}^{\infty} \left(1 - \frac{\lambda}{\lambda_n}\right). \quad \blacksquare$$

In general, as was shown, $D(\lambda)$ is an entire function of order 2. For a positive operator, however, it follows from (46) that $D(\lambda)$ is of order at most 1. We shall now show that by imposing stronger conditions on the kernel $K(x, y)$ more precise information can be gained regarding the order of the Fredholm determinant.

THEOREM 10. Suppose that the kernel $K(x, y)$ is continuous for $0 \leq x$, $y \leq 1$ and satisfies a Hölder condition

$$|K(x, y_1) - K(x, y_2)| \leq M |y_1 - y_2|^{\alpha} \tag{47}$$

for some $0 < \alpha \leq 1$. Then $D(\lambda)$ is an entire function of order at most $2/(1 + 2\alpha)$ and

$$D(\lambda) = e^{a\lambda} \prod_{n=1}^{\infty} \left(1 - \frac{\lambda}{\lambda_n}\right) e^{\lambda/\lambda_n}. \qquad (48)$$

PROOF.   We know that

$$D(\lambda) = \sum_{0}^{\infty} \frac{(-\lambda)^n}{n!} \int_0^1 \int_0^1 \cdots \int_0^1 K \begin{pmatrix} x_1 & x_2 & \cdots & x_n \\ x_1 & x_2 & \cdots & x_n \end{pmatrix} dx_1\, dx_2 \cdots dx_n$$

where

$$K \begin{pmatrix} x_1 & x_2 & \cdots & x_n \\ x_1 & x_2 & \cdots & x_n \end{pmatrix}$$

$$= \begin{vmatrix} K(x_1, x_1) & K(x_1, x_2) & \cdots & K(x_1, x_n) \\ K(x_2, x_1) & K(x_2, x_2) & \cdots & K(x_2, x_n) \\ \cdot & \cdot & & \cdot \\ \cdot & \cdot & & \cdot \\ \cdot & \cdot & & \cdot \\ K(x_n, x_1) & K(x_n, x_2) & \cdots & K(x_n, x_n) \end{vmatrix}$$

$$= \begin{vmatrix} K(x_1, x_1) - K(x_1, x_2) & K(x_1, x_2) - K(x_1, x_3) & \cdots & K(x_1, x_n) \\ K(x_2, x_1) - K(x_2, x_2) & K(x_2, x_2) - K(x_2, x_3) & \cdots & K(x_2, x_n) \\ \cdot & \cdot & & \cdot \\ \cdot & \cdot & & \cdot \\ \cdot & \cdot & & \cdot \\ K(x_n, x_1) - K(x_n, x_2) & K(x_n, x_2) - K(x_n, x_3) & \cdots & K(x_n, x_n) \end{vmatrix} \qquad (49)$$

the last being accomplished by standard determinant manipulations. Using the Hölder condition and the Hadamard inequality we have

$$\left| K \begin{pmatrix} x_1 & x_2 & \cdots & x_n \\ x_1 & x_2 & \cdots & x_n \end{pmatrix} \right| \leq n^{n/2} M^n \{|x_1 - x_2|\, |x_2 - x_3| \cdots |x_{n-1} - x_n|\}^{\alpha}. \qquad (50)$$

In order to use (50) it is necessary that a good estimate on the bracketed term in (50) be obtained. First we shall assume that

$$0 \leq x_1 \leq x_2 \leq \cdots \leq x_n \leq 1. \qquad (51)$$

Under this assumption we can show that

$$|x_1 - x_2|\, |x_2 - x_3| \cdots |x_{n-1} - x_n| \leq \frac{1}{(n-1)^{n-1}}. \qquad (52)$$

The latter follows by an inductive process. Clearly

$$0 \le |x_1 - x_2| = x_2 - x_1 \le x_2$$

$$0 \le |x_1 - x_2| |x_2 - x_3| = (x_2 - x_1)(x_3 - x_2) \le x_2(x_3 - x_2) \le \frac{x_3^2}{4}$$

Suppose now that

$$|x_1 - x_2| |x_2 - x_3| \cdots |x_{n-1} - x_n| \le \frac{x_n^{n-1}}{(n-1)^{n-1}}.$$

Then

$$|x_1 - x_2| \cdots |x_{n-1} - x_n| |x_n - x_{n+1}| \le \frac{x_n^{n-1}}{(n-1)^{n-1}} (x_{n+1} - x_n).$$

The last term can be maximized with respect to x_n, keeping x_{n+1} fixed, so that

$$|x_1 - x_2| \cdots |x_n - x_{n+1}| \le \frac{x_{n+1}^n}{n^n}.$$

Finally, if $x_{n+1} \le 1$, (52) is shown to be correct.

We wish to estimate

$$\int_0^1 \int_0^1 \cdots \int_0^1 K \begin{pmatrix} x_1 & \cdots & x_n \\ x_1 & \cdots & x_n \end{pmatrix} dx_1 \, dx_2 \cdots dx_n$$

by using (50). Equation (61) does not apply in the whole unit cube in the n dimensional integral. But at every point of the cube an inequality of the type (51) holds. By using the appropriate columns for our determinant manipulations in (49) we have

$$\left| K \begin{pmatrix} x_1 & \cdots & x_n \\ x_1 & \cdots & x_n \end{pmatrix} \right| \le n^{n/2} M^n \left\{ \frac{1}{(n-1)^{n-1}} \right\}^\alpha$$

throughout the cube so that

$$D(\lambda) = \sum_{n=0}^\infty c_n (-\lambda)^n$$

where

$$|c_n| = \frac{1}{n!} \left| \int_0^1 \int_0^1 \cdots \int_0^1 K \begin{pmatrix} x_1 & x_2 & \cdots & x_n \\ x_1 & x_2 & \cdots & x_n \end{pmatrix} dx_1 \, dx_2 \cdots dx_n \right|$$

$$\le \frac{n^{n/2} M^n}{n!} \left\{ \frac{1}{(n-1)^{n-1}} \right\}^\alpha.$$

By the use of (29) and Stirling's asymptotic formula for $n!$ we can now obtain an upper estimate on the order of $D(\lambda)$

$$\rho = \limsup_{n \to \infty} \frac{n \log n}{-\log |c_n|}$$

$$\leq \limsup_{n \to \infty} \frac{n \log n}{\log n! - n \log M - n/2 \log n + \alpha(n-1) \log(n-1)}$$

$$= \limsup_{n \to \infty} \frac{n \log n}{(\alpha + \frac{1}{2})n \log n} = \frac{2}{1 + 2\alpha}.$$

For $\alpha > 0$ we see that $\rho < 2$ and by the Hadamard factorization theorem

$$D(\lambda) = e^{a\lambda} \prod_{n=1}^{\infty} \left(1 - \frac{\lambda}{\lambda_n}\right) e^{\lambda/\lambda_n}. \quad \blacksquare$$

An important companion theorem is the following.

THEOREM 11. Under the hypotheses of theorem 10, but strengthened so that $\frac{1}{2} < \alpha \leq 1$, the order of $D(\lambda)$ is at most $2/(1 + 2\alpha) < 1$ and

$$D(\lambda) = \prod_{n=1}^{\infty} \left(1 - \frac{\lambda}{\lambda_n}\right) \tag{53}$$

PROOF. It was shown in theorem 10 that ρ is at most $2/(1 + 2\alpha)$ and if $\alpha > \frac{1}{2}$, we certainly have $\rho < 1$, so that by the Hadamard factorization theorem

$$D(\lambda) = \prod_{n=1}^{\infty} \left(1 - \frac{\lambda}{\lambda_n}\right). \quad \blacksquare$$

N.B. It can of course happen that $D(\lambda)$ reduces to a polynomial. Theorem 11 merely guarantees that $\sum (1/\lambda_n)$ converges under certain hypotheses, even if $D(\lambda)$ has an infinite number of zeros.

EXAMPLE 2. We consider the same integral equation as in Example 1, namely

$$\phi(x) - \lambda \int_{-1}^{1} e^{iH(x-y)^2} \phi(y) \, dy = 0.$$

Theorem 11 only tells us that $D(\lambda)$ is either a polynomial or else $D(\lambda)$ has an infinite series, but we will be able to preclude the possibility of $D(\lambda)$ being a polynomial. In the above case the kernel is differentiable so that the constant α in the Hölder condition is unity, and the order of $D(\lambda)$ is at most $\frac{2}{3}$. We shall show that, except for a countable set of H values, $D(\lambda)$ must be an infinite series so that $D(\lambda)$ has an infinite number of zeros. Let $H = ik$,

$k > 0$, so that

$$K\begin{pmatrix} x_1 & x_2 & \cdots & x_n \\ x_1 & x_2 & \cdots & x_n \end{pmatrix} = \begin{vmatrix} 1 & e^{-k(x_1-x_2)^2} & \cdots \\ e^{-k(x_1-x_2)^2} & 1 & \cdots \\ \cdot & \cdot & \\ \cdot & \cdot & \\ \cdot & \cdot & \end{vmatrix}$$

and clearly

$$\lim_{k \to \infty} K\begin{pmatrix} x_1 & x_2 & \cdots & x_n \\ x_1 & x_2 & \cdots & x_n \end{pmatrix} = 1$$

It follows that each coefficient in the series for $D(\lambda)$ is an entire function of H, that does not vanish identically. Hence for each coefficient in $D(\lambda)$ there are at most a countable number of H values for which it can vanish. Since the number of coefficients is also countable, then there is at most a countable number of H values for which $D(\lambda)$ reduces to a polynomial. In particular for $H = 0$

$$D(\lambda) = 1 - 2\lambda.$$

5. POSITIVE KERNELS

We had seen earlier that when an operator is compact, but not self-adjoint it need have no eigenvalues at all. A typical example of such an operator is, of course, a Volterra operator. In this section we shall investigate integral operators with positive kernels; that is operators with continuous kernels $K(x, y)$ for which $K(x, y) > 0$. These operators should not be confused with positive operators. The latter are necessarily self-adjoint and need not have positive kernels. It will be shown that operators with positive kernels necessarily have at least one positive eigenvalue. In order to prove this result we require the following function-theoretic lemma.

LEMMA.   Let

$$f(z) = \sum_{n=0}^{\infty} a_n z^n$$

be an analytic function and suppose that the above series has a finite radius of convergence R. Suppose that $a_n \geq 0$ for all n. Then the function $f(z)$ has a singularity at $z = R$.

PROOF.   It is well known that an analytic function must have at least one singularity on its circle of convergence. Suppose that $z = R$ is not a singular point of $f(z)$, in which case all points in a neighborhood of $z = R$ are also points of regularity. In this case $f(z)$ can be continued analytically outside the circle of convergence near $z = R$.

Now let

$$g(\zeta) = f(R - \epsilon + \zeta) = \sum_{n=0}^{\infty} a_n (R - \epsilon + \zeta)^n$$

$$= \sum_{r=0}^{\infty} \left(\sum_{n=r}^{\infty} a_n \binom{n}{r} (R - \epsilon)^{n-r} \right) \zeta^r.$$

For $0 < \epsilon < R$ the above series converges for all $|\zeta| < \epsilon$, and if $z = R$ is a point of regularity of $f(z)$, $g(\zeta)$ will have a radius of convergence greater than ϵ. In this case there will be a $\delta > \epsilon$, such that $g(\zeta)$ converges for $\zeta = \delta$. Then

$$\sum_{r=0}^{\infty} \left(\sum_{n=r}^{\infty} a_n \binom{n}{r} (R - \epsilon)^{n-r} \right) \delta^r$$

converges, and since all terms in the above are positive the series converges absolutely and therefore may be rearranged. Then

$$f(R - \epsilon + \delta) = \sum_{n=0}^{\infty} a_n (R - \epsilon + \delta)^n$$

converges. But this implies that the radius of convergence of $f(z)$ is at least $R - \epsilon + \delta > R$, which is a contradiction. It follows that $z = R$ is a singular point of $f(z)$. ∎

THEOREM 12. Let $K(x, y)$ be continuous and positive for all $0 \le x$, $y \le 1$. Then the integral operator associated with this kernel has at least one positive eigenvalue.

PROOF. If $K(x, y) > 0$, then

$$K_2(x, y) = \int_0^1 K(x, z) K(z, y) \, dz > 0$$

and by the corollary to theorem 8 we see that the operator K has at least one eigenvalue. More generally, it follows that $K_n(x, y) > 0$, and necessarily

$$\frac{D'(\lambda)}{D(\lambda)} = -\sum_{n=1}^{\infty} \lambda^{n-1} \int_0^1 K_n(x, x) \, dx$$

is a meromorphic function. The right side has a finite radius of convergence, say λ_0. All coefficients in this series are positive, and by the lemma $\lambda = \lambda_0$ is then a pole of $D'(\lambda)/D(\lambda)$. In this case $D(\lambda_0) = 0$ and λ_0 is an eigenvalue of K by theorem 3. ∎

THEOREM 13: Jentzsch's Theorem. Let $K(x, y)$ be continuous and positive on $0 \le x, y \le 1$. Let λ_0 be its smallest positive eigenvalue. All other eigenvalues are larger than λ_0 in absolute value and λ_0 is simple (i.e., there is only one independent eigenfunction associated with λ_0). The eigenfunction $\phi_0(x)$ associated with λ_0 is positive.

PROOF. We saw in theorem 12 that K has a positive eigenvalue λ_0. To show that the corresponding eigenfunction is positive we shall use the same technique employed previously in the proof of theorem 3. Equation 18 can be rewritten in the form

$$\frac{D(x, y; \lambda)}{D(\lambda)} = \lambda K(x, y) + \lambda \int_0^1 K(x, z) \frac{D(z, y; \lambda)}{D(\lambda)} \, dz. \tag{54}$$

From (16) we note that for $|\lambda| < \lambda_0$ (see also the proof of theorem 2)

$$\frac{D(x, y; \lambda)}{D(\lambda)} = \sum_{n=1}^{\infty} \lambda^n K_n(x, y), \qquad |\lambda| < \lambda_0 \tag{55}$$

where all coefficients in the above are positive.

As in theorem 3 we have

$$D(\lambda) = \sum_{n=r}^{\infty} d_n(\lambda - \lambda_0)^n, \qquad d_r \neq 0,$$

and

$$D(x, y; \lambda) = \sum_{n=s}^{\infty} d_n(x, y)(\lambda - \lambda_0)^n, \qquad d_s(x, y) \neq 0,$$

where

$$s \leq r - 1.$$

Then, using (55),

$$\sum_{n=1}^{\infty} \lambda^n K_n(x, y) = \frac{d_s(x, y)}{d_r(\lambda - \lambda_0)^{r-s}} + \cdots.$$

For $\lambda < \lambda_0$ we see that the left side is positive so that $d_s(x, y)$ must be of one sign. Inserting the above in (54), multiplying by $(\lambda - \lambda_0)^{r-s}$ and letting $\lambda \to \lambda_0$ we have

$$d_s(x, y) = \lambda_0 \int_0^1 K(x, z) d_s(z, y) \, dz.$$

We now let $y = y_0$ and set

$$\phi_0(x) = d_s(x, y_0).$$

Since $\phi_0(x)$ must be of one sign, we can assume without loss of generality that it is positive.

To complete the proof we suppose that λ_1 is another eigenvalue and

$$\lambda_1 \int_0^1 K(x, y)\psi(y) \, dy = \psi(x).$$

From the above we see that $\psi(x)$ is a continuous function, since $K(x, y)$ is continuous. Clearly every eigenfunction of such an operator is necessarily continuous. We also have

$$\lambda_0 \int_0^1 K(x, y)\phi_0(y) \, dy = \phi_0(x).$$

So far neither of these eigenfunctions has been normalized in any way whatever. Since $\phi_0(x)$ is positive we can multiply it by a suitable constant so that

$$|\psi(x)| \leq \phi_0(x), \qquad 0 \leq x \leq 1$$

and for at least one x_0, $|\psi(x_0)| = \phi_0(x_0)$. It follows that

$$\left| \frac{1}{\lambda_1} \psi(x) \right| \leq \int_0^1 K(x, y) \, |\psi(y)| \, dy$$

and then

$$\left| \frac{1}{\lambda_1} \psi(x) \right| - \frac{1}{\lambda_0} \phi_0(x) \leq \int_0^1 K(x, y)[|\psi(y)| - \phi_0(y)] \, dy \leq 0. \qquad (56)$$

Now we have

$$\left| \frac{1}{\lambda_1} \psi(x_0) \right| \leq \frac{1}{\lambda_0} \phi_0(x_0) = \frac{1}{\lambda_0} |\psi(x_0)|$$

so that $\lambda_0 \leq |\lambda_1|$.

We shall now consider the following two possibilities separately $|\psi(x)| \not\equiv \phi_0(x)$ and $|\psi(x)| \equiv \phi_0(x)$. In the former case we see that (56) is a strict inequality so that $\lambda_0 < |\lambda_1|$.

Finally, suppose that $|\psi(x)| \equiv \phi_0(x)$ and $\lambda_0 = |\lambda_1|$. We shall show that this eventuality cannot arise. Since both $\psi(x)$ and $\phi_0(x)$ are continuous functions, if $|\psi(x)| \equiv \phi_0(x)$ there exists a real function $\theta(x)$ such that $\psi(x) = e^{i\theta(x)}\phi_0(x)$. We also have $\lambda_1 = e^{i\alpha}\lambda_0$ for some real α. Then

$$\lambda_1 K\psi = \psi$$
$$\lambda_0 K\phi_0 = \phi_0$$

and

$$|\lambda_1 K\psi| = |\psi| = \phi_0 = \lambda_0 K\phi_0$$

Since $|\lambda_1| = \lambda_0$ we have

$$|K\psi| = K\phi_0 \qquad (57)$$

If we use the expressions

$$\psi = \cos \theta \phi_0 + i \sin \theta \phi_0$$

we see that

$$|K\psi|^2 = |K(\cos \theta \phi_0 + i \sin \theta \phi_0)|^2$$
$$= (K \cos \theta \phi_0)^2 + (K \sin \theta \phi_0)^2 = (K\phi_0)^2, \qquad (58)$$

by the use of (57). Now

$(K \cos \theta \phi_0)^2$

$$= \left[\int_0^1 K(x, y) \cos \theta(y)\phi_0(y) \, dy \right]^2 \leq \left[\int_0^1 K(x, y) |\cos \theta(y)| \, \phi_0(y) \, dy \right]^2$$

$$= \left\{ \int_0^1 K(x, y)[(1 - |\sin \theta(y)|)(1 + |\sin \theta(y)|)^{1/2} \phi_0(y) \, dy \right\}^2$$

$$\leq \int_0^1 K(x, y)(1 - |\sin \theta(y)|)\phi_0(y) \, dy \int_0^1 K(x, y)(1 + |\sin \theta(y)|)\phi_0(y) \, dy \quad (59)$$

by the Cauchy-Schwarz inequality. It follows from (59) that

$$(K \cos \theta \phi_0)^2 \leq [K\phi_0 - K |\sin \theta| \, \phi_0][K\phi_0 + K |\sin \phi| \, \phi_0]$$
$$= (K\phi_0)^2 - (K |\sin \theta| \, \phi_0)^2$$

and therefore

$$(K \cos \theta \phi_0)^2 + (K \sin \theta \phi_0)^2 \leq (K\phi_0)^2$$

An inspection of (58) and the above shows that in fact the equality holds in the above and therefore in (59) as well. It is known that the equality holds in the Cauchy-Schwarz inequality if and only if the functions concerned are linearly dependent. In this case we have now

$$(1 - |\sin \theta(y)|)^{1/2} = C(1 + |\sin \theta(y)|)^{1/2}$$

so that $\theta(y)$ has to be a constant so that $\psi(x) = k\phi_0(x)$ for some constant $k(|k| = 1)$. From this we can conclude that $\psi(x)$ is linearly dependent on $\phi_0(x)$ so that $\lambda_1 = \lambda_0$. All other eigenvalues must therefore exceed λ_0 in absolute value and λ_0 is a simple eigenvalue. ∎

EXERCISES

1. The integral equation

$$\phi(x) - \lambda \int_0^1 K(x, y)\phi(y) \, dy = f(x)$$

has a solution of the form

$$\phi(x) = f(x) + \int_0^1 \frac{D(x, y; \lambda)}{D(\lambda)} f(y) \, dy.$$

Solve in this form the following equations:

(a) $\phi(x) - \lambda \int_0^1 (xy^2 + yx^2)\phi(y) \, dy = f(x)$

(b) $\phi(x) - \lambda \int_0^1 (x^2 \sin \pi y + \cos \pi y)\phi(y) \, dy = f(x).$

2. Apply Hadamard's Factorization Theorem to
   (a) $\cos \sqrt{z}$
   (b) $\sin \pi z$
3. Prove that the trace of a matrix equals the sum of its eigenvalues.
4. Use theorem 8 to determine whether the following operators have at least one nonzero eigenvalue, where

$$\phi(x) - \lambda \int_a^b K(x, y)\phi(y) \, dy = f(x)$$

   (a) $K(x, y) = \sin \pi x \cos \pi y$ on $[0, 1]$
   (b) $K(x, y) = xy$ on $[0, 1]$
   (c) $K(x, y) = x + y$ on $[0, 1]$
   (d) $K(x, y) = x^2 + xy + y^2$ on $[0, 1]$
   (e) $K(x, y) = 1$ on $[0, 1]$.
5. Do the kernels in problem 4 satisfy the conditions of theorem 10? Find the eigenvalues and $D(\lambda)$.
6. Apply theorem 11 to the operators on $L_2[0, 1]$ associated with the following kernels.
   (a) $K(x, y) = e^{x-y}$
   (b) $K(x, y) = xy$
   (c) $K(x, y) = x + y$.
7. Do the following kernels satisfy theorem 12? If so, find one positive eigenvalue λ_0, using the method in the proof of theorem 12. The underlying space is $L_2[0, 1]$.
   (a) $K(x, y) = x^2 + y^2$
   (b) $K(x, y) = x^2 - y^2$
   (c) $K(x, y) = x^2 y + xy^2$
   (d) $K(x, y) = 1$
   (e) $K(x, y) = -1$
   (f) $K(x, y) = \sin 2\pi x \sin 2\pi y$.
8. Prove that if

$$\int_0^1 \phi(y) f(y) \, dy = 0$$

   for all continuous $f(y)$ and $\phi(y)$ is continuous, then $\phi(y) \equiv 0$.

7

NONLINEAR INTEGRAL EQUATIONS

SUMMARY

The previous chapters were primarily devoted to linear integral equations. As was shown for these, a substantial and powerful theory has been developed, but these techniques no longer apply to nonlinear integral equations. In this final chapter the Schauder Fixed Point Theorem will be proved and applied to a number of nonlinear equations, that cannot be treated by the methods discussed in earlier chapters.

1. INTRODUCTION

Basically the previous six chapters have been devoted to linear integral equations. The techniques discussed are generally not applicable to nonlinear cases, except for the fixed point theorems that were discussed in Chapter 2. There, it was shown that a contraction mapping acting on a Hilbert space had a unique fixed point. By means of this result certain existence theorems for linear as well as nonlinear integral equations were developed. However, rather severe restrictions had to be imposed on the equations in order for the resultant operators to be contraction operators. Theorem 4 of Chapter 1 provides such a result. Theorem 2 of that chapter was useful in lifting some of these restrictions when the integral operator was linear. It should be emphasized that these results led not only to existence, but also to uniqueness theorems.

The purpose of the present chapter is to develop a more general fixed point theorem, known as the Schauder fixed-point theorem. This theorem will enable us to prove existence theorems for nonlinear equations under rather

261

general conditions. This method does not, however, provide us with uniqueness results. The next section is devoted to a proof of the Schauder theorem. The reader who is willing to accept this theorem may wish to go on to the applications in subsequent sections.

2. THE SCHAUDER FIXED POINT THEOREM

To prove this theorem it will be necessary to derive a number of intermediate results.

THEOREM 1: Brouwer Fixed-Point Theorem. Let S be the closed unit sphere in an n dimensional real Euclidian space; that is, $S = [X \mid X$ in E_n and $\|X\| \leq 1]$. Let K be a continuous mapping of S into itself so that if $\|X\| \leq 1$, $\|K(X)\| \leq 1$. Then K has at least one fixed point; that is, there is at least one X in S such that $K(X) = X$.

If X has the components $(x_1, x_2, \ldots, x_n)$ we use the standard Euclidian norm $\|X\| = [\sum_i^n x_i^2]^{1/2}$. Note that the theorem guarantees the existence of at least one fixed point, but makes no assertion regarding its uniqueness.

PROOF. We shall first assume that $K(X)$ has continuous first derivatives with respect to all x_i. Subsequently, we will show how this restriction can be lifted. Consider now the expression $\psi(a, X) = X + a(X - K(X))$ where a is a real parameter and X is in S. Every point $\psi(a, X)$ lies on the line connecting X and $K(X)$. In particular $\psi(0, X) = X$ and $\psi(-1, X) = K(X)$. For $-1 < a < 0$, $\psi(a, X)$ is between X and $K(X)$.

We shall now make the assumption that $X \neq K(X)$ for all X in S. On the basis of this assumption we will be led to a contradiction so that $X = K(X)$ for at least one X, thereby establishing the theorem. If now $X \neq K(X)$ the line given by $\psi(a, X)$, $-\infty < a < \infty$ must intersect the surface of S at precisely two points. Since for $-1 < a < 0$, $\psi(a, X)$ lies between X and KX, there must be precisely one non-negative value of a denoted by $a(X)$, such that $\|\psi(a(X), X)\| = 1$. That value $a(X)$ can be derived analytically as the larger of the two roots of the quadratic equation

$$\|\psi(a, X)\|^2 = \|X + a(X - K(X))\|^2$$

$$= \|X\|^2 + 2a(X, X - K(X)) + a^2 \|X - K(X)\|^2 = 1. \quad (1)$$

In fact $a(X) = 0$ if and only if $\|X\| = 1$.

From (1) it follows (since $X - K(X) \neq 0$) that the larger of the two solutions $a(X)$ will have the same differentiability properties as $K(X)$. Now $\psi(a(X), X)$ is a continuous mapping of all points of S onto the boundary of S and if $K(X)$ has continuous first derivatives it follows that $\psi(a(X), X)$ is not only continuous, but has continuous first derivatives.

Consider now the function $f(t, X) = X + ta(X)(X - K(X))$ where t is a real parameter and $0 \leq t \leq 1$. Evidently $f(0, X) = X$ and $f(1, X) = \psi(a(X), X)$ so that $f(t, X)$ is a point on the line connecting X and the boundary point $\psi(a(X), X)$. Therefore $f(t, X)$ maps the sphere S into itself and has also continuous first derivatives, and $f(0, X)$ is the identity mapping and $f(1, X)$ maps the sphere onto the boundary of S. If $\|X\| = 1$ we see that $a(X) = 0$ so that $f(t, X) = X$ and on the boundary $f(t, X)$ is the identity. More generally let $S(t)$ denote the set of points $\{f(t, X)\}$ where $\|X\| \leq 1$, and let $V(t)$ denote the volume of $S(t)$. By a standard theorem* the *volume* $V(t)$ is given by

$$V(t) = \int_{\|X\| \leq 1} \left| \frac{\partial f}{\partial X} \right| dv \qquad (2)$$

where $|\partial f / \partial X|$ denotes the Jacobian of $f(t, X)$. Evidently

$$V(0) = \int_{\|X\| \leq 1} dv > 0 \qquad (3)$$

since $\partial f / \partial X|_{t=0} = I$ and $|I| = 1$, and $V(0)$ is the volume of the unit sphere. Also

$$V(1) = \int_{\|X\| \leq 1} \left| \frac{\partial f}{\partial X} \right|_{t=1} dv = 0 \qquad (4)$$

since $f(1, X)$ maps all points onto the boundary and the volume of the boundary must vanish.

We shall now show that $V(t)$ is in fact a constant, which contradicts (3) and (4). This contradiction will then provide us with the proof of the theorem. Since $f(t, X)$ is linear in t, $V(t)$ is necessarily a polynomial in t by (2). $f(t, X)$ will be shown to be a $1-1$ mapping of the sphere S onto itself for all $0 \leq t < \epsilon$ for some positive ϵ. Consider the equation

$$F(t, X) \equiv f(t, X) - Y = X + ta(X)(X - K(X)) - Y = 0 \qquad (5)$$

where $\|Y\| \leq 1$. We can show that for every Y in S (5) has a unique solution; at least for sufficiently small t. To prove this we shall invoke the implicit function theorem. First we observe that

$$F(0, Y) = 0 \qquad (6)$$

so that a solution exists at $t = 0$. To show that a unique solution exists near $t = 0$, we have to show that the functional derivative of $F(t, X)$ does not vanish near $t = 0$. But

$$\frac{\partial F}{\partial X}(0, Y) = I \qquad (7)$$

---

* See, for example, W. Maak, *An Introduction to Modern Calculus*, Holt, Rinehart, and Winston, New York, 1963, Sec. 11-4.

and since $F(t, X)$ has continuous first derivatives we see that

$$\left| \frac{\partial F}{\partial X}(0, Y) \right| = 1$$

and

$$\left| \frac{\partial F}{\partial X}(t, Y) \right| > 0, \quad 0 \leq t < \epsilon. \tag{8}$$

It follows that (5) has a unique solution for every Y in S. Therefore, $S(t) = S$ for small t and $V(t) = V(0)$. But a polynomial that is constant over some interval is identically constant. Then we have a contradiction to (3) and (4) and $X = K(X)$ for at least one X. Thus K has at least one fixed point.

This concludes the proof under the assumption that $K(X)$ has continuous first derivatives. To remove that restriction we make use of the Weierstrass approximation theorem. The latter asserts that any continuous function can be approximated arbitrarily closely over a closed and bounded domain by polynomials. Then we can find a sequence of differentiable mappings $\{K_n\}$ (where in fact each component is a polynomial) such that $\| K(X) - K_n(X) \| \leq 1/n$ for X in S. But for each K_n there exists X_n such that $K_n(X_n) = X_n$ by the preceding proof.* It follows that $\| K(X_n) - K_n(X_n) \| = \| K(X_n) - X_n \| \leq 1/n$. By the Bolzano-Weierstrass theorem it must be possible to extract from $\{X_n\}$ a convergent sequence $\{X_{n_k}\}$ such that $X_{n_k} \to \tilde{X}$. Then

$$\| K(\tilde{X}) - \tilde{X} \| = \| K(\tilde{X}) - K(X_{n_k}) + K(X_{n_k}) - X_{n_k} + X_{n_k} - \tilde{X} \|$$
$$\leq \| K(\tilde{X}) - K(X_{n_k}) \| + \| K(X_{n_k}) - X_{n_k} \| + \| X_{n_k} - \tilde{X} \|. \tag{9}$$

By the continuity of K and the fact that $X_{n_k} \to \tilde{X}$ the first and last terms on the right of (9) can be made small for large n_k. The middle term is at most $1/n_k$. It follows that

$$\| K(\tilde{X}) - \tilde{X} \| < \epsilon, \quad n_k > N(\epsilon)$$

and since the left side is independent of n we have $K(\tilde{X}) = \tilde{X}$ so that K has at least one fixed point. ∎

It is easy to see that the Brouwer fixed-point theorem can be applied not only to continuous mappings of the sphere S into itself, but to more general domains as well. Let $\tilde{S}$ be any set that can be put into a 1–1 correspondence with the unit sphere S by a continuous function f, such that f^{-1} exists and is also continuous. Let K be a continuous mapping of $\tilde{S}$ into itself, and for every X in S, $f(X)$ denotes a point in $\tilde{S}$. We can now define a continuous mapping of ψ of S into itself by

$$Kf(X) = f\psi(X). \tag{10}$$

* It is not obvious that K_n maps S into itself. But K_n can be so constructed that it maps S into the set $S_n = [X \,|\, \| X \| \leq 1 - 1/n]$ and also $\| K(X) - K_n(X) \| \leq 1/n$. Then the fixed point theorem for differentiable mappings can be applied to K_n.

Since f has a unique and continuous inverse, (10) defines ψ as a continuous mapping on S. By the Brouwer fixed-point theorem ψ has a fixed point, say $\tilde{X}$. Then

$$Kf(\tilde{X}) = f\psi(\tilde{X}) = f(\tilde{X}) \tag{11}$$

so that K has the fixed point $f(\tilde{X})$ in $\tilde{S}$.

In the preceding proof we assumed that E_n was a real Euclidian space. In fact the Brouwer fixed-point theorem is valid in complex spaces as well. That conclusion follows from the above proof immediately, because every n-dimensional complex Euclidian space is equivalent to a $2n$-dimensional real space.

In order to state and prove the Schauder fixed-point theorem we have to introduce some new terminology.

DEFINITION.   A set S in a linear space is said to be convex, if for any two points X, Y in S the set of points $tX + (1 - t)Y$, $0 \leq t \leq 1$, also belongs to S.

A simple consequence of this definition is the following. If $\{X_1, X_2, \ldots, X_n\}$ is a finite point set in a convex set S, then $\sum_i^n t_i X_i$ is also in S, if all $t_i \geq 0$ and $\sum_i^n t_i = 1$.

DEFINITION.   A set S in a Hilbert space is said to be compact if every infinite sequence of points in S, say $\{X_n\}$, has a convergent subsequence, that converges to a point in S.

For example, in a Hilbert space H all points in the unit sphere $\|X\| \leq 1$ form a convex set. On a finite dimensional Euclidian space every closed and bounded set is necessarily compact by the Bolzano-Weierstrass theorem. Next we require the following lemma.

LEMMA.   Let S be a compact set in a Hilbert space. For every n there exists a finite set of points $X_1, X_2, \ldots, X_{N(n)}$ such that

$$\min_{1 \leq i \leq N(n)} \|X - X_i\| < 1/n, \tag{12}$$

where X is an arbitrary point in S. In other words, every X in S is within a distance of $1/n$ from at least one of the X_i.

PROOF.   The proof is by contradiction. Start with any point X_1 in S. We can find a second point X_2 in S such that $\|X_2 - X_1\| \geq (1/n)$. In this fashion we construct an infinite sequence $X_1, X_2, \ldots, X_k, \ldots$, and $\|X_k - X_i\| \geq (1/n)$, $i = 1, 2, \ldots, k - 1$. If the theorem is false it must be possible to construct such a sequence. Since we have $\|X_i - X_j\| \geq (1/n)$ for $i \neq j$ the sequence will not converge and also can have no convergent subsequence.

But S is compact so that some subsequence must converge. It follows that the above construction must stop after a finite number of steps, yielding a point set in S with the required properties. ∎

We are at last in a position to state and prove the following key theorem.

THEOREM 2: Schauder Fixed-Point Theorem. Let S be a closed convex and compact set in a Hilbert space, and K a continuous mapping of S into itself. Then K has at least one fixed point; that is, there is at least one X in S such that $K(X) = X$.

PROOF. The outline of the proof is as follows. We shall construct an infinite sequence $\{K_n\}$ of mappings of S into itself such that each has a fixed point. From these we shall be able to extract a subsequence that converges to a fixed point of K.

For every n we can select a sequence $X_1, X_2, \ldots, X_{N(n)}$ in S, satisfying the conditions of the previous lemma, since S is compact. Consider $K(X_i)$ and find $X_{j(i)}$ so that

$$\min_{1 \leq k \leq N(n)} \|K(X_i) - X_k\| = \|K(X_i) - X_{j(i)}\| < \frac{1}{n}. \tag{13}$$

If the above minimum is attained for more than one X_k, we choose one specific one.

We now consider the subset of S, say S_n, consisting of all X of the form

$$X = \sum_{i=1}^{N(n)} t_i X_i, \qquad t_i \geq 0, \qquad \sum_{i=1}^{N(n)} t_i = 1. \tag{14}$$

Since S is convex S_n is a subset of S. On S_n we can now define a mapping K_n by

$$K_n(X) = \sum_{i=1}^{N(n)} t_i X_{j(i)} = \sum_{i=1}^{N(n)} t_{i'} X_i \tag{15}$$

and evidently $K_n(X)$ is a continuous mapping of S_n into itself. The third term on the right of (15) indicates merely that we can rearrange the sum in the middle term so as to sum over $\{X_i\}$ instead of $\{X_{j(i)}\}$.

Next we wish to apply the Brouwer fixed-point theorem to show that K_n has a fixed point. To do so we view the point X in (14) as equivalent to the point $T = (t_1, t_2, \ldots, t_{N(n)})$ in an $N(n)$ dimensional Euclidian space. The mapping K_n, as shown in (15) maps T into $T' = (t_{1'}, t_{2'}, \ldots, t_{N'(n)})$. Let $\tilde{S}$ be the set of all points T in $E_{N(n)}$ with all $t_i \geq 0$ and $\sum_{i=1}^{N(n)} t_i = 1$. This set $\tilde{S}$ is evidently equivalent to the unit sphere and the mapping $T \to T'$ induced by K_n is evidently continuous, and maps $\tilde{S}$ into $\tilde{S}$. Accordingly, there is a fixed point $T^{(n)} = (t_1^{(n)}, t_2^{(n)}, \ldots, t_{N(n)}^{(n)})$ such that

$$X^{(n)} = \sum_{i=1}^{N(n)} t_i^{(n)} X_i \tag{16}$$

and

$$K_n(X^{(n)}) = X^{(n)}. \tag{17}$$

Let X be any point in S_n. Then for at least one X_i we have $\|X - X_i\| \leq (1/n)$. Since the mappings K and K_n are continuous we can assert that

$$\|K(X) - K(X_i)\| < \epsilon, \qquad n > N(\epsilon)$$

$$\|K_n(X) - K_n(X_i)\| < \epsilon, \qquad n > N(\epsilon)$$

where $\epsilon > 0$ can be made arbitrarily small for sufficiently large $N(\epsilon)$. It follows that

$$\|K(X) - K_n(X)\| \leq \|K(X) - K(X_i)\| + \|K(X_i) - K_n(X_i)\|$$

$$+ \|K_n(X_i) - K_n(X)\| \leq \epsilon + \frac{1}{n} + \epsilon, \tag{18}$$

by using the estimate (13) and the definition of $K_n(X_i)$. Using (17) and (18) we see that

$$\|K(X^{(n)}) - X^{(n)}\| < 2\epsilon + \frac{1}{n}. \tag{19}$$

Since S is compact there exists a convergent subsequence of $\{X^{(n)}\}$ namely $\{X^{(n_k)}\}$ converging to $\tilde{X}$. Then

$$\|K(\tilde{X}) - \tilde{X}\| = \|K(\tilde{X}) - K(X^{(n_k)}) + K(X^{(n_k)}) - X^{(n_k)} + X^{(n_k)} - \tilde{X}\|$$

$$\leq \|K(\tilde{X}) - K(X^{(n_k)})\| + \|K(X^{(n_k)}) - X^{(n_k)}\| + \|X^{(n_k)} - \tilde{X}\|$$

$$< \epsilon + 2\epsilon + \frac{1}{n_k} + \epsilon, \qquad n_k > M(\epsilon) \tag{20}$$

by invoking both the continuity of K and the convergence of $\{X^{(n_k)}\}$. Since the left side is independent of n_k we see that $K(\tilde{X}) = \tilde{X}$ so that K has a fixed point. ■

Theorem 2 holds in more general spaces than Hilbert spaces. An inspection of the proof shows that all we used was the fact that the space was complete with respect to the norm. The inner product was not brought into play at all. Complete normed linear spaces are known as Banach spaces and the Schauder fixed-point theorem evidently is true in such spaces as well.

There is a modification of theorem 2 that turns out to be far more useful in applications than theorem 2. In many applications the set S in question is closed and convex, but not necessarily compact. The image of the set S under the mapping K, that is, $K(S)$, is often compact. To deal with such cases we require the following theorem.

THEOREM 3. Let S be a closed and convex set in a Hilbert space, and K a continuous mapping of S into itself. Suppose that the set $K(S)$ is compact. Then K has at least one fixed point in S.

N.B. Theorem 3 also holds in Banach spaces.

PROOF.   The proof is almost identical to the one of theorem 2. We select a sequence $X_1, X_2, \ldots, X_{N(n)}$ in $K(S)$ using the previous lemma. (In the proof of theorem 2 the sequence was selected in S, but now we work in the compact set $K(S)$.) Since X_i is in $K(S)$ we can find an $X_{j(i)}$ such that

$$\| K(X_i) - X_{j(i)} \| < \frac{1}{n}. \tag{21}$$

From this point on we proceed as before. We consider all X as defined in (14) and define $K_n(X)$ for these X as in (15). The set S_n generated is convex, and therefore still in S. That each mapping K_n has a fixed point follows exactly as in the proof of theorem 2. We then have as in (19)

$$\| K(X^{(n)}) - X^{(n)} \| < \frac{1}{n}. \tag{22}$$

The set $\{K(X^{(n)})\}$ lies in the compact set $K(S)$ and must therefore contain a convergent subset, say $\{K(X^{(n_k)})\}$ such that $\lim_{n_k \to \infty} K(X^{(n_k)}) = \tilde{X}$. But

$$
\begin{aligned}
\| \tilde{X} - X^{(n_k)} \| &= \| \tilde{X} - K(X^{(n_k)}) + K(X^{(n_k)}) - X^{(n_k)} \| \\
&\leq \| \tilde{X} - K(X^{(n_k)}) \| + \| K(X^{(n_k)}) - X^{(n_k)} \| \\
&< \epsilon + \frac{1}{n_k} \quad \text{for} \quad n_k > M(\epsilon)
\end{aligned}
$$

so that $\lim_{n_k \to \infty} X^{(n_k)} = \tilde{X}$. Therefore, $K(X^{(n_k)}) \to K(\tilde{X}) = \tilde{X}$ so that K has a fixed point. ■

3. APPLICATIONS

We shall consider an integral equation of the form

$$\phi(x) - \lambda \int_0^1 K(x, y)\psi(y, \phi(y)) \, dy = 0 \tag{23}$$

and examine a number of conditions under which we can guarantee the existence of a solution.

THEOREM 4. Suppose that $K(x, y)$ is continuous for all x, y in $[0, 1]$ and that $\psi(y, t)$ is continuous for all y in $[0, 1]$ and all t and that

$$\int_0^1 |\psi(y, \phi(y))|^2 \, dy \leq A^2 \|\phi\|^2. \tag{24}$$

Suppose that $\psi(y, t)$ also satisfies the Lipschitz condition

$$|\psi(y, t_1) - \psi(y, t_2)| \leq B |t_1 - t_2| \tag{25}$$

where B is independent of y, and let

$$|K(x, y)| \leq C. \tag{26}$$

Then (23) has a unique solution in $L_2[0, 1]$ provided

$$|\lambda| < \frac{1}{BC}. \tag{27}$$

PROOF.   Define the operator T by

$$T\phi = \lambda \int_0^1 K(x, y)\psi(y, \phi(y))\, dy \tag{28}$$

so that a solution of (23) is a fixed point. T is a bounded operator since

$$\|T\phi\| = |\lambda| \left\{ \int_0^1 \left| \int_0^1 K(x, y)\psi(y, \phi(y))\, dy \right|^2 dx \right\}^{\frac{1}{2}}$$

$$\leq |\lambda| \left\{ \int_0^1 \int_0^1 |K(x, y)|^2\, dx\, dy \int_0^1 |\psi(y, \phi(y))|\, dy \right\}^{\frac{1}{2}}$$

$$\leq |\lambda|\, CA \,\|\phi\| \tag{29}$$

using (24) and (26). Similarly we can show that

$$\|T\phi_1 - T\phi_2\| \leq |\lambda|\, BC \,\|\phi_1 - \phi_2\| \tag{30}$$

by the use of (25) so that T is continuous. If $|\lambda|\, BC < 1$, T is a contraction operator and by theorem 1 of Chapter 2 has a unique fixed point. Since $K(x, y)$ is continuous in x the solution satisfying (23) everywhere will in fact be also continuous. ∎

We shall now derive a more general theorem using theorem 3.

THEOREM 5.   Consider Eq. (23) with the following conditions on $K(x, y)$ and $\psi(y, t)$. Then $K(x, y)$ is continuous for all (x, y) in $[0, 1]$ and

$$|\psi(y, t)| \leq B$$

and for every $\epsilon > 0$ we can find a $\delta(\epsilon)$ such that

$$\int_0^1 |\psi(y, \phi_1(y)) - \psi(y, \phi_2(y))|^2\, dy < \epsilon \quad \text{if} \quad \|\phi_1 - \phi_2\| < \delta(\epsilon), \tag{31}$$

where ϕ_1 and ϕ_2 are in $L_2[0, 1]$. Then (23) has at least one solution in $L_2[0, 1]$ provided

$$|\lambda| \leq \frac{1}{BC}. \tag{32}$$

PROOF.   Let S be the set of all functions $\phi(x)$ in $L_2[0, 1]$ for which $\|\phi\| \leq 1$. Clearly S is closed and convex. To verify the convexity we note that, if ϕ_1 and ϕ_2 are in S, then

$$\|t\phi_1 + (1 - t)\phi_2\| \leq t \|\phi_1\| + (1 - t) \|\phi_2\| \leq 1$$

if $0 \leq t \leq 1$. S is closed since $L_2[0, 1]$ is complete and if $\{\phi_n\}$ is a sequence in S, having a limit ϕ then $\|\phi\| \leq 1$ and ϕ is in S. Let T be defined as in (28). Clearly

$$\|T\phi\| \leq |\lambda| BC \leq 1$$

so that T maps S into itself. By the use of (31) we see also that

$$\|T\phi_1 - T\phi_2\| \leq |\lambda| C\left\{\int_0^1 |\psi(y, \phi_1(y)) - \psi(y, \phi_2(y))|^2 \, dy\right\}^{\frac{1}{2}} < \epsilon$$

$$\text{if} \quad \|\phi_1 - \phi_2\| < \delta(\epsilon)$$

so that T is a continuous mapping. We note that $T\phi$ is a continuous function of x and that

$$|T\phi| \leq |\lambda| \int_0^1 |K(x, y)| \, |\psi(y, \phi(y))| \, dy \leq |\lambda| BC, \qquad (33)$$

so that the set $T(S)$ is uniformly bounded. Now

$$\left|\lambda \int_0^1 K(x_1, y)\psi(y, \phi(y)) \, dy - \lambda \int_0^1 K(x_2, y)\psi(y, \phi(y)) \, dy\right|$$

$$\leq |\lambda| B \int_0^1 |K(x_1, y) - K(x_2, y)| \, dy < \epsilon \quad \text{if} \quad |x_1 - x_2| < \delta(\epsilon) \quad (34)$$

since $K(x, y)$ is continuous. Then $\delta(\epsilon)$ in (34) is independent of the choice of ϕ in S. Now we see that the set $T(S)$ is not only uniformly bounded, but also equicontinuous. We can now use Arzelà's theorem (Chapter 3, Section 3) to show that $T(S)$ is compact. According to that theorem every bounded, equicontinuous set of functions has a convergent subsequence. But if that is the case, $T(S)$ is compact, so that by theorem 3, T has a fixed point and (23) has at least one solution in $L_2[0, 1]$. As in the previous theorem, if that solution satisfies (23) everywhere, and since $K(x, y)$ is continuous the solution must be continuous. ■

EXAMPLE 1.   Consider

$$\phi(x) - \lambda \int_0^1 \frac{x^2 + y^2}{1 + |\phi(y)|} \, dy = 0. \qquad (35)$$

Here $K(x, y) = x^2 + y^2$, so that $C = 2$ in (26). $\psi(y, t) = [1/(1 + |t|)] < 1$ so that $B = 1$ in (31). Now if $|\lambda| \leq \frac{1}{2}$, (35) must have a continuous solution.

While the Schauder fixed-point theorem does provide existence proofs under more general conditions than the contraction mapping principle it is weaker in some other respects. It, first of all, provides no information regarding uniqueness. Secondly, the contraction mapping principle asserts that the unique fixed point can be located by a successive approximation scheme. While it may be difficult in practice to find the limit of such an infinite sequence it does lead to approximations to the solution. The Schauder fixed-point theorem leads to pure existence results. There are many different theorems that can be formulated. We shall now discuss some other modifications of theorem 5.

THEOREM 6.   Consider the equation

$$\phi(x) - \lambda \int_0^1 K(x, y)\psi(y, \phi(y))\, dy = 0 \tag{36}$$

where $K(x, y)$ and $\psi(y, t)$ are continuous functions of their variables. Let S be the set of functions ϕ in $L_2[0, 1]$ for which $\|\phi\| \leq M$. Suppose that

$$|K(x, y)| \leq C, \qquad 0 \leq x, y \leq 1 \tag{37}$$

$$\int_0^1 |\psi(y, \phi(y))|^2\, dy \leq B^2 \qquad \text{for all } \phi \text{ such that } \|\phi\| \leq M \tag{38}$$

and for every positive ϵ we can find $\delta(\epsilon)$ such that

$$\int_0^1 |\psi(y, \phi_1(y)) - \psi(y, \phi_2(y))|^2\, dy < \epsilon \qquad \text{if} \quad \|\phi_1 - \phi_2\| < \delta(\epsilon).$$

Then (36) has at least one solution in S for

$$|\lambda| \leq \frac{M}{CB}. \tag{39}$$

PROOF.   Define the operator T by

$$T\phi = \lambda \int_0^1 K(x, y)\psi(y, \phi(y))\, dy. \tag{40}$$

For ϕ in S

$$|T\phi| \leq |\lambda| \left\{ \int_0^1 |K(x, y)|^2\, dy \int_0^1 |\psi(y, \phi(y))|^2\, dy \right\}^{1/2} \leq |\lambda|\, CB \tag{41}$$

and

$$|T\phi(x_1) - T\phi(x_2)| \leq |\lambda|\, B \left\{ \int_0^1 |K(x_1, y) - K(x_2, y)|^2\, dy \right\}^{1/2} < \epsilon,$$

$$\text{if} \quad |x_1 - x_2| < \delta(\epsilon). \tag{42}$$

Evidently S is closed and convex, T is continuous, and if $|\lambda|\,CB \leq M$, T maps S into itself. From (41) and (42) we see that the set $T(S)$ is uniformly bounded and equicontinuous. By Arzelà's theorem $T(S)$ is now compact so that T has a fixed point in S. That point is a solution of (36). ∎

The conditions of the previous theorems have required that the kernel $K(x, y)$ be continuous. In theorem 7 of Chapter 3 we saw that a linear integral operator with a square integrable kernel was compact. Similarly we can show that in the previous theorems the kernels need not be continuous, but only square integrable in order to be able to derive the same conclusions regarding the existence of solutions of equations like (36).

THEOREM 7.   Consider Eq. (36) with the same conditions on $\psi(y, t)$ as in theorem 6. Let $K(x, y)$ be such that

$$\int_0^1 \int_0^1 |K(x, y)|^2 \, dx \, dy < C^2 < \infty. \tag{43}$$

Then (36) has at least one solution in S for

$$|\lambda| \leq \frac{M}{CB}. \tag{44}$$

PROOF.   Let $K_n(x, y)$ be a sequence of continuous kernels such that

$$\lim_{n \to \infty} \int_0^1 \int_0^1 |K(x, y) - K_n(x, y)|^2 \, dx \, dy = 0 \tag{45}$$

and without loss of generality we can assume that

$$\int_0^1 \int_0^1 |K_n(x, y)|^2 \, dx \, dy \leq C^2 \tag{46}$$

by use of (43). We can define a sequence of integral operators

$$T_n\phi = \lambda \int_0^1 K_n(x, y)\psi(y, \phi(y)) \, dy. \tag{47}$$

S is the set of all ϕ in $L_2[0, 1]$ for which $\|\phi\| \leq M$. Evidently S is closed and convex. We also see that

$$\|T\phi\| \leq |\lambda| \left\{ \int_0^1 \int_0^1 |K(x, y)|^2 \, dy \, dx \int_0^1 |\psi(y, \phi(y))|^2 \right\}^{\frac{1}{2}}$$

$$\leq |\lambda|\,CB \leq M \quad \text{if} \quad |\lambda| \leq \frac{M}{CB}$$

and also $\|T_n\phi\| \leq M$, so that T and all T_n map S into itself. These mappings are also continuous and if $T(S)$ were compact we would be done. That all $T_n(S)$ are compact follows immediately from the proof of theorem 6. Use of

(45) shows that

$$\|T\phi - T_n\phi\| \le |\lambda| \left\{ \int_0^1 \int_0^1 |K(x, y) - K_n(x, y)|^2 \, dy \, dx \int_0^1 |\psi(y, \phi(y))|^2 \, dy \right\}^{\frac{1}{2}}$$

$$\le \frac{M}{CB} \cdot B \left\{ \int_0^1 \int_0^1 |K(x, y) - K_n(x, y)|^2 \, dy \, dx \right\}^{\frac{1}{2}} < \epsilon$$

$$\text{if} \quad n > N(\epsilon). \quad (48$$

Equation (48) is clearly uniform for all ϕ in S.

Let $\{\phi_n\}$ be any sequence in S. We can select a subsequence $\{\phi_{n(1)}\}$ such that $\{T_1\phi_{n(1)}\}$ converges. From that subsequence we can extract a new subsequence $\{\phi_{n(2)}\}$ so that $\{T_2\phi_{n(2)}\}$ converges, and so on. In this fashion we obtain a chain of subsequences

$$\{\phi_n\} \supset \{\phi_{n(1)}\} \supset \{\phi_{n(2)}\} \supset \cdots \supset \{\phi_{n(k)}\} \supset \cdots \quad (49)$$

such that the sequence $\{T_i\phi_{n(k)}\}$ converges for all $i = 1, 2, \ldots, k$. Finally, we take the sequence $\{\phi_{n(m)}\}$ which is a subsequence of every $\{\phi_{n(k)}\}$ except for a finite number of elements and clearly $\{T_k\phi_{n(m)}\}$ converges for every k. Now

$$\|T\phi_{n(n)} - T\phi_{m(m)}\|$$
$$= \|T\phi_{n(n)} - T_k\phi_{n(n)} + T_k\phi_{n(n)} - T_k\phi_{m(m)} + T_k\phi_{m(m)} - T\phi_{m(m)}\|$$
$$\le \|T\phi_{n(n)} - T_k\phi_{n(n)}\| + \|T_k\phi_{n(n)} - T_k\phi_{m(m)}\|$$
$$+ \|T_k\phi_{m(m)} - T\phi_{m(m)}\|. \quad (50)$$

The first and third terms on the right of (50) can be made small for large k by the use of (48). The middle term becomes small for large n and m since the sequence $\{T_k\phi_{n(n)}\}$ converges. It follows that

$$\|T\phi_{n(n)} - T\phi_{m(m)}\| < 2\epsilon, \quad n, \quad m > M(\epsilon)$$

and this implies that the sequence $\{T\phi_{n(n)}\}$ is a Cauchy sequence so that $T(S)$ is compact and T must have a fixed point. ∎

EXAMPLE 2. Consider

$$\phi(x) - \frac{2}{\pi} \int_0^\pi \sin x\phi^2(y) \, dy = 0. \quad (51)$$

The preceding theorems guarantee at least one solution. We see however that both $\phi_1 = 0$ and $\phi_2 = \sin x$ satisfy (51) so that we certainly don't have unique solutions.

Essentially, every theorem that guarantees the existence of eigenvalues and eigenvectors for linear operators is a fixed-point theorem. Theorem 12 of Chapter 4 guarantees the existence of at least one positive eigenvalue for an integral operator with a positive kernel. This theorem too, can be proved by an application of the Schauder fixed-point theorem.

THEOREM 8. Let $K(x, y)$ be a continuous, positive function for all $0 \leq x, y \leq 1$. Consider the operator

$$K\phi = \int_0^1 K(x, y)\phi(y) \, dy. \tag{52}$$

K has at least one positive eigenvalue, corresponding to a positive eigenfunction.

PROOF. Here we shall work in the Hilbert space consisting of all real functions $\phi(x)$ for which $\int_0^1 |\phi(x)|^2 \, dx < \infty$, namely the real space $L_2[0, 1]$. We can suppose without loss of generality that $0 < m \leq K(x, y) \leq M$ since $K(x, y)$ is taken to be positive and continuous. Now let S be the set of all functions in our Hilbert space for which $\int_0^1 |\phi(x)|^2 \, dx \leq 1$ and $\phi(x) \geq 0$, almost everywhere. S is clearly closed and convex.

Since $K(x, y) \geq m > 0$ we have

$$K\left[\phi + \frac{1}{n}\right] = \int_0^1 K(x, y)\left(\phi(y) + \frac{1}{n}\right) dy \geq \frac{m}{n}, \qquad \phi \in S, \tag{53}$$

where n is a positive integer. From (53) we see that

$$\left\|K\left[\phi + \frac{1}{n}\right]\right\| = \left\{\int_0^1 \left(\int_0^1 K(x, y)\left(\phi(y) + \frac{1}{n}\right) dy\right)^2 dx\right\}^{\frac{1}{2}} \geq \frac{m}{n}. \tag{54}$$

Consider now the mapping of S into itself defined by

$$T_n\phi = \frac{K[\phi + (1/n)]}{\|K[\phi + (1/n)]\|}. \tag{55}$$

Evidently the denominator is bounded away from 0 by use of (54). Also $\|T_n\phi\| = 1$ and for ϕ in S we clearly have $T_n\phi \geq 0$, so that T_n maps S into itself. It is also easy to see that T_n is a continuous mapping. In order to apply the Schauder fixed point theorem we still have to verify that the image of S under T_n is compact. This will, as in previously discussed cases, be a consequence of Arzelà's theorem.

First we observe that

$$\left|K\left(\phi + \frac{1}{n}\right)\right| \leq \left\{\int_0^1 K^2(x, y) \, dy \int_0^1 \left(\phi + \frac{1}{n}\right)^2 dy\right\}^{\frac{1}{2}} \leq 2M \tag{56}$$

since $K(x, y) \leq M$ and ϕ is in S so that the numerator of (55) is uniformly bounded. Similarly

$$\left|K\left(\phi + \frac{1}{n}\right)\right| \geq \left|K\frac{1}{n}\right| \geq \frac{m}{n}$$

so that

$$|T_n\phi| \leq \frac{2nM}{m}$$

so that the set of functions $\{T_n\phi\}$, $\phi \in S$ is uniformly bounded, and also continuous. Similarly they are equicontinuous so that $T_n(S)$ is compact and T_n has a fixed point so that for some ϕ_n we have

$$T_n\phi_n = \phi_n. \tag{57}$$

The parameter n was introduced in order to be sure that the denominator of (55) is bounded away from 0. We shall now show that we can select a subsequence of $\{\phi_n\}$ that converges to an eigenfunction of K, corresponding to a positive eigenvalue.

By combining (55) and (57) we obtain

$$K\left[\phi_n + \frac{1}{n}\right] = \lambda_n\phi_n \tag{58}$$

$$\lambda_n = \left\|K\left[\phi_n + \frac{1}{n}\right]\right\| \tag{59}$$

the latter being a consequence of the fact that $1 = \|T_n\phi_n\| = \|\phi_n\|$ by (57). Since $\|\phi_n + (1/n)\| \leq \|\phi_n\| + \|1/n\| \leq 2$ we see that the set $\{\phi_n + (1/n)\}$ is uniformly bounded in $L_2[0, 1]$. The linear integral operator K is a compact operator, and therefore we can extract a subsequence $\{\phi_{n_k} + (1/n_k)\}$ such that the sequence $\{K[\phi_{n_k} + (1/n_k)]\}$ converges to some function, say ψ.

We shall now show that the sequence $\{\lambda_{n_k}\}$ also converges to some positive limit. From (58) we see that

$$\phi_n = \frac{1}{\lambda_n}K\left[\phi_n + \frac{1}{n}\right] \geq \frac{1}{\lambda_n}K\left[\frac{1}{n}\right] \geq \frac{m}{n\lambda_n} \tag{60}$$

and also that $\phi_n(x)$ is a positive continuous functions on $[0, 1]$. Let

$$\min_{0 \leq x \leq 1}\phi_n(x) = \beta_n > 0$$

so that

$$\beta_n \geq \frac{m}{n\lambda_n} \tag{61}$$

We can refine these estimates by using (58) again so that

$$\phi_n = \frac{1}{\lambda_n}K\left[\phi_n + \frac{1}{n}\right] \geq \frac{1}{\lambda_n}\left[m\beta_n + \frac{m}{n}\right]$$

and it follows that

$$\beta_n \geq \frac{1}{\lambda_n}\left[m\beta_n + \frac{m}{n}\right]. \tag{62}$$

The last estimate shows that $\lambda_n \geq m$ and is therefore bounded away from the origin. Equation (59) now shows that $\lambda_{n_k} = \|K[\phi_{n_k} + (1/n_k)]\| \to \|\psi\|$, and we shall denote that limiting value by $\lambda_0 \geq m$.

Finally we see from (58) that $\phi_{n_k} \to (1/\lambda_0)\psi$. Accordingly $K[\phi_{n_k} + (1/n_k)] \to K(1/\lambda_0)\psi$ and also $K[\phi_{n_k} + (1/n_k)] \to \psi$ so that

$$K\left[\frac{1}{\lambda_0}\psi\right] = \psi$$

or more explicitly

$$\int_0^1 K(x, y)\psi(y)\,dy = \lambda_0\psi(x). \tag{63}$$

The left side of (63) is continuous and positive so that $\psi(x)$ is continuous and positive. ▪

EXAMPLE 3. Consider the equation

$$\phi(x) - \lambda\int_0^1 (1 + xy)\phi(y)\,dy = 0 \tag{64}$$

where the kernel is evidently positive. The operator is degenerate and we can look for a solution in the form

$$\phi(x) = a + bx. \tag{65}$$

Insertion of the above in (64) leads to the linear algebraic system

$$(1 - \lambda)a - \frac{\lambda}{2}b = 0$$

$$-\frac{\lambda}{2}a + \left(1 - \frac{\lambda}{3}\right)b = 0 \tag{66}$$

The determinant of the above system is given by $1 - \frac{4}{3}\lambda + \frac{1}{12}\lambda^2$ and its zeros are $8 \pm \sqrt{52}$. The latter are the eigenvalues of (64). Both happen to be positive. By theorem 8 at least one has to be positive. The corresponding eigenfunctions are given by

$$\phi_+ = 1 - \frac{\sqrt{13} + 2}{3}x$$

$$\phi_- = 1 + \frac{\sqrt{13} - 2}{3}x \tag{67}$$

Only ϕ_- is positive as guaranteed by the theorem. This is consistent with theorem 13 of Chapter 6, where the smaller eigenvalue has a positive solution associated with it.

EXERCISES

1. Let $K(x, y)$ be a real, continuous, positive kernel for all $0 \leq x, y \leq 1$, such that $K(x, y) \geq 0$. Suppose that $\int_0^1 K(x, z)K(z, y)\,dz$ is positive. Show that the integral operator associated with $K(x, y)$ has at least one positive eigenvalue.

2. Show that

$$\phi(x) - \lambda \int_0^1 (x^2 + y^2) \sin \phi(y)\,dy = 0$$

has solutions in $L_2[0, 1]$ for suitably restricted values of $|\lambda|$.

3. Show that

$$\phi(x) - \lambda \int_{-\infty}^{\infty} e^{-\sqrt{x^2+y^2}} e^{-|\phi(y)|}|\,dy = 0$$

has solutions in $L_2[-\infty, \infty]$ for suitable restricted values of $|\lambda|$.

BIBLIOGRAPHY

Bôcher, Maxime, An Introduction to the Study of Integral Equations. Cambridge University Press. 1913. Reissued by Hafner Pub. Co., New York.

Carrier, George F., Max Krook, and Karl E. Pearson, *Functions of a Complex Variable; Theory and Technique*, McGraw-Hill, New York, 1966.

Courant, Richard, and David Hilbert, *Methods of Mathematical Physics*, Vol. I, Interscience, New York, 1953.

Frank, Philipp, and Richard V. Mises, *Die Differential und Integral gleichungen der Mechanik und Physik*, 2 vol. Dover, New York. 1961.

Gakhov, F. D., *Boundary Value Problems*, Pergamon Press, New York, 1966.

Goursat, E., *A Course in Mathematical Analysis*, Vol. III, Part 2, "Integral Equations, Calculus of Variations," Dover, New York, 1964.

Hamel, Georg, *Integralgleichungen, Einführung in Lehre und Gebrauch*, Springer-Verlag, Berlin, 1937.

Hellinger, E., and O. Toeplitz, *Integralgleichungen und Gleichungen mit Unendichvielen Unbekannten*, Chelsea Publishing Co., New York, 1953.

Hilbert, David, *Grundzüge einer Allgemeinen Theorie der Linearen Integralgleichungen*, Chelsea Publishing Co., New York, 1953.

Hoheisel, Guido, *Integral Equations*, Ungar Publishing Co., New York, 1968.

Hopf, E., *Mathematical Problems of Radiative Equilibrium*, Cambridge University Press, Cambridge, 1934.

Kneser, A., *Die Integralgleichungen und ihre Anwendungen in der Mathematischen Physik*, 2nd ed., Vieweg, Braunschweig, 1922.

Kourganoff, V., *Basic Methods in Transfer Problems*, Dover, New York, 1963.

Kowalewski, Gerhard, *Einfuhrung in die Determinantentheorie einschliesslich der Fredholmschen Determinanten*, 3rd ed., Chelsea Publishing Co., New York, 1948.

Kowalewski, G., *Integralgleichungen*, W. de Gruyter, Berlin, 1930.

Krasnosel'skii, M. A., *Topological Methods in the Theory of Nonlinear Integral Equations*, Pergamon Press, New York, 1964.

Lalesco, T., *Introduction à la Theorie des Equations Integrals*, Hermann, Paris, 1912.

Lovitt, William Vernon, *Linear Integral Equations*, Dover, New York, 1950.

Mikhlin, S. G., *Integral Equations, and their Applications to Certain Problems in Mechanics, Mathematical Physics and Technology*, Pergamon Press, New York, 1957.

Mikhlin, S. G., *Multidimensional Singular Integrals and Integral Equations*, Pergamon Press, New York, 1965.

Morse, Philip M., and Herman Feshbach, *Methods of Theoretical Physics*, McGraw-Hill, New York, 1953.

Muskhelishvili, N. I., *Singular Integral Equations*, Noordhoff, Holland, 1953.

Noble, B., *Methods Based on the Wiener-Hopf Technique, for the Solution of Partial Differential Equations*, Pergamon Press, New York, 1958.

Paley, Raymond E. A. C., and Norbert Wiener, "Fourier Transforms in the Complex Domain," *Amer. Math. Soc. Coll. Pub.* Vol. XIX, 1934.

Petrovskii, I. G., *Lectures on the Theory of Integral Equations*, Graylock Press, Rochester, N.Y., 1957.

Pogorzelski, W., *Integral Equations and their Applications*, Vol. I, Pergamon Press, New York, 1966.

Riesz, Frigyes and Bela Sz.-Nagy, *Functional Analysis*, Ungar Publishing Co., New York, 1955.

Schmeidler, Werner, *Integralgleichungen mit Anwendungen in Physik und Technik*, Akad. Verlagsgesell., Leipzig, 1955.

Smirnov, V. I., *A Course of Higher Mathematics*, Vol. IV, *Integral Equations and Partial Differential Equations*, Pergamon Press, New York, 1964.

Smithies, F., *Integral Equations*, Cambridge University Press, London, 1958.

Sneddon, Ian N., *Fourier Transforms*, McGraw-Hill, New York, 1951.

Stakgold, Ivar, *Boundary Value Problems of Mathematical Physics*, 2 vols. Macmillan, New York, 1968.

Titchmarsh, E. C., *Introduction to the Theory of Fourier Integrals*, Clarendon Press, Oxford, 1948.

Tricomi, F. G., *Integral Equations*, Interscience, New York, 1957.

Widom, Harold, *Lectures on Integral Equations*, Van Nostrand-Reinhold, New York, 1969.

Zaanen, Adriaan Cornelis, *Linear Analysis, Measure and Integral, Banach and Hilbert Space, Linear Integral Equations*, North-Holland, Amsterdam, 1964.

INDEX